Computer Numerical Control Simplified

Computer Numerical Control Simplified

by

Michael Fitzpatrick
Instructor
Sno-Isle Skills Center
Everett, Washington

GLENCOE PUBLISHING COMPANY

Send all inquiries to:
Glencoe Publishing Company
15319 Chatsworth Street
Mission Hills, California 91345

Printed in the United States of America

ISBN 0-02-676410-5

1 2 3 4 5 6 7 8 9 10 93 92 91 90 89

ACKNOWLEDGMENTS

I would like to express my appreciation to the following individuals and companies. Their cooperation has made this book possible.

Technical Advisory Committee

Mr. Kelly Sloan — C/NC Instructor, Oklahoma City, Oklahoma
Mr. Harley Gillett — Industrial Training Consultant, Pontiac, Michigan
Mr. Warren Carman — Machinist and C/NC Instructor, New Jersey

Technical Readers

Mr. Kelly Sloan
Mr. Warren Carman
Mr. Karl Hommer — Machinist and C/NC Instructor, Renton, Washington
Mr. Don Olson

Illustrations

The Boeing Company
Component Products
Eldec Corporation
Electronic Industries Association
Kennametal, Inc.
Massachusetts Institute of Technology
Mazak Machine Tools
Mr. Steve Monson — CAD Instructor, Everett, Washington
X-Act Machining

Programs and Information

DYNA Mechtronics
Fanuc-General Numeric Controls USA
Kennametal, Inc.
Metatron Corporation
Point Control Company

My special thanks to Steve Monson, for his preparation of the CAD drawings and for his special help in writing Chapter 25.

INTRODUCTION

Modern industry is in a period of rapid technological change. The new technologies will place a serious demand on you, the student machinist. There was once a time when scientific and technical knowledge was doubling every one hundred years. Then, as we developed greater technical and industrial abilities, our knowledge of technology doubled every fifty years. More recently it has doubled every five years. Keeping up with advances in a certain field of technology is a challenging task. Computer numerical control (C/NC) is only one of these expanding technologies. The student and teacher alike face a challenge to stay on the leading edge of C/NC developments.

Machines that are directed by a code of letters and numbers are called numerical control machines, a term that is usually shorted to NC. Computer numerical control developed when the microcomputer revolution made inexpensive computer chips available for NC controls. When referring to the general subject of programmed machine tools, which includes NC and CNC, I will use the term C/NC. Often, there is little difference between an NC and a CNC principle. The term C/NC will, then, be used to refer to the whole field.

The organization of information in *Computer Numerical Control Simplified* is shown in the table of contents. As you will notice, the information is organized into units, chapters, and sections.

Each unit deals with a general topic in computer numerical control. The chapters in the unit deal with specific topics as they apply to the general field of computer numerical control. The sections within the chapters develop and explore topics in fairly specific detail.

The information presented in this text is core information for all C/NC programs. The general application of information is discussed, along with its particular application to equipment that may be available to you. Thus, you first study the universal application of the C/NC information. This information has a general application industry-wide. You will then complete related activities in your lab. The coordination of these two approaches will enable you to obtain a solid grasp of C/NC information.

Taken one step at a time, C/NC procedures are not difficult. It is important to remember that the correct performance of a C/NC procedure depends on the careful and accurate completion of each of its steps. If one of the steps is incorrectly performed, the procedure will be incorrect. This need to be attentive to technical detail should not frighten you. The equipment used in C/NC is complex. However, most C/NC procedures are not difficult if they are approached methodically, in a step-by-step manner.

Skill-building and information retention are promoted through the use of three types of learning reinforcement activities. These are titled Activities, Reference Points, and In Your Lab. By highlighting essential information, each of these supplements the instruction presented in the text. The *Activities* present questions and problems that will allow you to gauge your grasp of key points. The *Reference Points* will alert you to items of special interest. The sections entitled *In Your Lab* will prompt you to relate text information to the equipment and procedures used in your lab.

The student of C/NC who wants to be prepared for future developments in this field must fully understand the basics. Many aspects of C/NC are changing rapidly. However, many procedures have not changed or have changed only slightly. This text covers the newer developments, as well as the basics. Learn the basics well. A sound knowledge of the basics will present you with more career options. Such knowledge will also enable you to improve your employment opportunities by taking more technical courses. If you learn the theory of machining and the operation of many kinds of machines, you will have learned the basics. Then, when given the opportunity to operate different equipment, you will find it easy to adapt because of your background knowledge of the basics. This is called transfer learning. This text will prepare you to transfer your knowledge of one subject in C/NC to another subject in C/NC. You will then be starting with a solid foundation in one of the most exciting careers in modern industry.

TABLE OF CONTENTS

UNIT 1

Industrial Foundation of Programmed Machine Tools

By studying Unit 1 you will gain an understanding of the scope and nature of computerized numerical control (C/NC) in metal manufacturing. You will learn the history of C/NC and its place in industry today. You will also learn about the careers available in C/NC. This necessary background information will give you an overall perspective of the field.

CONTENTS

CHAPTER 1

PROGRAMMED MACHINE TOOL HISTORY

INTRODUCTION

Twenty-Five Years Ago

You might think of a computerized numerical control machine tool as an example of an emerging technology. However, even though microcomputer advances have made many improvements in the field, you should understand that, programmed machine tools (NC) have been used in American industry for more than twenty-five years.

Machines operating from punched tape or punched cards were used in industry in the early 1960s. These machines were different from those available today. Some, however, are still in use today, making acceptable parts at a low cost. Fig. 1-1.

OBJECTIVES

After studying this chapter, you will be able to:

- Outline the early history of programmed machine tools.
- Define the evolution of machine tools from standard equipment through NC to C/NC.
- Identify the basic parts of a C/NC drive.

Fig. 1-1. This three axis, tape-driven horizontal milling machine has been in service since 1965. The Boeing Company

1-1 The History of Numerical Control

THE YEARS OF GROWTH

In 1945, at the end of World War II, several events led to an experiment that changed metal manufacturing.

1945

In the early 1940s, the need to produce military products, such as airplanes, accelerated technical research. As a result, the products being produced at the end of the war were too complex in shape or too closely toleranced for practical manufacturing.

To assist in engineering calculations, a computer was developed at the University of Pennsylvania. The ENIAC (Electrical Numerical Integrator and Calculator) as it was called, was a huge mass of tubes and wires. It was difficult to program and very slow by today's standards, but it was a *computer*.

Later, numerical control machines and computers were used to develop today's computer numerical control machines (C/NC). The Parsons Corporation, the United States Air Force, and the Massachusetts Institute of Technology each played a role in this development.

1946

In 1946, the Parsons Corporation tried to find accurate ways to make complicated aircraft parts. In an effort to generate an accurate rotor blade for a helicopter, they experimented with complicated tables of coordinates and manual machines. To generate the compound curves, they placed one human operator per axis handle, and called each move out in turn. This was slow and prone to errors. The Parsons Corporation then turned its attention to automatically generating these shapes. It seemed that an automatic method was possible.

1949

John Parsons then set up a demonstration of his ideas for the Air Force. With a demonstration, Parsons convinced the Air Force to award a research contract.

1952

Shortly thereafter, Parsons set up a subcontract with the Servomechanisms Laboratory of the Massachusetts Institute of Technology (MIT). After three years, MIT built the first NC milling machine. In 1952, a Cincinnati Hydrotel vertical spindle milling machine ran the first true NC-produced parts. See Fig. 1-2. The electrical cabinet took up more floor space than the machine. This was, however, the beginning that would change machining forever.

Fig. 1-2. This Cincinnati Hydrotel is thought to be the first true NC machine.

The prototype numerical control machine developed by MIT used a punched tape to generate movements of three axes. This machine was capable of making curved shapes, quickly, accurately, and reliably.

1955

In 1955, the Air Force granted 35 million dollars to produce 100 NC machines. These were used to make military aircraft.

PRIVATE INDUSTRY AND NUMERICAL CONTROL

1960

In 1960, machine manufacturers began to make NC equipment that many companies could afford. In 1960, equipment was on the market at a price that allowed many shops to purchase their first NC machine.

Although the first NC machine developed at MIT was a three-axis contouring machine, most NC machines used in private industry were simple two-axis machines. They could position a drilling table accurately. This was because the cost of NC machines was high. Managers were yet to be convinced that NC machining was to become the backbone of production.

Today, over half of all manufacturers and machine shops in the world have some form of C/NC. At the Chicago International Machine Tool Show, most of the displays are devoted to computer-aided manufacturing (CAM), which includes C/NC. In terms of dollars, over 75 percent of all new machine tools delivered to U.S. shops are programmable. This trend is expanding.

In a single day, it is now possible to deliver a machine, set it up, and run it without problems. The electronics are reliable; the controls and software are user friendly; the machine teaches the user and assists them in their progress to mastery. C/NC is no longer a special subject to be taught to a machinist at the end of his or her study of the craft. Numerical Control is a basic part of the craft—an entry-level skill.

1-2 Applications of C/NC

C/NC IN INDUSTRY

Although commonly found in machine shops, any manufacturing situation may use some form of programmed equipment. Listed below you will find some creative uses of C/NC. With the addition of microprocessors, the list is growing because of the flexibility of programming and quick turnaround time. Work not well suited to NC operations could be done quickly and inexpensively with computers assisting.

C/NC is common on all standard machine shop and inspection equipment. It is also found in cabinet shops, sheet metal shops, welding shops, print shops, and lumber mills. In fact, it is found in any type of manufacturing where quick error-free repetition is needed. Drafting machines and robots are, to some extent, C/NC-programmed machines.

Perhaps one of the more interesting applications of C/NC is in bending automotive exhaust tubing. Large inventories are a problem for shops that sell tail pipes to fit many automobiles. With an NC bender, all that is required for the shop is a set of programs and a rack of unbent tubing. The tubing is bent as needed, saving space and inventory.

The first NC machines were very expensive. For a shop to buy an NC machine, the work had to be repetitive or complicated to justify the high cost of the equipment. Also, the NC machine was rather inflexible. The programs were difficult to edit, and the turnaround time was relatively long. *Turnaround* refers to the changing time required from one setup to another. Today, however, NC has changed.

Long-run production is easily completed by a programmed machine. However, a modern C/NC is so adaptable that it is now economically feasible to produce a single part often faster than on many standard machines. Very high density microelectronics and intelligent software combine to form fourth generation controls. Parts far too complex for any standard machining method can now be made by programming computers and intelligent controllers.

1-3 Evolution of Programmed Machine Tools

The development of programmed machine tools has affected the machine, the controller, and the software and firmware.

THE MACHINE

Of the three areas, the machine has evolved the *least*. Machinery was well developed when the NC machine was invented. Of course, there have been improvements on the early equipment modified by MIT. The following machine improvements were needed to control machines by NC programs, instead of operators.

A Complete NC Drive System

Four component parts make up a complete C/NC drive system. Each part is different from a part in a standard machine due to the special needs created by programmed control. These parts are:

- Ball screws.
- Feedback device.
- Precision drive motors.
- Master control unit (MCU).

A new system was needed to locate machine slides and tools accurately with minimum mechanical backlash. This new system is called the NC drive.

A nut and ground screw are used to move a slide on a standard lathe or mill. These have a given amount of clearance to allow smooth operation. This screw and nut system has too much *backlash* (clearance) for the accuracy needed in NC machining. A system with less backlash was needed.

The most common device for eliminating backlash is the ball screw. Fig. 1-3. Ball screws not only eliminate backlash, they also reduce mechanical friction and wear. Backlash is controlled when the central preload pushes both directions on the round thread. Ball screws produce closer repeatability and resolution for C/NC machines.

Two qualities important in any C/NC machine are repeatability and resolution. The ability to locate a machine slide or tool within a given tolerance time after time is called *repeatability.* Accuracies of less than .0005″ are common.

The *resolution* of a C/NC machine is the smallest movement the machine is capable of making. Resolution is a combination function of the machine and control hardware. The control must be able to output small movement commands while the axis drives are able to obey the commands. Standard machines needed better drive motors to meet NC repeatability and resolution.

In addition to ball screws, special drive motors were needed to decrease the resolution. Two common types of motors are used today: stepper motors and servo motors.

A *stepper motor* reliably moves a given amount per pulse. For example, it may move 1/300 of a motor revolution with one pulse of electricity. The distance the axis moves is based upon the number of electrical pulses the control generates. The speed of the motor is determined by how fast the pulses are sent to the motor by the controller. Stepper motors were first used for light machining. However, improvements have now made them suitable for use as drivers on medium-duty C/NC machines.

Servo motors differ from standard motors in that they are highly controllable. Electric or hydraulic servos are available, but direct current electric servos are the more common. Servo motors allow complete control of:

- Motor speed.
- Reversing time.
- Acceleration and deceleration.
- Distance rotated (given a known amount of input energy).

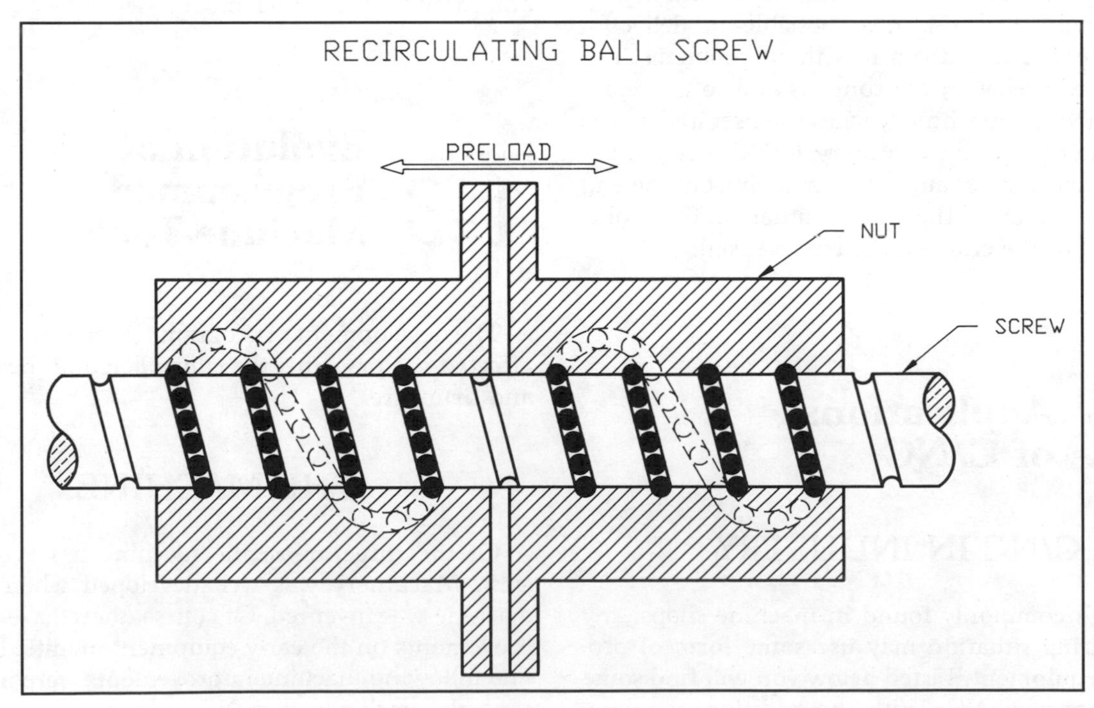

Fig. 1-3. Ball screws eliminate backlash and friction in C/NC drives, by preloading two matching nuts. The load pushes in both directions against the screw.

Machines equipped with ball screws and precision drive motors could locate accurately. However, there was no allowance for changing loads on the drive, such as a cutter entering the work and leaving the work. There was a need for a message to be sent from the machine axis to the controller that the command was being carried out accurately.

Feedback

Feedback is the ability of a machine to send a message back to the control unit. This message tells the control that it has moved the amount requested. We will study this concept again when we look at closed and open loop controls. There are several different feedback systems.

Indirect Devices

Some feedback sending units are connected to the drive motors, and sense the amount of rotation of the screw. Fig. 1-4. These are categorized as *indirect devices*. They are also called transducers, tachometers, resolvers, or encoders.

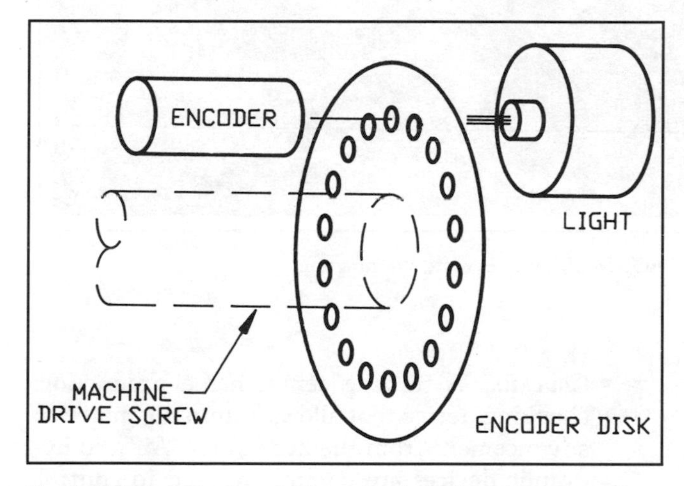

Fig. 1-4. An encoder is an indirect feedback device. The optical sensor receives pulses of light through the slotted plate as the drive screw rotates.

Direct Feedback

Direct feedback devices called *scales* are connected to the machine slides and sense distance traveled. These are usually optical. They sense a rule or scale laid out on the machine slide. A similar readout is found on many standard Bridgeport mills. The position of the table is determined by the reading of a scale on the machine. The difference is that the NC signal is sent to the controller for comparison as well as a position readout for the operator.

A diagram of a complete C/NC drive is shown in Fig. 1-5. This system is used on both NC and C/NC machines.

New Developments

The invention of NC prompted the development of new types of machinery. This new machinery was needed because of the following:

- More demand on the equipment.
- Higher production needs.

Equipment demand. To justify their high expense, C/NC machines often had to be used twenty-four hours a day. Therefore, they had to be stronger. Both sliding and rotating bearings were needed that could withstand continuous use and still deliver the accuracy needed. Castings and the general machine needed to be stronger, more rigid, and produce less vibration. Spindles had to go faster with less vibration.

Higher production need. Because NC could do more work simultaneously, more machine functions were added. The improved machines could come much closer to completing a machined part in one loading. Less handling time meant faster production.

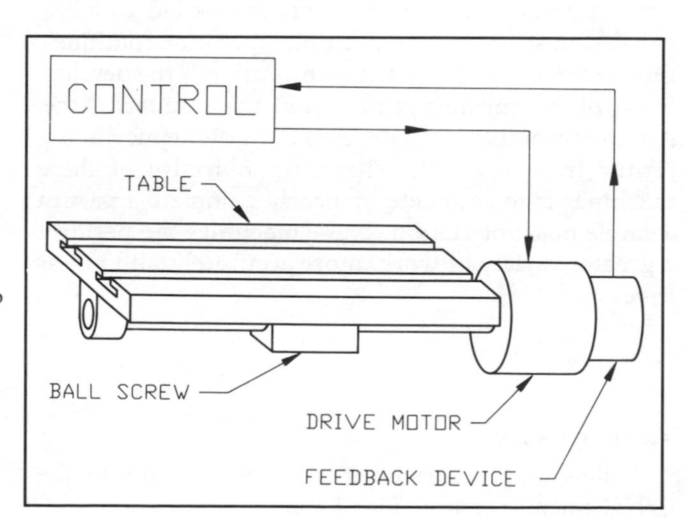

Fig. 1-5. These components form a complete C/NC drive to provide repeatability and resolution.

Fig. 1-6. A modern C/NC turning center. X-ACT Machining, Everett, Washington

Multiple Functions

Because NC control allowed more than one function, and the controller could handle simultaneous operations, machine manufacturers responded with bigger, faster, stronger, and multiple-operation machines. One example of such machine evolution is the development of the turning center from the standard lathe. Another example is the development of the machining center from the mill. The main objective of these machines is to complete or nearly complete a part in a single machine station. These machines can perform a greater variety of work, more accurately, and in less time.

Turning Center

Following is a brief list of improvements in the C/NC turning center (Fig. 1-6).

- *Chucking.* A turning center has a production chucking device that allows automatic material advancement from the head stock. Air and hydraulic devices are commonly used to control the pressure on the part.
- *Multiple tools.* A tool turret with many tool stations.
- *Extra tool slides.* More than one tool slide may be added to a turning center. Often, each slide can operate independently.
- *Programmable spindles.* The spindle motor is a servo. The controller knows exactly where the part is in terms of rotation.
- *Secondary operations.* While not always on a turning center, limited mill and drill functions are often included to allow single-station part completion.
- *Chip removal.* The high volume of chips produced requires a system to continuously remove the waste.

Machining Center

Following is a brief list of improvements in the machining center (Fig. 1-7):

- *Multiple tools.* A tool storage system and automatic tool changing device are necessary.
- *Multiple spindle.* Machining centers may have more than one tool spindle.
- *Programmable auxiliary axes.* While not always found on a machining center, they may be equipped with extra rotating or sliding axes.
- *Material handling and secondary operations.* To complete parts in a single station, machining centers often include universal pallets to position, load, and unload parts with no operator intervention. All sides of a part may be machined in a single program.

To complete the Unit 1 Activity, you will need to understand the following history of NC and machine evolution.

You should be able to identify the individuals and events that led to the development of NC in industry. You should be able to give important dates.

You should understand the overall concept of a C/NC drive and the general evolution of machines. Study Fig. 1-5. You will be asked to identify the major components in the Unit 1 Activity. You should be able to define the following words and concepts:

- Ball screw.
- Drive motor.
- Feedback device.
- Resolution.
- Repeatability.
- Turning and machining centers.

Fig. 1-7. A modern C/NC machining center is a direct descendant of the first milling machines. The Boeing Company

THE CONTROLLER

A major area of evolution has been the controller. Control capabilities are the difference between NC and C/NC. We will compare NC to C/NC. The major difference between an NC and C/NC machine is that the C/NC machine has a microprocessor coupled with the NC controller. The C/NC microprocessor has:

- The capability of managing data and making calculations.
- The capability of making decisions based on data input.
- Flexibility in the form of random access memory (RAM).

Refer to Fig. 1-8.

NC Compared with C/NC

NC is the operation of a machine from data stored in a program. The data is a combination of numbers and letters that command machine movements and functions such as "spindle on" to a machine from a *master control unit* (MCU). The MCU directs the machine movements.

An NC machine simply follows directions from a program; it is a slave. There is no flexibility nor authority over the work being done. The program is external to the control such as a punched tape. The program is fixed—all information must be in the program.

Computer numerical control (C/NC) means that data stored within the MCU memory *directs* the operation of a machine. The microprocessor allows the machine to exercise authority over the operation. A C/NC control can make calculations and decisions.

A C/NC machine is more flexible than an NC machine. Similar to NC, a C/NC follows program directions. In addition, it assumes responsibility for calculations and decisions and contains a memory in which the program resides. This means that the program can be changed easily. The memory that can be edited is called random access memory (RAM). A C/NC can detect and help correct problems and communicate with the operator and with external devices such as robots and central programming computers.

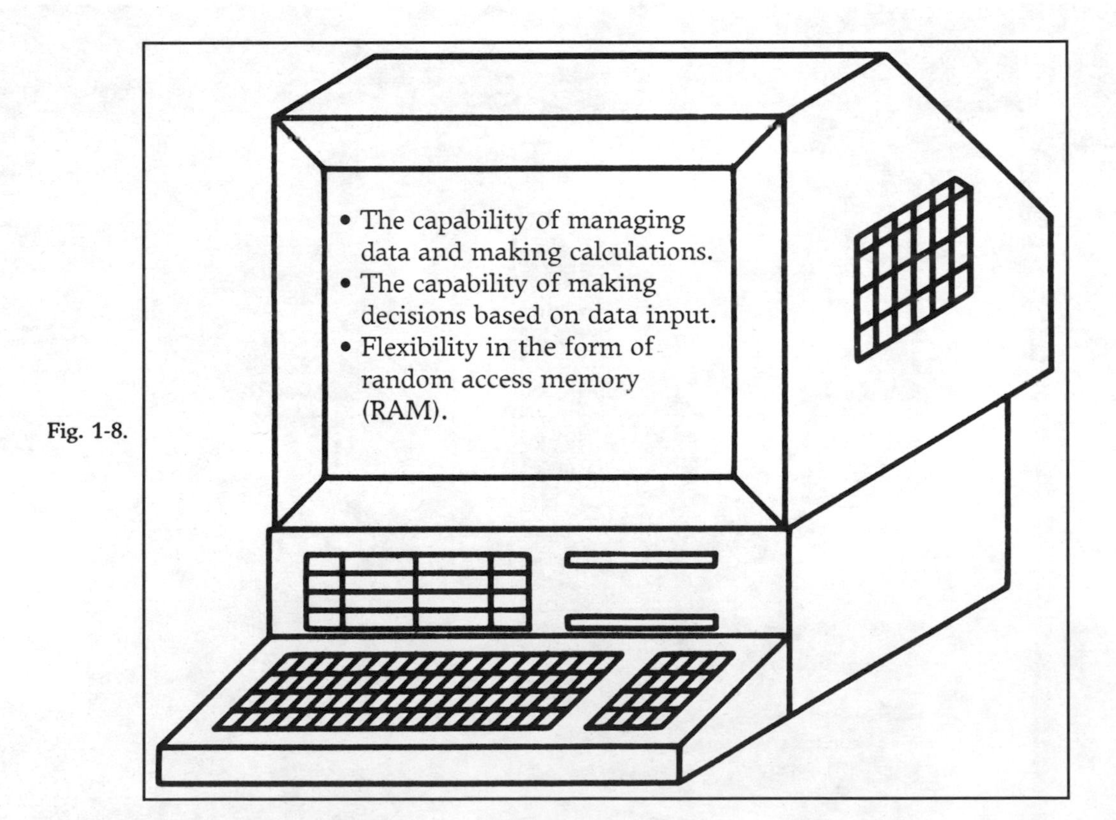

Fig. 1-8.

NC, CNC, and C/NC—all three terms will be used in this book. NC means simple numerical control, while CNC means computer numerical control. C/NC refers to either subject. The abbreviation C/NC is used to denote the entire subject of programmed machine tools.

Table 1-A compares numerical control (NC) with computer numerical control (CNC).

Table 1-A. Numerical Control (NC) compared with Computer Numerical Control (C/NC)

NC	CNC
Program Storage and Program Operation	
Programs are stored on some external medium such as a punched tape or cards. The tape must be run each time the machine is cycled.	Programs may be stored on some medium such as floppy disk, punched tape or cassette tape but they are downloaded into the MCU memory and operated from there. The internal memory holds the program.
Memory	
An NC machine has *no memory*. The program is held on the storage medium.	Random access memory (RAM) is part of the controller. RAM allows easy editing of programs and also allows on-board programming. This is manual data insert (MDI).
Programming Ability	
An NC machine must be programmed from an external device. The program can not be created on the control. There are attachments that allow tape punching at the machine but they are not part of the control. Complicated shapes had to be described as a long set of coordinates for each movement of the tool.	CNC may be programmed at an external device similar to an NC but also from an external computer, or at the control. This program may then be uploaded to a computer disk, or put on punched tape. CNC allows flexibility in programming and editing. Complicated shapes can be programmed by describing the shape to the control. The control then generates the tool path.
Control Authority	
Has no authority. NC does interpret much data beyond a limited scope. NC is a play-back slave.	Has authority over operation. CNC controls can compensate for size, determine tool wear, inspect parts, communicate with external computers, robots, and the operator. Modern CNC machines can display the tool path on a screen (CRT) and detect errors in the program.

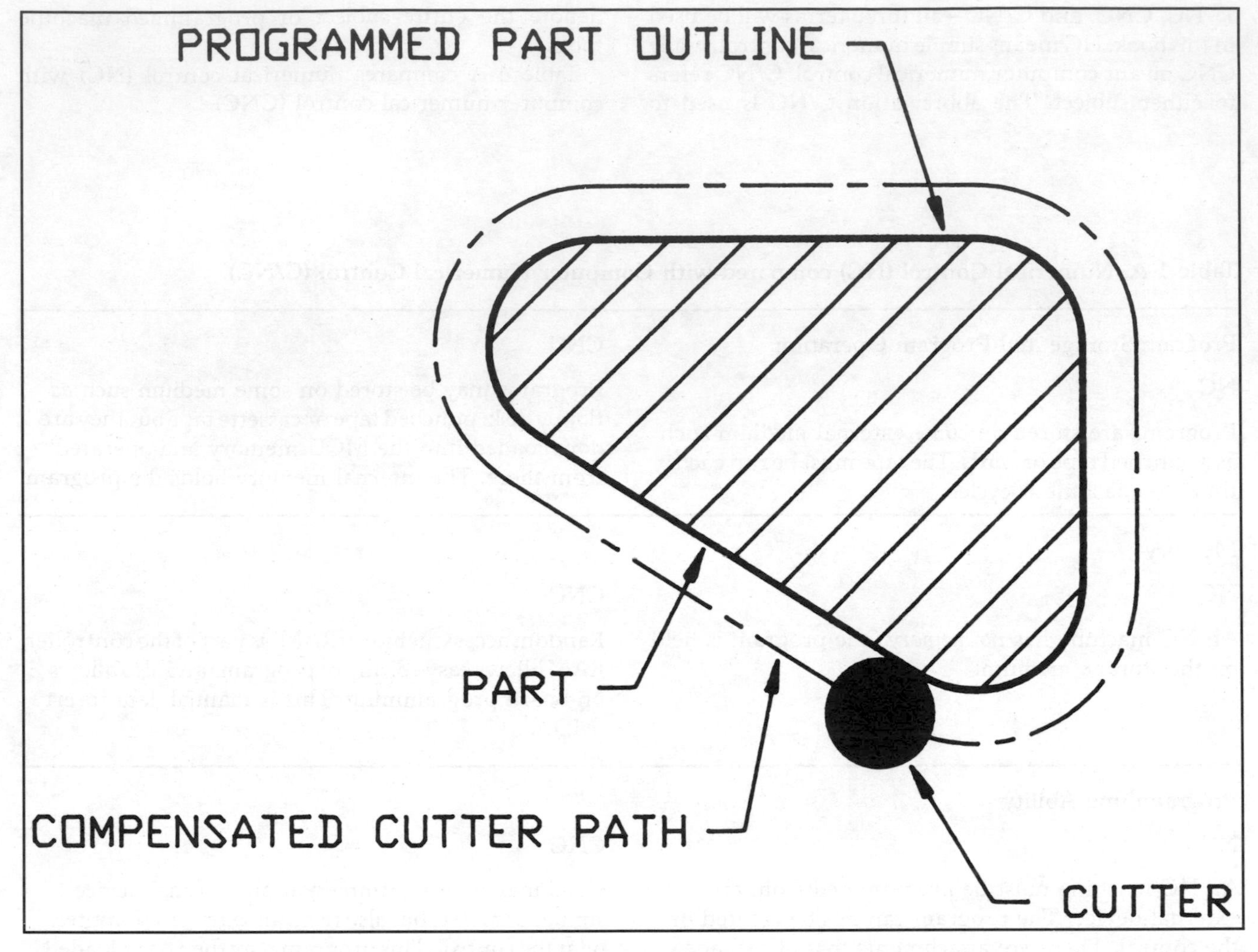

Fig. 1-9. Computer tool radius compensated cutter path. The input program is the actual part shape. The generated path is the cutter centerline to produce the part.

Tool Compensation

Tool compensation is one of the most useful aspects of CNC. We will look further into tool compensation in later units.

Inputting the Part Shape

To input the part shape, the programmer writes the part shape into the program. Then he requests the computer to adjust for the actual path the tool should follow to compensate for the size of the tool. On a milling machine, the control can allow for a new diameter or length of the cutter. On a lathe, the radius on the nose of the cutter may be changed. Yet the control can still recalculate the path the tool is to take to generate the original part. This eliminates calculations. Fig. 1-9.

If the tool becomes dull and must be replaced, it need not be the same size. Once the controller is informed of the changed size, it can compensate (adjust) the tool path for the new tool size or shape.

An NC machine program is locked to a cutter of a certain size or shape. If that cutter is replaced, it must be replaced with an exact duplicate. If it is not, the part will be bigger or smaller than intended due to the size of the new tool.

SOFTWARE AND FIRMWARE EVOLUTION

Of equal importance to controller evolution are the operating instructions and capabilities built into the control. *Firmware* suggests intelligence you cannot change, while *software* is changeable. Either term is used to indicate the intelligence level of the control. From previous discussions we have learned NC controls had little intelligence because they were a controller only and did not have the microprocessor as part of their system.

NC Intelligence

The NC intelligence was mostly in the program supplied by the programmer. An NC control had a buffer area for the next command. It could follow customizable pre-planned cycles, but that is the limit of NC intelligence. The program had to be specific for each move and operation. Perhaps the biggest single difference was in machining curved shapes. If a curved shape was to be made on an NC, each point for the path of the cutter had to be programmed. Many points were needed to make the curve smooth. Fig. 1-10. This led to long and difficult programs. Errors were common. To avoid these errors, offline programming languages were developed to generate the tool path data. The most common of these languages was *automatic programmed tooling language* (APT). Much math or an offline computer was necessary to produce an NC tape.

NC Curves

Curves were difficult to program in NC. Refer to the NC curve shown in Fig. 1-10. This shows how an NC control can be used to machine a complex shape. It is possible to approximate a circle (or a line of any shape) with a series of short straight lines. If the lines are close enough together, the curve will be very close to a true curve. An NC curve had to have each point in the program. This created long sets of point coordinates. In programming an NC curve, the programmer determines the amount of tolerable divergence from a true curve. Then the APT computer or the programmer calculates points that are close enough to give the correct results.

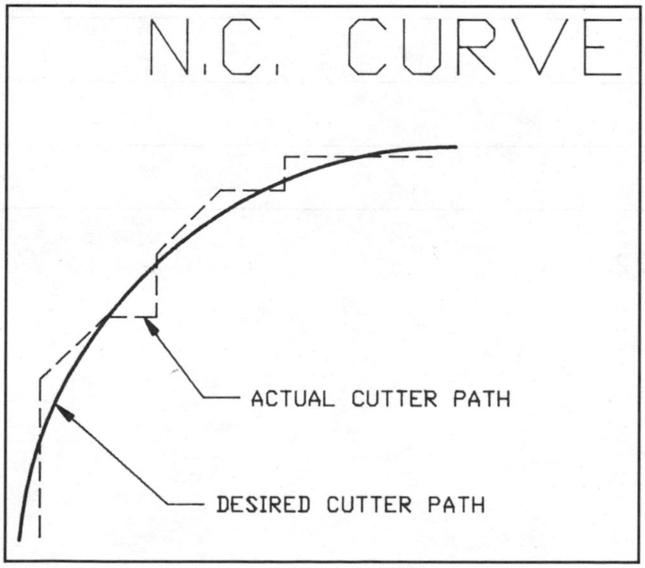

N.C. CURVE

ACTUAL CUTTER PATH

DESIRED CUTTER PATH

A CNC machine, however, is capable of producing a smooth, exact curve. The programmer inputs only the parameters of the curve. The *parameters* are the conditions of the curve's shape. Thus, the CNC needs the radius, length, center point, and start/stop points or number of degrees in the curve.

CNC Intelligence

On the CNC controller, it was natural to let the computer assume some of the work required of the programmer and the offline computer. CNC made several features available. The following ten features are found only on a CNC machine. These features are possible because of the microprocessor.

1. Random access memory.
2. Tool path compensation.
3. On-line programming and editing MDI.
4. Full geometry programming—true curves, not approximations. Program information describes the shape. The microprocessor generates the machine axis moves perfectly.
5. Tool compensation for size, shape, length, and wear.
6. Constant surface speed on lathe cuts. The rim speed is maintained no matter what diameter is being cut. On a face cut, as the tool moves inward, the spindle speeds up in direct proportion.
7. Part scaling. The computer is requested to generate a bigger or smaller version of the part program. This is often used in mold work where a shrink factor must be included in the mold.
8. Graphic representations of the program. Controls are capable of displaying the program on a screen before running it. This is a safety factor and eliminates errors before actual tryout.
9. Program error analysis. Many controllers are capable of finding certain program errors and alerting the user. The safe work envelope is designated to help avoid cutter-to-machine contact.
10. Advanced control authority. This allows the following in-process inspection; part loading is triggered by the control to a robot; tools are replaced by wear factors or a timed number of parts; tools and work fixtures are set up by probes in the machine, rather than by an operator using an indicator.

Because each of these operations requires a computer, each of the above is possible only with a CNC machine controller.

Fig. 1-10. To produce a curve on an NC machine, many short line segments had to be programmed. These short lines approximated the desired shape.

1. Can you list the four major differences between NC and CNC? Can you describe these differences?

2. Can you list the three areas of machine evolution?

3. Can you list and describe the ten features available only on C/NC?

4. The Unit 1 activity will be found at the end of Chapter 2.

CAREERS IN C/NC

As your skills grow, you will want to take advantage of the many careers available in C/NC. Is programming, setup work, or operating what you want to do? How will you plan? What skills are needed?

OBJECTIVES

After studying this chapter, you will be able to:

- Understand the career opportunities and job titles in computer numerical control.
- Identify the tasks that must be done to put a job into production.
- Describe the jobs in the C/NC process.

2-1 The C/NC Process from Idea to Finished Product

To understand the careers available in industry today, we will examine a C/NC job from start to finish. By looking at the whole process, the work that must be done, and the job titles for this work, you should understand the options available to you in C/NC.

Each of the following tasks must be done to manufacture a finished product. The organization of the tasks may differ from shop to shop. In small shops, a single person may do most of the jobs. In larger shops, some jobs may be done by a computer.

MACHINE SHOP C/NC PROCESS

The work needed to complete a C/NC job fall into three main phases: creating the idea, planning, and production. Fig. 2-1. Each one of these tasks requires distinct job skills.

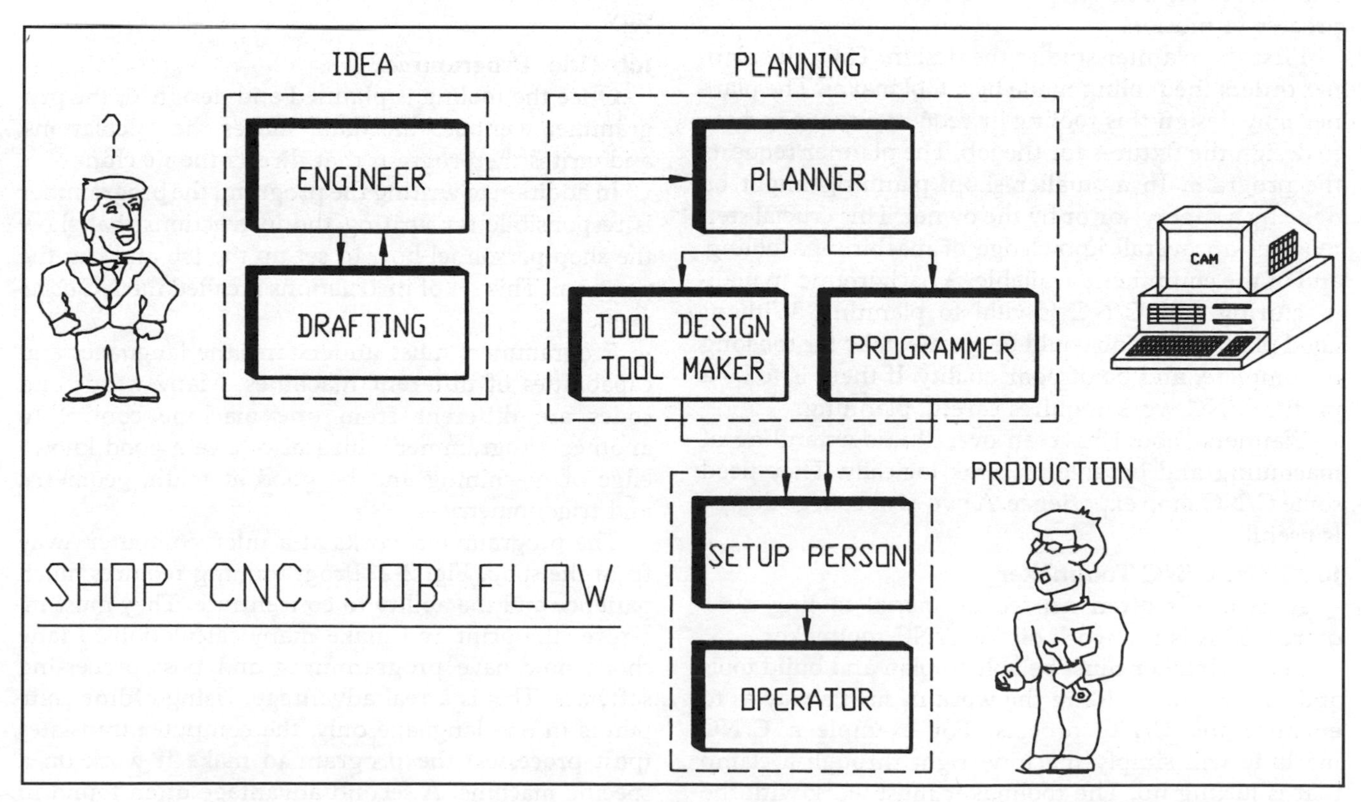

Fig. 2-1. The three phases of a C/NC job are the creation of the idea, planning and the production of the part.

Creating the Idea

Job Title: Mechanical Design Engineer

Most products require a design and a print before they can be programmed or made. The mechanical engineer is responsible for the design.

The engineer must know materials and mechanical design. He or she must be aware of the needs of the shop, customer, company, and programmer. The engineer must know the capabilities of the machines available.

After designing the product, the engineer may prepare the final drawing. He or she might also ask a *drafter* or *engineering aid* to do this. Or perhaps the engineer may do the final drawing also. The original plan could be drawn manually or on a CAD unit (computer-aided drafting). If the drawing is on CAD, it can be turned directly into a C/NC program by the computer. We will study this in Unit 7. Most of the drawings in this book were drawn on a CAD unit.

The mechanical engineer need not be a machinist, but he or she must understand machining. This job usually requires a four-year degree in engineering. Engineers must have a good background in mathematics, physics, and mechanical principles.

Planning the Job for the Shop

Job Title: Shop Planner

The shop planner determines which machines will be used in the work and how the work will be held and machined. The shop planner sets the whole shop process in motion.

First the planner studies the design. Then the planner orders the tooling made by a tool maker. The planner may design this tooling or request a tool engineer to design the fixtures for the job. The planner requests the program. In a smaller shop, planning might be done by a supervisor or by the owner. This crucial step requires an overall knowledge of machining, tooling, and of the equipment available. A background in manufacturing and C/NC is vital to planning. Without good planning, a job could be unsafe, take far too long to complete, and be of poor quality. If there is to be a profit, C/NC work requires careful planning.

Planners should have an overall understanding of machining and be able to think logically. They need some C/NC shop experience. A two-year college degree is useful.

Job Title: C/NC Toolmaker

A fixture is often needed or special tooling is required. This is the work of the C/NC toolmaker.

The toolmaker must be able to plan and build tools and fixtures for holding the work in such a way as to enhance the C/NC process. For example a C/NC machine will simply machine right through a clamp that is jutting up. The toolmaker must work with the planner and programmer to build holding fixtures that

Fig. 2-2. A programmer at an offline microcomputer. The program is written, stored, and edited on this unit away from the shop environment. The Boeing Company

work with the program. The toolmaker must understand C/NC and be highly skilled in machining. The tooling itself might be made on a C/NC machine in the tooling department.

A four- or five-year apprenticeship or a company training program is necessary to be a toolmaker. Related training in math and printmaking also are necessary.

Job Title: Programmer

Once the tooling is planned and designed, the programmer compiles the data, makes the calculations, and writes the program that directs the machine.

In addition to writing the program, the programmer is responsibile for writing the instructions that show the shop personnel how to set up the job and run the program. This set of instructions is called the *NC document.*

Programmers must understand the languages and capabilities of different machines. Many words and codes are different from one machine control to another. Programmers must also have a good knowledge of machining and be good at math, geometry, and trigonometry.

The programmer works at a microcomputer away from the shop. Fig. 2-2. Programming requires much patience and the ability to concentrate. They must interpret the print and make many calculations. Many shops now have programming and post processing software. This is a real advantage. Using offline computers in one language only, the computer translates (post processes) the program to make it work on a specific machine. A second advantage often found in this type of software allows the programmer to enter a description of the part's shape and features. The computer then makes the calculations to generate the actual program. This is called interactive graphic programming. It is also found on advanced C/NC controls in the shop.

Community college or vocational technical schools teach programming. Obtaining on-the-job training is also a good way to become a programmer. This is most useful after some experience in C/NC machine operation and setup.

Production—Making the Product

Job Title: Machine Setup Person

A machine setup person gets the job going in the shop and corrects errors in tooling, the program, or the fixtures.

With the tooling, raw material, and program ready, the setup person reads the print, the work order, and the NC document. Then, he or she sets up the fixture and the cutting tools and coordinates the machine. The setup person loads the program into memory and produces a first part. The setup person must also maximize production. *Maximizing* means to get everything going as fast as possible without lowering quality or safety standards or exceeding conservative use of tools. Another responsibility of the machine setup person is to train the operator.

A C/NC setup person must have a complete knowledge of machining and tooling. He or she must have a working understanding of programs. Depending on the shop policy, they may have to edit programs on the machine or communicate the needed editing to the programmer. A setup person must be a machinist or C/NC machine operator with extensive C/NC experience.

Job Title: Machine Operator

The machine operator assumes the responsibility for the machine process once it is running smoothly. Operators must load and unload parts, monitor the machine and the parts produced, make sure the coolant is flowing, and change tools as needed. The operator may have to burr or do other secondary operations while the machine is running. Setup and operation may be performed by the same person in some shops. If both responsibilities are assigned, this person is then called a *C/NC journeyman.*

Most students enter the work force as machine operators. It is a starting place where you will gain the knowledge and experience to move into the titles above. You should have a vocational education in machining. This can be obtained in a vocational high school, area occupational center, community college, or vocational technical school. You might also enter a company training program.

Changing Jobs in C/NC

Jobs in C/NC are changing. As industry moves further into CIM (computer integrated manufacturing), jobs and careers will change because the process will change. This will create new career titles, while others will be replaced. In Unit 7 we will look at the C/NC process in a CIM factory, where the work is the same but many of the jobs are done by a computer.

The secret of survival in a changing world is to adapt to the changes. In C/NC you may not be doing the same type of work in five years. If you understand the information presented in this book and continue to receive training, you will be able to adapt to the changes.

1. Can you outline the three main work areas required to produce a C/NC machined product?

2. Can you describe the jobs within these areas?

3. What is your goal for this class? Does it involve any of the following?

   ☐ To learn about C/NC in preparation for engineering training.

   ☐ To obtain vocational training for an entry-level job.

   ☐ To explore possible careers.

   ☐ To obtain information on a specific career.

   ☐ To obtain vocational training for job promotion.

   Your goals may overlap. For example, you may have checked more than one category. Do any of your goals require other education or on-the-job experiences? If they do, outline your plans to reach the above goals. Ask yourself the following questions:

   - What training will be needed?
   - What job experience is necessary?
   - Where and how will you get these?
   - How long will it take to obtain the necessary training?
   - What will the training cost?

____

____

____

____

____

____

Activities

To check your understanding of Chapters 1 and 2, complete the following:

1. The first NC machine was operated at MIT in 1952.

   Is this statement ☐ True or ☐ False?

   ____

   ____

2. List the three original partners that produced the first NC machine?

   ____

   ____

   ____

3. Programmed machine tools have been commercially available in industry for

   ____________________ years?

4. What two reasons promoted the development of an NC machine?

____

____

5. List the three major areas of evolution from *Standard Equipment* to *C/NC*.

____

____

____

6. Define machine tool *repeatability*.

____

____

7. What is a common tolerance for repeatability on modern C/NC machines?

____

8. Define *resolution*.

____

____

9. Circle the letter of the list that has all the components of a complete C/NC drive system?

   A) Drive motor, ball screw, feedback device, tool changer.

   B) Controller, drive motor, ball screw, feedback device.

   C) Feedback device, resolution, tool changer, ball screw.

   D) Repeatability, drive motor, backlash, controller.

10. A feedback device sends a signal from the controller to the drive motor, to determine the position of the motor and the slide it is driving.

    Is this statement ☐ True or ☐ False?

    If it is false, what will make it true?

____

11. Servo and stepper motors differ from standard motors in that they are highly

    ____________________.

12. Microprocessors in the MCU are common in CNC but rarely found in NC machine controllers.

    Is this statement ☐ True or ☐ False?

    If it is false, what will make it true?

____

____

13. An NC machine is capable of limited authority over the program, but has no flexibility.

Is this statement ☐ True or ☐ False?

If it is false, what will make it true?

____

____

14. C/NC machines can use punched tape.

Is this statement ☐ True or ☐ False?

If it is false, what will make it true?

____

____

15. Circle only the features found on *both* NC and CNC machines.

A) Ballscrews
B) Constant surface speed
C) Accurate resolution
D) On-line editing of programs
E) Graphic part representation
F) Feedback devices
G) Full geometry programming
H) Tool compensation
I) Random access memory
J) Part scaling
K) Tape reading ability
L) Program error analysis
M) Multiple machine slides
N) Stronger bearings

16. For each of the dates, list an important event in machine tool evolution.

1945—Event: ______________________________

1949—Event: ______________________________

1952—Event: ______________________________

1960—Event: ______________________________

17. Three *planning* jobs must be completed to put a C/NC job into production. List the job titles.

____

____

____

18. List three tasks that must be completed by a C/NC setup person or lead person.

____

____

____

UNIT 2

Machine Movements

In this unit, you will learn the universal system of machine axes and points and coordinates. These subjects are the foundation of programming and operating C/NC machinery.

Machine setup, programming, and operation of C/NC equipment require a solid understanding of axes and coordinates. You will write simple programs at the end of Unit 2. The major body of data in C/NC programs is axes and point coordinates.

CONTENTS

CHAPTER 3

THE AXIS SYSTEM

Axes is the plural of *axis*. They both mean a reference for motion and position. An axis is a central line to which all C/NC movement is compared or referenced. The axis system is a worldwide standard for machine movement.

OBJECTIVES

After studying this chapter, you will be able to:

- Identify standard and auxiliary axes or lathes and mills using the "right hand rule."
- Diagram axis frames using the EIA axis identification characteristics.
- Apply the "rule of thumb" to determine rotary axis direction.

3-1 What Is a Machine Axis?

MOTION AND DIRECTION

Axes are used to identify motion. All C/NC machines require some method of identifying which motion is needed in the program. For example, on a vertical milling machine, if the programmer requires motion left or right, the X axis is called upon, while in and out motion of the table would require the Y axis.

Direction is determined by sign value. Once the axis is identified, the sign of the axis, plus or minus, determines the direction of its movement: left or right, in or out, up or down, clockwise or counterclockwise.

Refer to the milling machine in Fig. 3-1. The X axis of this milling machine table is the left and right movement. If the programmer needs a tool movement to the right, he or she would call out a "plus X movement." A movement to the left would be "minus X." Movement of the tool inward (away from the operator) is plus Y, and tool-up would be plus Z.

Refer to the C/NC lathe shown in Fig. 3-2. A C/NC lathe has two common axes. They are the X and Z axis. Plus X movement of the tool would be the cross slide moving away from the work. A plus Z movement would also be away from the work, with the saddle moving. Note that the Z axis is perpendicular to the X axis.

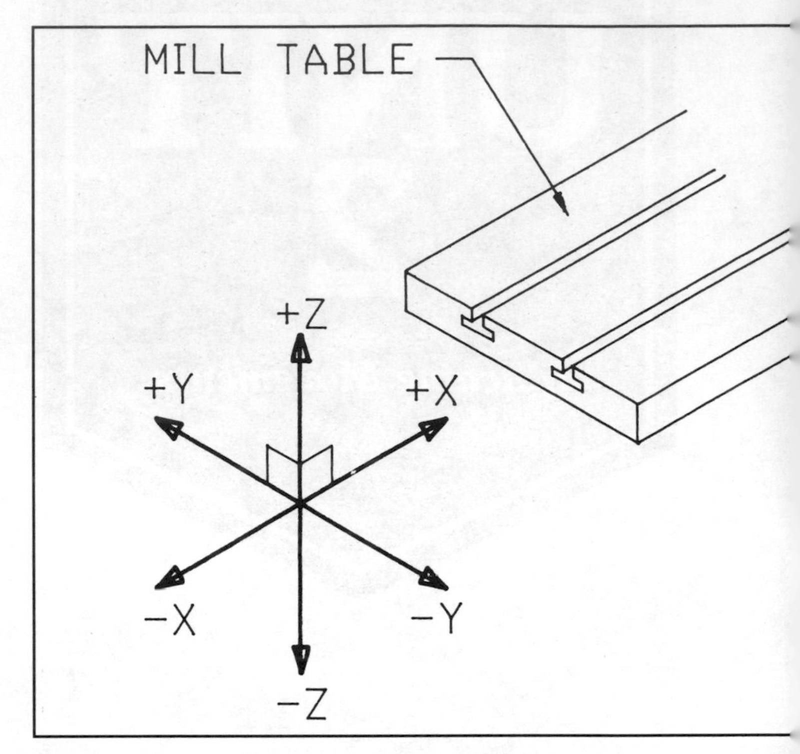

Fig. 3-1. The three axes of a vertical milling machine. The X axis is usually parallel to the floor on most C/NC machines.

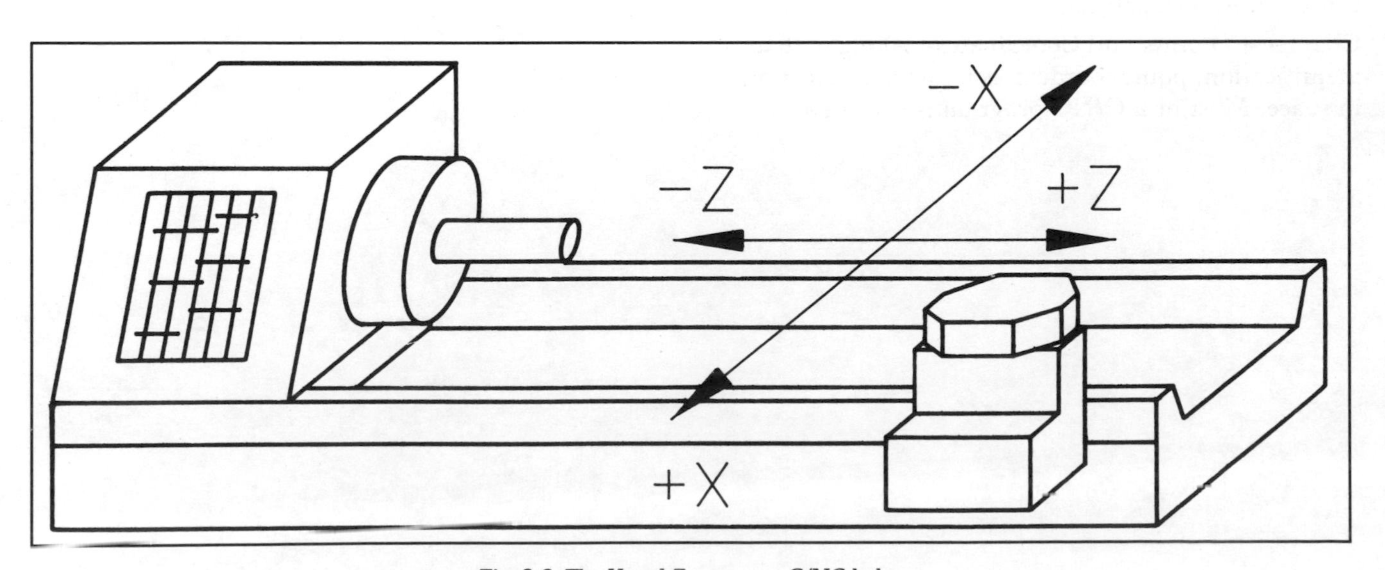

Fig. 3-2. The X and Z axes on a C/NC lathe.

RELATIVE TOOL MOVEMENT

Always think of the tool as moving. When programming or setting up a C/NC machine, view the axis movement as though the tool is moving. This is called *relative tool movement.*

As an example, refer to Fig. 3-3. After drilling hole A on this NC drilling machine, we wish to drill hole B 2.5″ farther to the right on the part.

The programmer would call out a plus X movement of 2.5″. Would the tool actually move 2.5″ to the right? Actually, the table moves to the left. Thus, the tool position *relative to the part,* is shifted to the right. We ignore the table movement and think about the tool as moving. From this point on we will always envision the tool as the moving object. This is a standard in C/NC programming. If the tool moves or the machine slide moves, always think of the tool as the moving object. By following this rule, programming is simplified. Also, there is no confusion as to the movement that will be produced.

For another example, refer to Fig. 3-1. Using the concept of relative tool movement, we see that the axis directions on a vertical mill are:

Tool to the right	= plus X
Tool left	= minus X
Tool away from front	= plus Y
Tool toward	= minus Y
Tool up from table	= plus Z
Tool down	= minus Z

Refer to Fig. 3-2.

Crossslide tool toward work	= minus X
Crossslide away from work	= plus X
Saddle away from chuck	= plus Z
Saddle toward chuck	= minus Z

3-2 The Axis System

THE AXIS FRAMEWORK

Note that the three axes on the mill (Fig. 3-1) and the two on the lathe (Fig. 3-2) are at 90° to each other. This is called an *orthogonal axis frame. Orthogonal* means "at 90°." The standard axes on most C/NC machines are orthogonal. Fig. 3-4.

RIGHT-HAND RULE

You can identify the axis framework on most C/NC machines by the right-hand rule illustrated in Fig. 3-5.

If the thumb of your right hand points along the positive X axis, the first finger will point out the positive Y axis and the second finger will identify the positive Z axis.

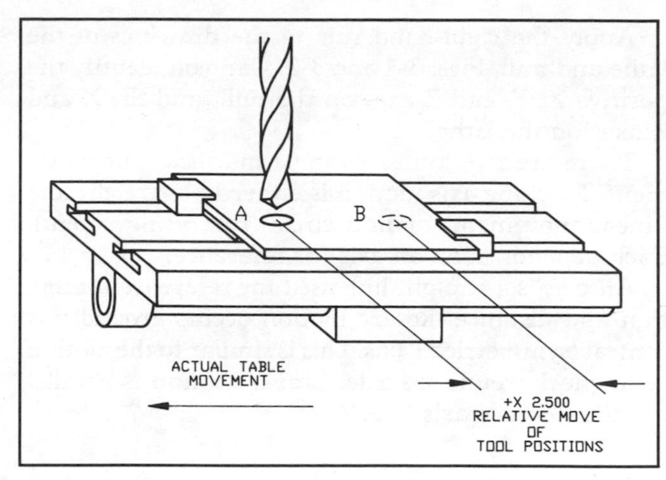

Fig. 3-3. Relative tool movement means that we always think of the tool as the moving object. Hole B is to the right if you think of the tool as the moving object.

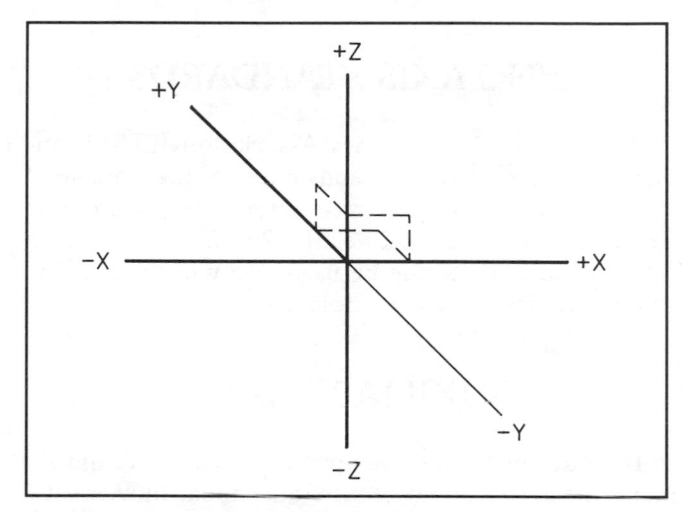

Fig. 3-4. An orthogonal axis framework.

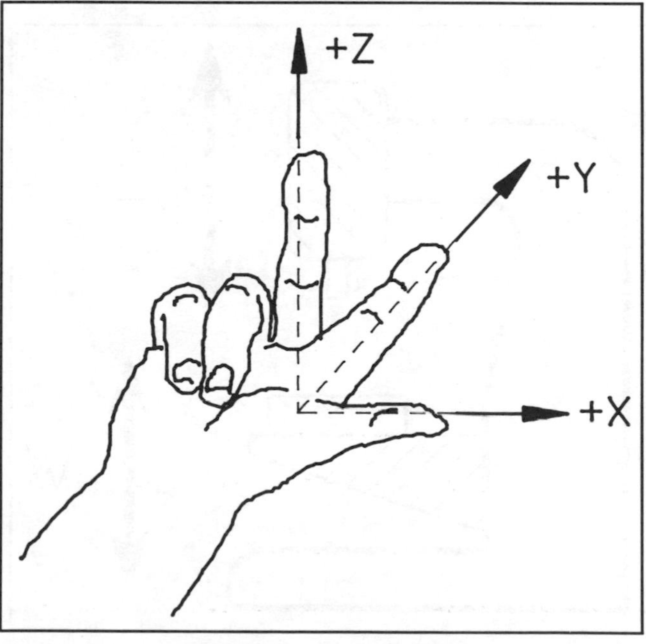

Fig. 3-5. The right-hand rule shows you the relationship of the primary linear axes of C/NC machines.

Apply the right-hand rule to the drawings of the lathe and mill. Figs. 3-1 and 3-2. Can you identify the positive X, Y, and Z axes on the mill, and the X and Z axes on the lathe?

There are two kinds of single machine axis movement. Machine axis motion is either in a straight line (linear movement) or in a circle (rotary movement). Each depends upon an axis for reference.

An *axis* is a straight line used for reference for motion and distance. Rotary motion occurs around this central (symmetrical) axis. This is similar to the motion of a wheel around an axle. Linear motion is parallel to the reference axis line.

3-3 EIA Axis Identification

C/NC AXIS STANDARDS

The Electronic Industries Association (EIA) is an organization that recommends many of the standards in C/NC work. They have recommended standards for C/NC axes of motion. The EIA-267-B lists fourteen different axes that may be used. We will look at the characteristics of nine of these axes.

AUXILIARY AXES

In addition to the three primary axes, X,Y, and Z, many machines might require additional movements or functions. A lathe may have more than one cutting tool slide or a programmable spindle. A mill may have a rotating toolhead, a rotary table, or optional slides. These are called auxiliary axes or secondary axes. The EIA standards show us how to identify standard and auxiliary axes.

Primary Linear Axes X, Y, Z

In naming the axes on any machine, first identify the Z axis. The Z axis is parallel to the main spindle. It is the axis that brings the main spindle to the work such as a milling machine quill. On a lathe, the Z axis brings the tool to the work.

Remember, whatever the actual movement is, we think of the tool as the object that travels.

X axis. The X axis is the longest travel at 90° to the Z axis. The X axis is usually horizontal.

Y axis. The Y axis is mutually perpendicular to both Z and X.

Secondary Linear Axes U, V, W

U, V, and W are parallel to the primary axes X, Y, and Z. U is parallel to X, V to Y, and W to Z.

Special note on U and W axes. On some lathe controllers, the U and W axes are also used to call out independent jumps called incremental moves. We will study incremental movement in Chapter 4. A "U" movement is an incremental movement in the X axis direction, and a "W" movement is in the Z axis.

Secondary mill axis example. Refer to Fig. 3-6. A vertical mill might have a moving quill and a programmable knee. These two slides are parallel. The quill then would be the Z axis, while the knee movement would be called out as a W axis. These axes do not perform the same function, yet they are parallel.

Secondary lathe axis example. Refer to Fig. 3-7. The saddle on a standard C/NC lathe is called the Z axis. What would a programmable tailstock be called?

This movement would be parallel to the Z axis but be a secondary slide. It would be denoted as the W axis.

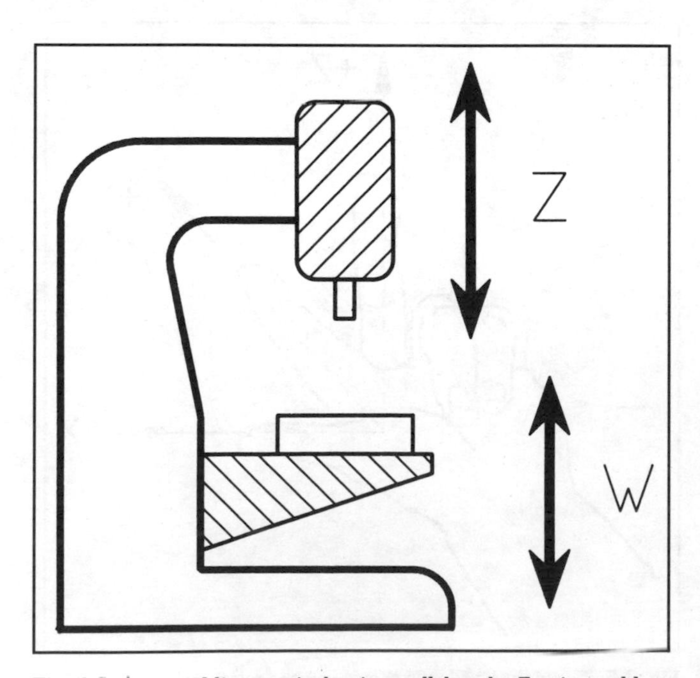

Fig. 3-6. A second linear axis that is parallel to the Z axis would be the W axis.

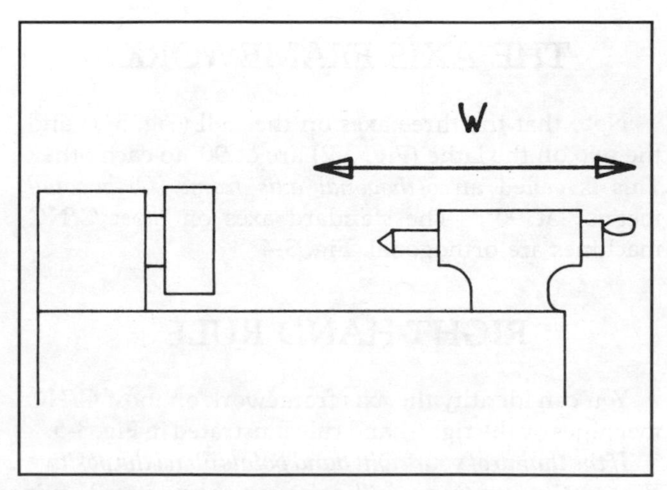

Fig. 3-7. Since the tool saddle would be the Z axis on a C/NC lathe, the tailstock spindle is called the W axis.

Primary Rotary Axes A, B, C

The three rotational axes rotate around the primary linear axes. The A axis rotates around a line parallel to the X axis, B around Y, and C around Z. Fig.3-8.

Lathe rotary axis example. Refer to Fig. 3-9. Many C/NC lathes have a programmable spindle. This is a rotary axis. Around which linear axis does the spindle rotate? In other words, around the centerline of which axis does the spindle rotate? And, the centerline of the spindle is parallel to which axis? The answer is the Z axis. This means that a programmable lathe spindle would be the C axis.

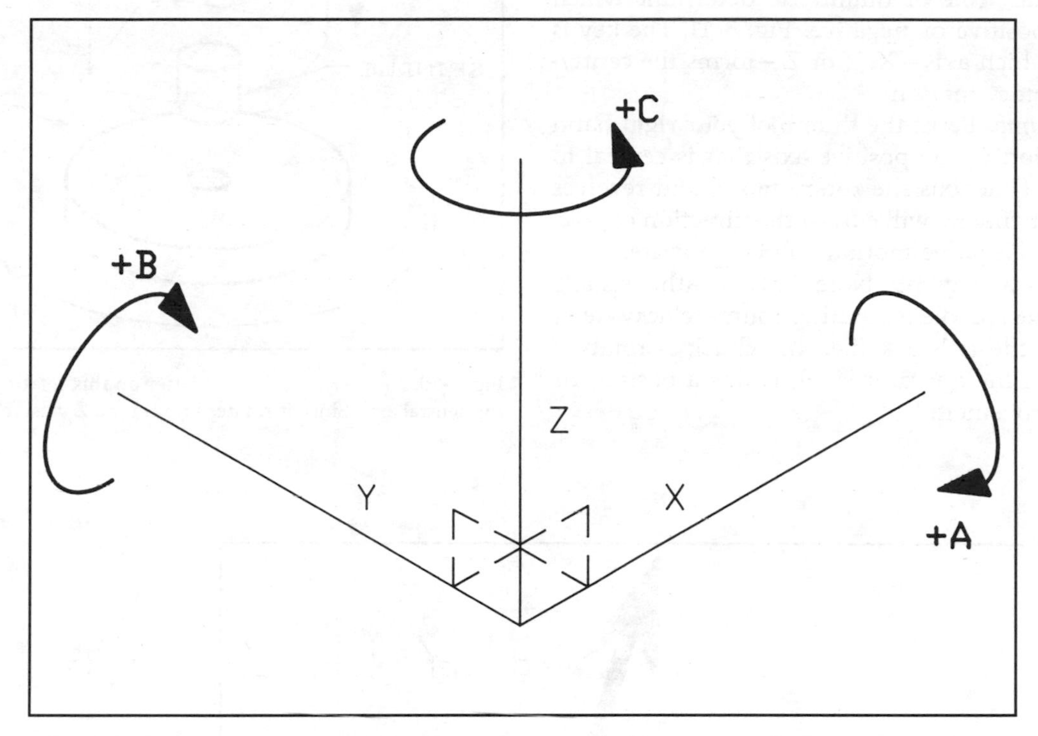

Fig. 3-8. The A axis rotates around the X axis, the B around the Y axis and the C around the Z axis.

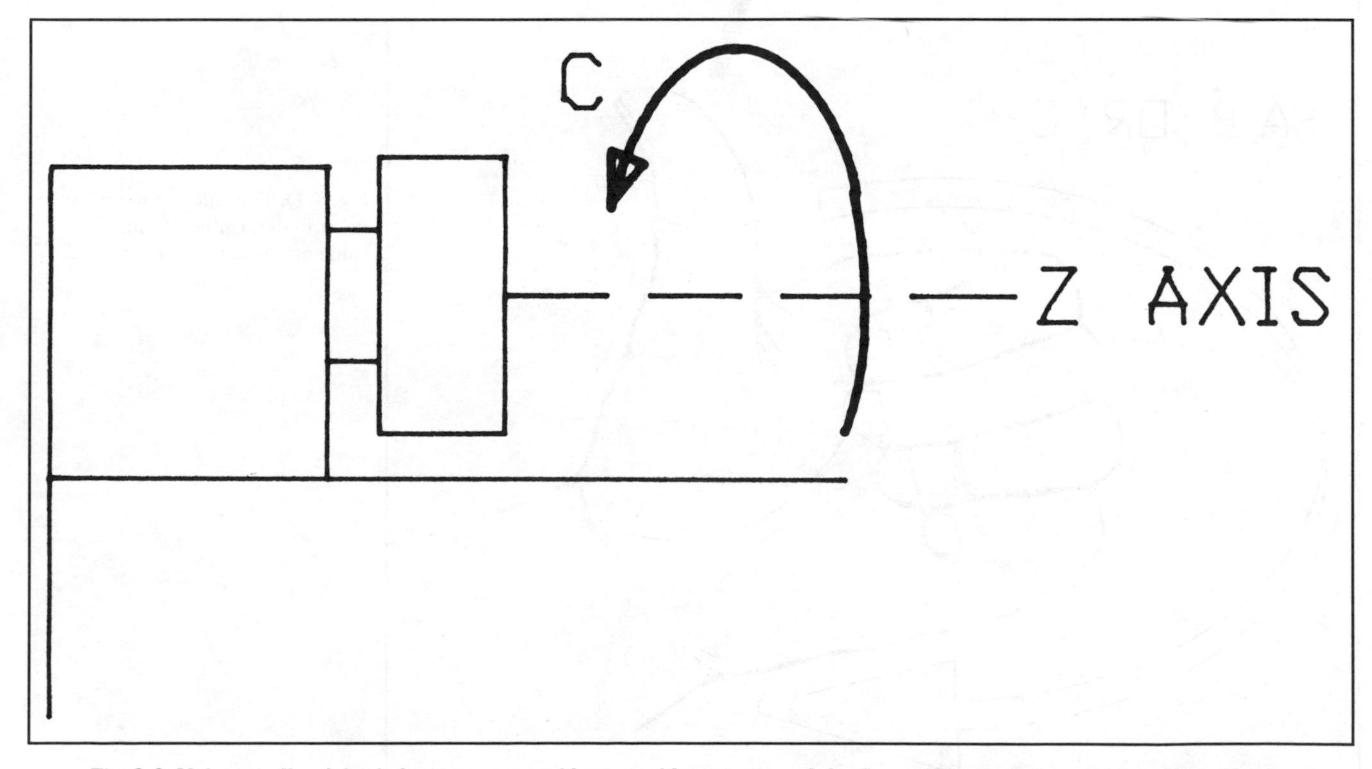

Fig. 3-9. If the spindle of this lathe is programmable, it would rotate around the Z axis. Therefore, it would be called a C axis.

Mill rotary axis example. Refer to Fig. 3-10. Suppose a milling machine has a horizontal rotating table that is programmable. What is the letter callout for this axis? The rotating table is called a C axis because it rotates around the Z axis of the mill.

Direction of rotary axes. When dealing with a rotary axis, apply the "rule of thumb" to determine which direction is positive or negative. Fig. 3-11. The key is identifying which axis—X, Y or Z—forms the centerline of the rotary motion.

Rule of thumb. Point the thumb of your right hand in the direction of the positive axis that is central to the rotation. (The axis the rotary movement revolves around.) Your fingers will curl in the direction of positive rotation. Negative motion will be opposite.

Look again at Fig. 3-9. Note that the lathe has a C axis. When the spindle is rotating counterclockwise as viewed from the tailstock (i.e., the chuck is rotating downward on the operator side), is this a positive or negative C movement?

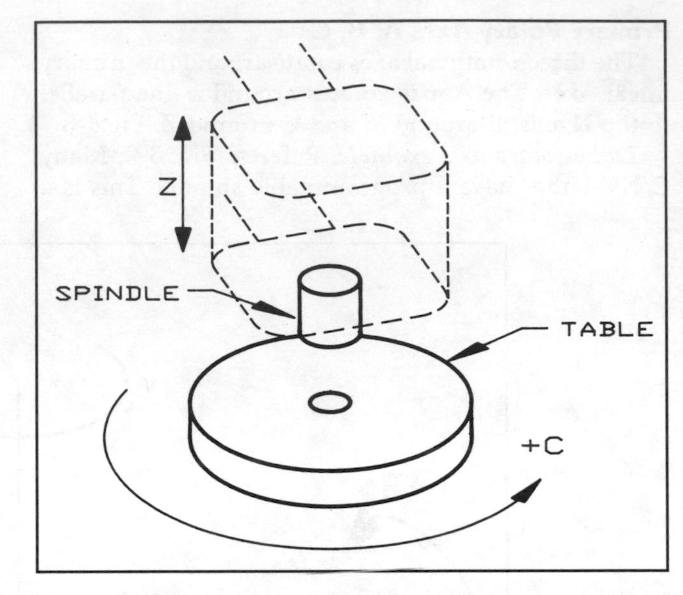

Fig. 3-10. To determine the axis letter on this rotating table, find the central axis. Since it rotates around the Z axis, it is a C axis.

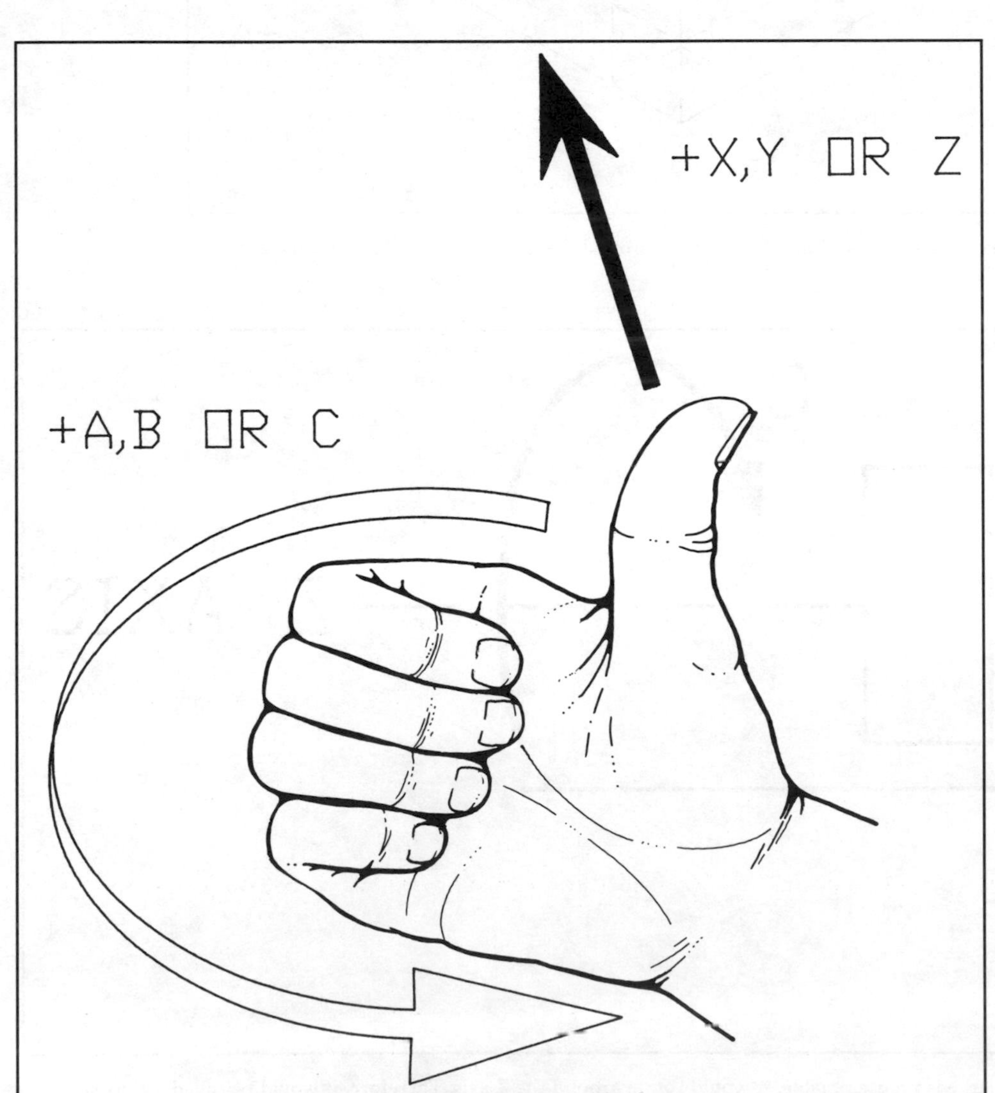

Fig. 3-11. The rule of thumb helps to determine the sign value of a rotary move in C/NC.

You must understand the following material to be a competent C/NC machinist:

- Axis reference for motion.
- Direction of motion.
- Primary and secondary axes.
- Rotary axes and the "rule of thumb"
- The "right hand rule" for axis identification.
- The EIA axis designations.

1. Refer to Fig. 3-12. Fill in the axes and their direction sign (plus or minus). Use the "right hand rule." Remember, the arrows in the frame indicate relative tool movement. The actual machine movement may be opposite.

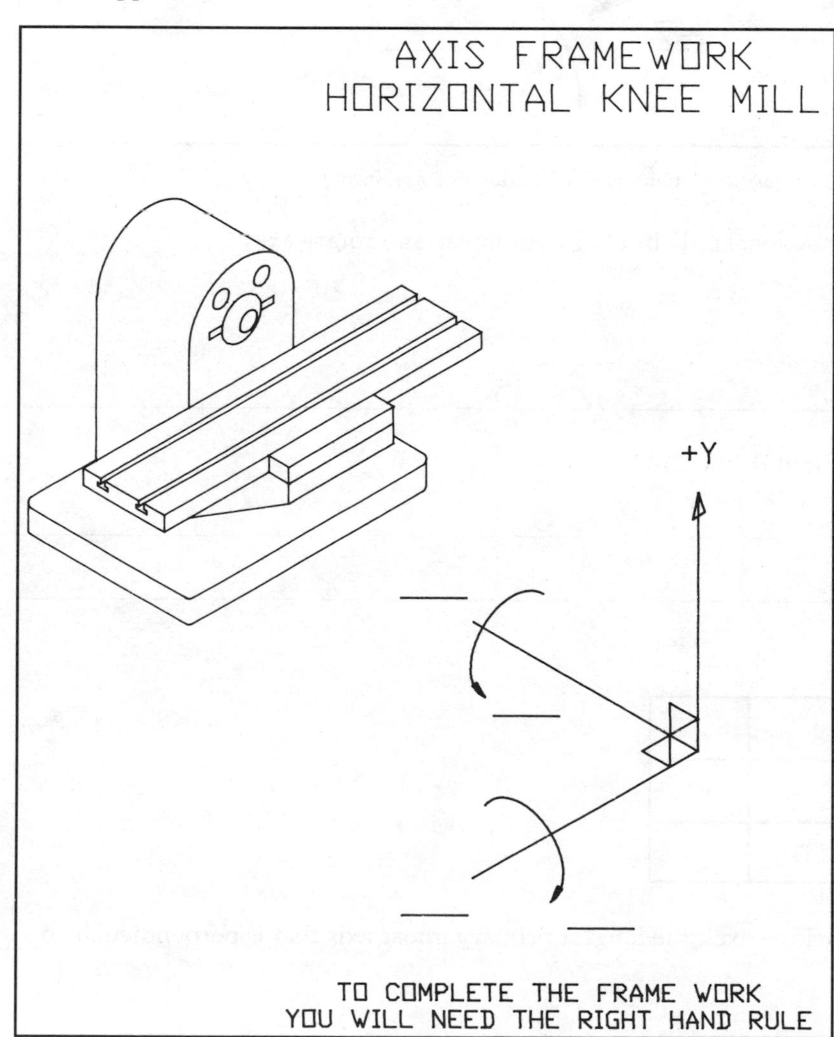

Fig. 3-12. Fill in the axis letters and direction signs. Use the Right Hand Rule.

2. According to EIA standards, which axis rotates around the Y axis?

____

3. Refer to Fig. 3-13. Identify the rotary axis letter. Indicate the sign of the arrow head direction. Put a plus or minus by the axis name. Use the "rule of thumb."

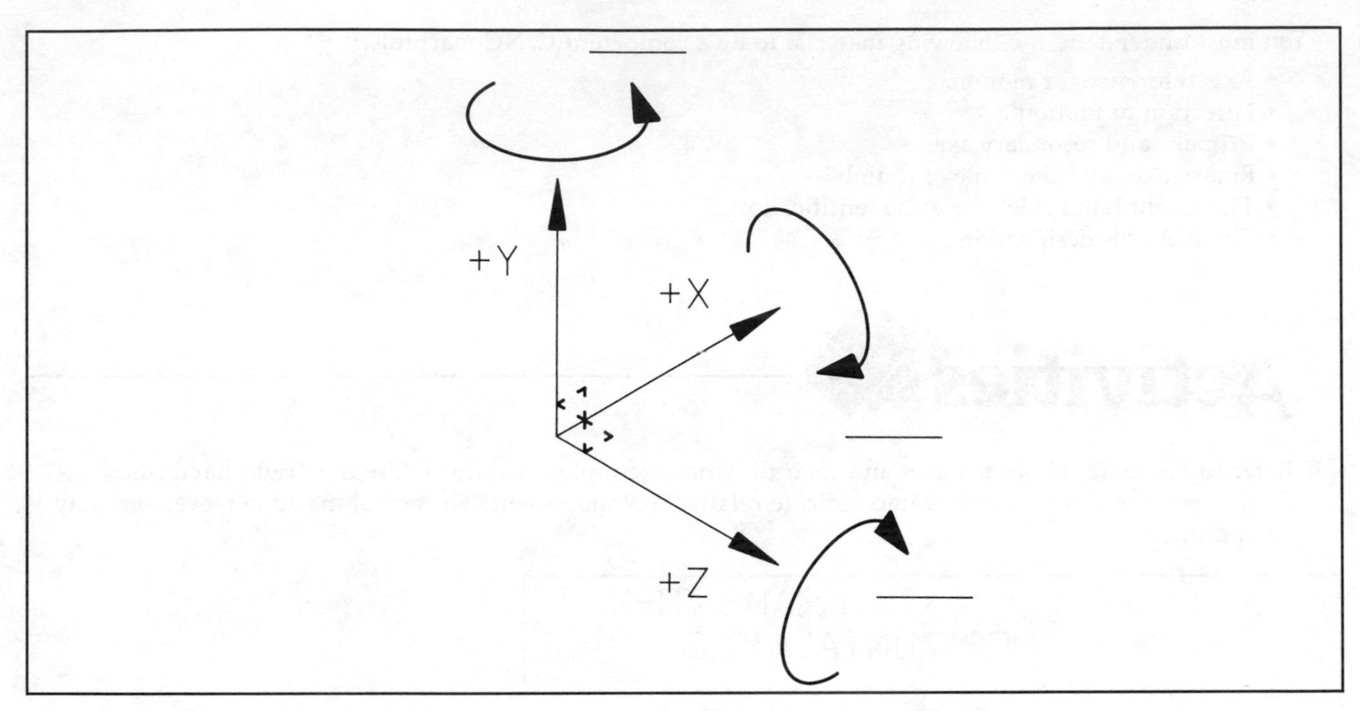

Fig. 3-13. Identify the rotary axis letter and the direction sign as shown.

4. The axis frame on most machines is orthogonal only if it has both linear and rotary axes.

   Is this statement ☐ True or ☐ False?

   If it is false, what will make it true?

   ____

5. What are the *secondary* linear axes as listed by the EIA?

   ____

   ____

6. Fill in the following chart.

Primary linear axes	X		
Secondary linear axes	U		W
Primary rotary axes	A	B	

7. According to the EIA standards for machine axes, the longest primary linear axis that is perpendicular to the Z axis is the

   ______________________ axis.

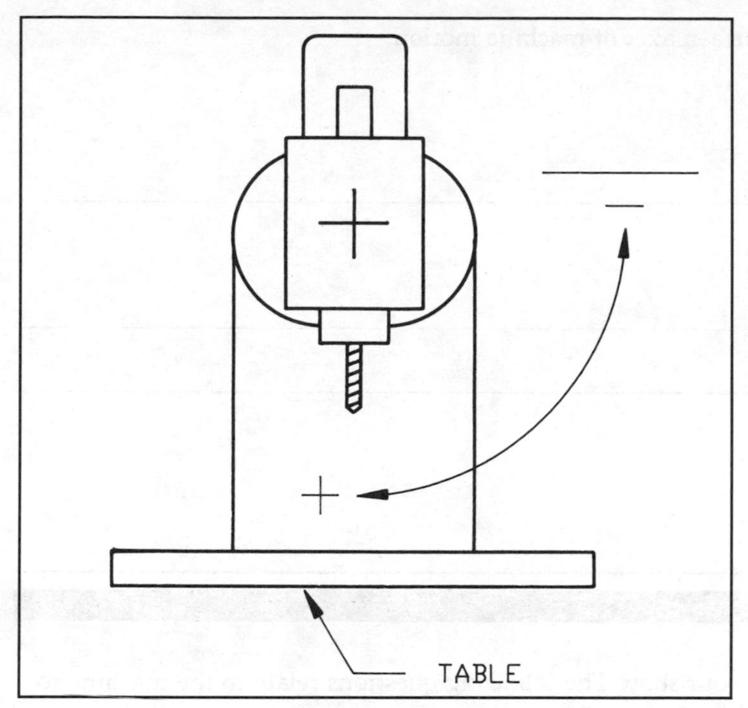

Fig. 3-14. Indicate the sign of the rotary motion shown.

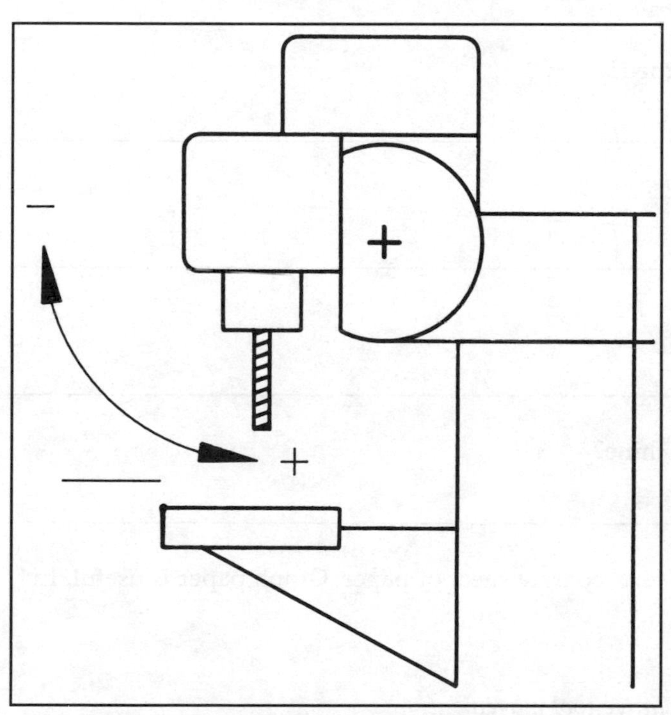

Fig. 3-15. Identify the rotary motion shown and the direction sign.

8. On some vertical milling machines, the spindle head can rotate in two directions. If these motions are programmable, they would be rotary axes.

   A) Refer to Fig. 3-14. What is the rotary axis callout for the motion that moves the head left and right as viewed from the front? Indicate the sign of the motion shown.

   ____

   B) Refer to Fig. 3-15. What is the axis callout and the sign of the motion shown that rotates the head forward and back as viewed from the side?

   ____

9. According to the EIA standards, there are fourteen axes of machine motion.

   Is this statement ☐ True or ☐ False?

   If it is false, what will make it true?

   ____

10. Define *axis* as used in C/NC work.

   ____

   ____

In Your Lab

Your instructor will assign you to a machine in your shop. The following questions relate to the machine to which you have been assigned.

11. What is the longest linear axis on the machine assigned?

   ____

12. How many axes are programmable on the machine?

   ____

13. What is the shortest travel axis on the machine?

   ____

14. Are there any programmable rotary axes on the machine?

   ____

15. Draw a sketch of the axis frame on this machine. Use a separate sheet of paper. Graph paper is useful. In making your sketch:

   A) Indicate 90° relationships.

   B) Label the positive and negative directions for *relative tool movement.*

   C) Label all rotary axes.

POINTS AND COORDINATES

Programs use coordinates to identify specific locations on the part or tooling. A coordinate is a combination of a letter and numbers. The letter is the axis being used for reference. The numbers are the distance. The major part of any C/NC program is simply sets of coordinates that identify points.

OBJECTIVES

After studying this chapter you will be able to:

- Describe the characteristics and difference between position and reference points.
- Select coordinate points using absolute Cartesian values.
- Select coordinate points using incremental Cartesian values.
- Locate and select coordinate values for reference points.

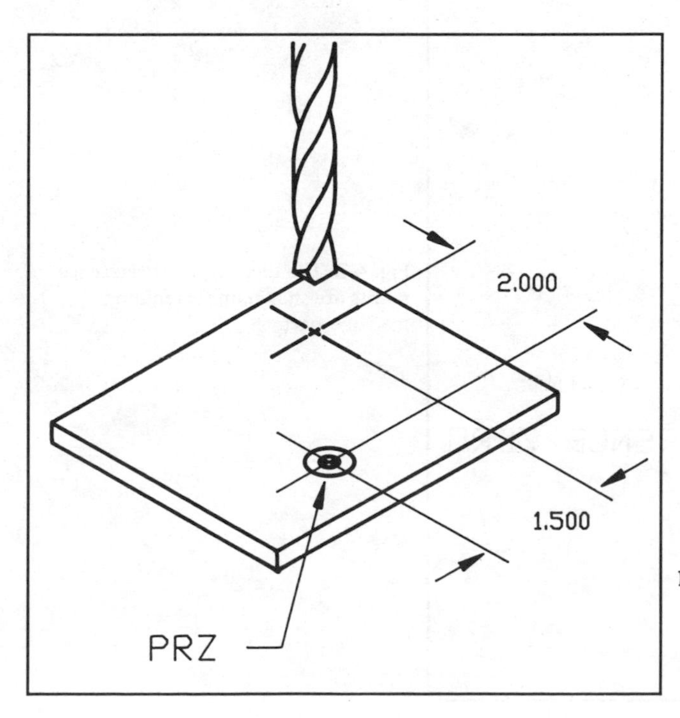

4-1 Coordinates and Significant Points

Coordinates identify significant points. The largest part of any program is sets of *coordinates* that identify single locations on the part, the holding fixture, or tooling. These locations, called *significant points*, are often referred to as simply "points."

Coordinates identify unique locations in space. Coordinates are sets of letters and numbers that identify how far along each axis the significant point lies. If the coordinates are properly used, the point identified cannot be confused with any other point. For example, X1.5, Y2.0.

Each coordinate identifies first the reference axis and then the distance from the reference point. The pair of coordinates above identifies a point in the X-Y plane.

Example. Refer to Fig. 4-1. This is a set of coordinates (X1.5, Y2.0) for a two-axis machine. The coordinates identify a point that lies 1.5″ to the right of the program reference zero (PRZ) point in X distance, 2″ on the plus side of the Y axis PRZ. The program reference zero point is the start point for coordinates, the origination of coordinate values in the program. The coordinates for the PRZ are X0.0, Y0.0, Z0.0. Values that refer to the PRZ for their numeric values are called *absolute.*

If a coordinate is given for X, Y, and Z on a mill or X and Z on a lathe, the point is locked in. It is unique. The controller can exactly identify the point the programmer has selected.

There are two types of significant points. Points in a program may be used as positioning points or reference points.

Positioning points identify a location to which the cutting tool is moved. This may be a cutting pass or a rapid move to get to the location such as positioning over a hole to be drilled.

Reference points identify a point to be used for reference. This point is used to calculate other points. The cutter may or may not actually be moved to this location. The PRZ is the prime reference point.

In machining a circle, a reference point is needed at the center of an arc. The controller must know where the center lies. The center is the point the arc pivots around. The tool would not be moved to the reference point, but the controller would use it to calculate the circular move.

Fig. 4-1. A pair of coordinates identify a position in a flat plane.

4-2 Reference Point Identification

Reference points are used in three ways. Fig. 4-2. They are used as *program reference zero, local reference,* and *machine reference zero.*

We will look at the characteristics and uses of each type of reference point in turn:

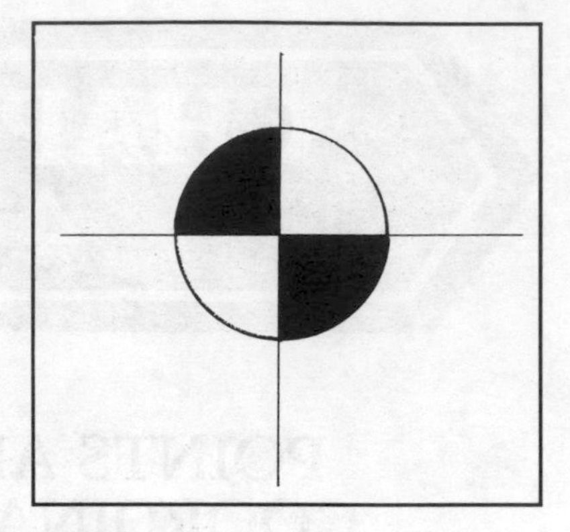

Fig. 4-3. This is the symbol for the program reference zero point PRZ.

PROGRAM REFERENCE ZERO

Program reference zero (PRZ) is a position chosen by the programmer. Fig. 4-3. PRZ is the mathematical center of the grid for coordinates. Because PRZ is used to identify all other points in the program, it must be chosen first, before any programming can be started.

The PRZ may be located anywhere within the machine travel on or off the actual part. However, it must lie within the physical limits of the machine. This is called a *full floating reference point.* Older NC machines have a fixed location for PRZ. Thus, the part and tooling had to be coordinated to the point. This was slow and inconvenient. Normally, there is only one PRZ per program.

There is a situation in which there might be multiple PRZs. This could occur when more than one part of exactly the same shape and size is to be run. For example, there might be two vises on a mill table with the same part. The PRZ would be shifted a given distance in the program and the next part in the next vise would be run. The first would be reloaded while the second was running. Two PRZs would allow a single program control both vises with a shifted reference point.

The program reference point (PRZ) is also called the program reference, program zero, set point, program home, zero point, part zero, datum point, and zero. Each of these items refers to the reference point from which all other coordinates originate in the program. The actual PRZ location is optional and is chosen by the programmer.

Later, we will discuss the proper placement and setup of the PRZ. The shape of the part and the kind of machining to be done are two major factors in selecting the PRZ location. The PRZ must be carefully chosen by the programmer and accurately coordinated by the setup person. An incorrectly chosen or coordinated PRZ will cause scrap parts.

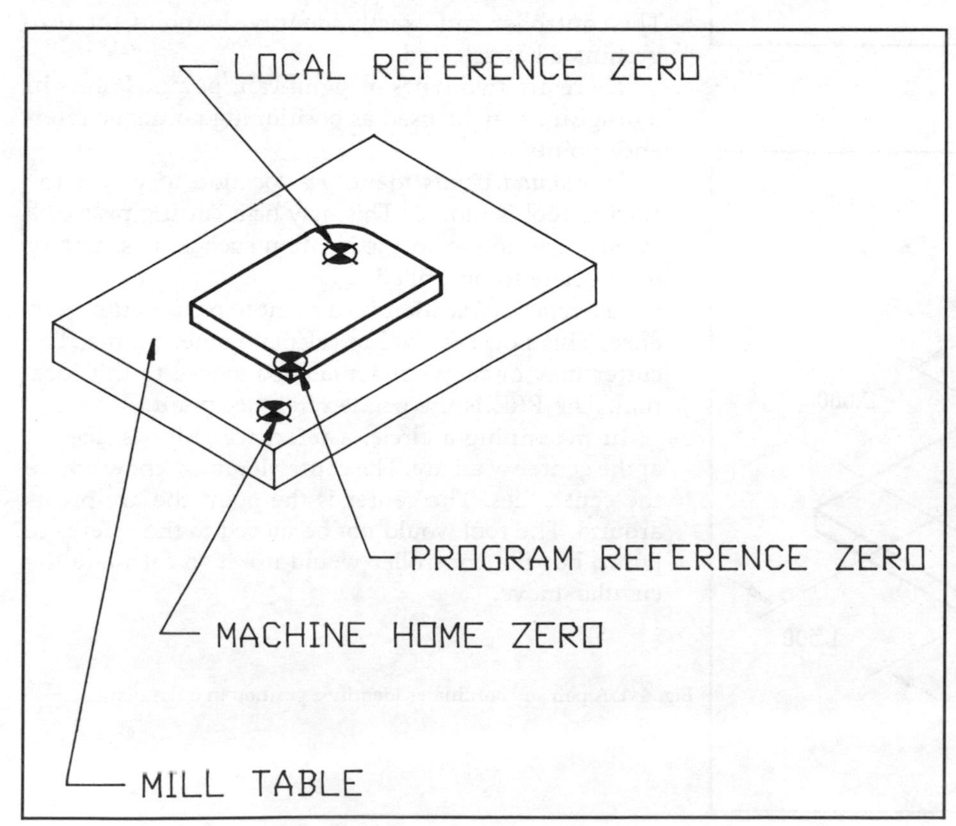

Fig. 4-2. The three types of reference points are shown on this milling machine part.

For a PRZ example, refer to Fig. 4-4. On this simple lathe job, the programmer has placed the PRZ on the centerline for the X axis and just touching the stock in the Z axis. The setup person must ensure that the tool is moved to this position and that the axis registers are reset at X0.0, Z0.0. This will establish the physical PRZ on the machine. Fig. 4-5.

SETTING UP A PRZ—COORDINATING

Refer to Fig. 4-5. When setting up a machine for a part run, the setup person must coordinate the physical machine to the program location for PRZ. To establish the PRZ, the setup person:

1. Reads the instructions written by the programmer to determine the correct location for the PRZ.
2. Manually moves the tool or machine exactly to this position on the part or tooling as indicated in the instruction.
3. Zeros the machine axis registers and resets all axis values at zero. This tells the controller that it is now parked on the PRZ. If commanded to move to X0, Y0, Z0, the machine would always return to that same exact point—the established PRZ.

This process is called *coordinating* the machine. In coordinating, the tool must physically be at the PRZ. It must be set exactly where the program requires before starting the program run.

TOOLS USED FOR COORDINATING

Dial indicators, center finders, edge finders, feeler gages, and electronic probes are commonly used in coordinating. Also, the tool itself may be used to "touch off" the work by slowly moving the cutting tool until it just touches the work. Once contact is made, the operator may then move the tool the correct amount away from the work or fixture and then reset the axis registers to zero. If you are to be a C/NC operator or setup person, much of your job responsibility will be in coordinating the setup.

LOCAL REFERENCE ZERO (LRZ)

A *local reference zero* (LRZ), or local reference point, is a position used temporarily for reference, much the same as PRZ. Well-chosen local reference points can simplify the math required to machine complicated parts. In the previous example, the center point of a circle was identified as a local reference point. It will be used until the circle is completed. Then the reference point will be cancelled. Other LRZ points might be used as a reference for several coordinates before cancelling. They are used as a handy math tool any time the programmer chooses.

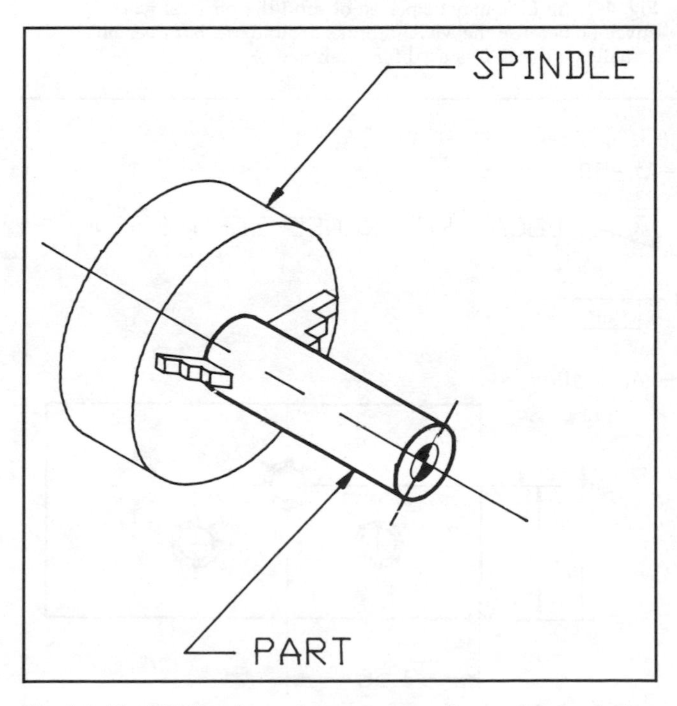

Fig. 4-4. The PRZ has been chosen on centerline at the tip of this lathe job.

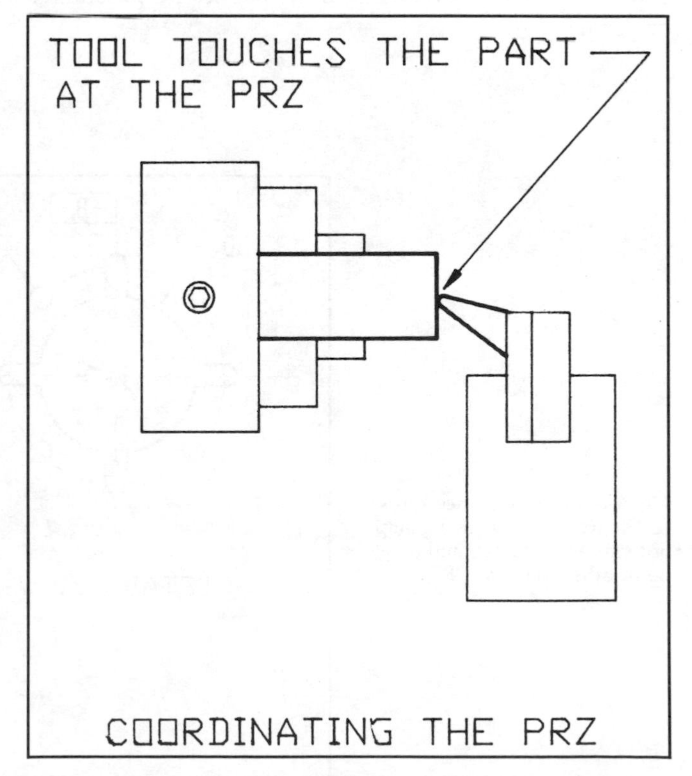

Fig. 4-5. The setup person must coordinate his or her tools to the exact point, as indicated on the setup document.

The center of an arc or circle is a local reference (LRZ). The center position must be known, yet the cutter need not actually move there. It must only refer to the point. The arc must be concentric to the center point, so it must be called out in the program. A center LRZ is usually established when the coordinates for the arc are given. We will look further into this in circular interpolation and programming in Chapter 15.

Local reference points are temporary in the program. They are used only to machine a certain area of the part. They are then cancelled. There may be any number of local reference zeros (LRZ) in the program. They may be located anywhere within the machine travel limits. LRZs are not present in all programs due to the shape of the part. Also, some controls do not effectively use LRZs. Local references are more common in mill programs than in lathe programs. This is because more complex shapes are encountered in milled work.

LRZ Example

The instrument panel in Fig. 4-6 is a good example of a need for an LRZ. Note that the cutouts and related hole patterns are each referenced from the center of the pattern. An individual pattern would be easy to program, but a series of the same pattern would be complicated if each coordinate point in the pattern was referred back to the PRZ in the left corner of the part. The programmer could set an LRZ at the center of each pattern, then for that pattern only, refer to the LRZ. This would save time and reduce the possibility of error.

On some jobs, an LRZ may be located beyond the physical limits of the machine. For example, an arc is needed whose radius is too large to place the center within the machine work envelope. An LRZ is placed at 36″ outside the actual machine. The work envelope is the physical limits of the machine. On many controllers, it is possible to place the local reference beyond the envelope because the controller uses this point for reference, not to actually move there. The controller "thinks around the point," although the point is away from the actual machine. Fig. 4-7.

Refer to Fig. 4-2 to see the relationship of the three types of reference points.

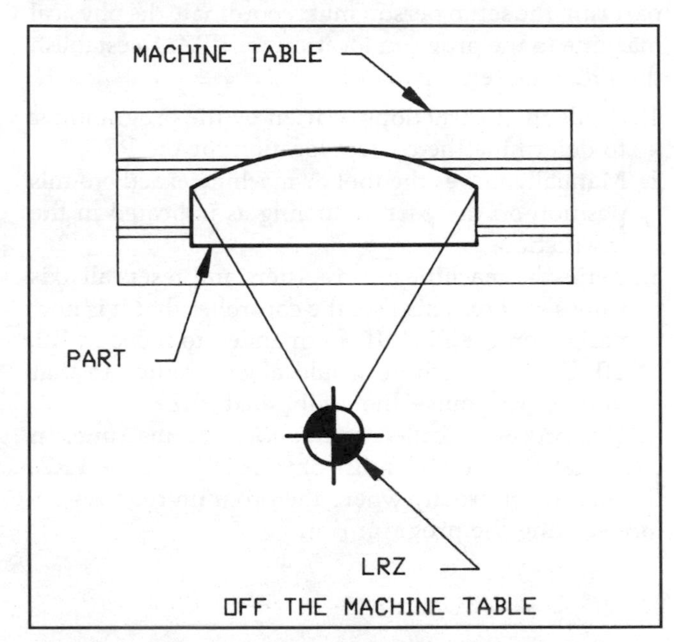

Fig. 4-7. An LRZ may be placed beyond the physical work envelope because the MCU "thinks around" it, but does not actually go there. It is used for reference.

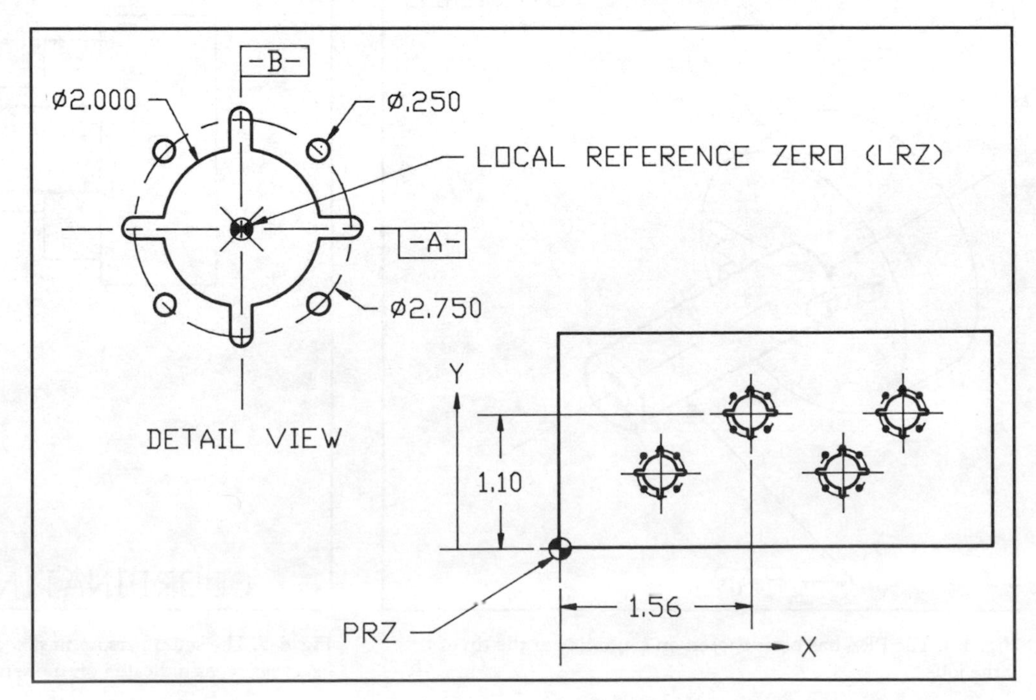

Fig. 4-6. Local reference zeros (LRZ) are useful when a group of features center around a point other than the PRZ.

MACHINE HOME REFERENCE

The *machine home reference* (M/H) is also called machine home zero. This point is used for setup. It is seldom used in programs. Machine home is a constant position. It is established by moving against limit switches on each axis, usually at the extreme limit of each axis travel. Machine home reference is not present on all machines. Machine home is used:

- For setup and initial coordination of the machine.
- When the machine is first turned on as a wakeup coordinating position.
- As a tool change position.

Machines that use M/H will automatically park at that position when commanded to do so. From there it becomes simpler to set up the PRZ, to change tools, or to begin again after a crash.

The machine home position is sometimes used in a program as the PRZ location. This might be done to squeeze every bit of possible travel out of the machine for a part near the capacity of the machine. This practice is not recommended because the cutter must then come very close to the limits of machine travel. M/H is often used as a safe parking position for tool changing.

PLANES FORMED BY AXES

To learn the concept of coordinates, we will limit further examples to a two-dimensional flat plane. We will not look at three-axis programming for a while. All examples and work problems will be given in either the X-Y (vertical mill) or X-Z (lathe) plane. This plane is signified by the axes that lie within it.

The programming plane for the lathe is the imaginary X-Z surface. For the mill, it is the X-Y surface. The mill is easy to visualize, because it is parallel to the table surface. The lathe X-Z plane is formed by the saddle and cross slide movements of the tool.

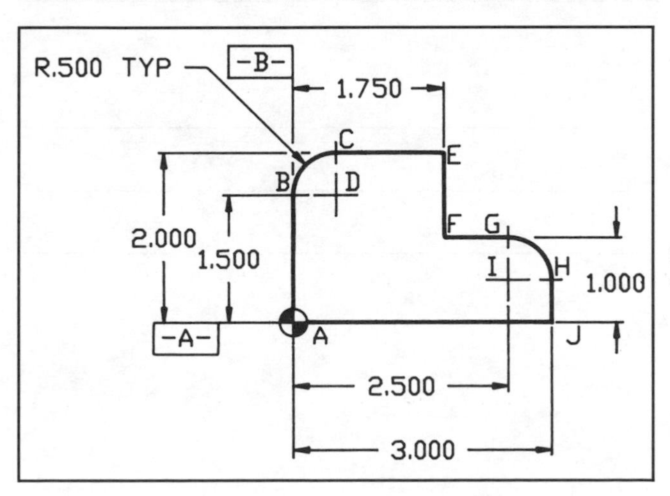

Fig. 4-8. This milling machine part drawing may be identified with coordinates.

You should realize that X-Y examples are interchangeable with X-Z. They are the same concept and the ideas may be interchanged. The axis letter Y could be exchanged for the letter Z with no difference in the lesson or concept being shown as far as points and coordinates.

Point Example

Refer to Fig. 4-8. On the milling machine part drawing shown, points A, B, C, E, F, G, and H are places we wish the cutter to go. Points A, D, and I are reference points. Point A does double duty. It is the program reference point as well as a *position point*. It is the reference starting point for the part program.

To make this part, we would need to start at point A, go around the part, then return to point A to complete the shape. Often the PRZ is chosen as a location on the part.

Coordinate Example

Below is a set of coordinates for the significant points on Fig. 4-8. These coordinates are *absolute*, which means they refer to the PRZ for their distance values. The sequence number is simply a utility number to find your place in a program. Given some rearrangement and extra information such as feed rate and spindle on, the table below could be a program.

SEQUENCE NUMBER	POINT	X COORDINATE	Y COORDINATE
1.	A	X0.0	Y0.0
2.	B	X0.0	Y1.5
3.	C	X0.50	Y2.0
4.	D	X0.50	Y1.5 Ref Point
5.	E	X1.75	Y2.0
6.	F	X1.75	Y1.0
7.	G	X2.50	Y1.0
8.	H	X3.00	Y.50
9.	I	X2.50	Y.50 Ref Point
10.	J	X3.00	Y0.0
11.	A	X0.0	Y0.0 Return to Start

Null Entries

In sequence number 2 in the previous example, the X coordinate was X0.0. The previous coordinate was also X0.0. The tool was already at X0.0. Thus, the movement required only the Y axis motor. The X axis motor did not need to move in order to position at point B. This is called a *null coordinate*. On many C/NC controls, you may omit null coordinates in a program, or you may leave them in for information. Either way, the controller will be able to understand that you have identified point B. Thus, sequence 2 could have simply read Y1.5, because the X component could be omitted.

Remember, on most CNC control controls, if a position point is called out and a certain axis is already at that position, the coordinate for that axis can be omitted. If you expect no axis movement, you may omit the coordinate. The coordinate can also be left in for clarity.

At this point, you should be able to answer "yes" to the following questions.

1. Can you identify the two types of significant points?

___

2. Can you describe the three types of reference points?

___

___

___

3. Do you understand machine home?

___

___

4. Can you describe a two-axis plane?

___

___

___

1. There was no need for local references on Appendix A Drawing 2. Why?

2. Refer again to Fig. 4-8 and the coordinate chart. Which entries (position points only) could be omitted as null?

3. When might multiple PRZs be used in a C/NC part run?

4. Which type of reference point may be positioned beyond the machine work envelope?

5. The two general types of significant points are:
   A) Zero and full-floating.
   B) Reference and absolute.
   C) Coordinate and reference.
   D) Position and reference.
   E) Machine home and local.

6. A setup person coordinates a machine by moving the tool to the exact location as called for in the program in relation to the part. He or she then resets the axis registers to zero. Is this statement ☐ True or ☐ False? If it is false, what will make it true?

7. Reference points are used in three ways in a program. Match each description below with the reference point that will be used. **Reference Points** A = PRZ B = LRZ C = MH

_______ Uses a never-changing point for initial machine turn-on, or as a tool-changing position.

_______ The prime reference chosen before the program is written.

_______ A reference point that may be used more than once in a program.

_______ A reference point that may be used more than once in a program for multiple parts of the same shape.

_______ A reference point that can be cancelled in a program.

_______ A reference point that cannot be cancelled in a program.

_______ A reference point which, in a special situation, may be used more than once in a program.

8. The programming plane on a C/NC lathe is X-Y. Is this statement ☐ True or ☐ False? If it is false what will make it true?

4-3 Coordinate Identification Systems

In C/NC work, there are two general coordinate systems used to identify a significant point on the part. These systems are the polar coordinate system and the Cartesian coordinate system. At present, the Cartesian coordinate system is the most common.

POLAR COORDINATE SYSTEM

Polar coordinates are based on the radius distance and angle from the reference point. Because polar coordinates eliminate a great deal of math, the latest controllers are adding polar coordinates to their capability. Not all C/NC machines have polar coordinates—they are less common. So, we will look first at Cartesian coordinates. All C/NC machines use Cartesian coordinates. We will study polar coordinates in a later chapter.

CARTESIAN COORDINATE SYSTEM

The Cartesian coordinate system works from a square reference grid similar to graph paper. Reference lines run at 90° to each other. Each is parallel to a primary axis. The coordinate examples given earlier are Cartesian coordinates.

There are two distinct methods to identify a point on a Cartesian grid. One method is absolute coordinate values. The other method is incremental coordinate values (also called relative values). The difference lies in the point used as the coordinate reference. We will look first at absolute coordinate values.

4-4 Absolute Value Cartesian Coordinates

The numbers assigned to absolute coordinates always refer to their distance from the PRZ. As mentioned, previous coordinate examples were absolute value Cartesian coordinates. Each value indicated the distance of the point from the PRZ.

EXAMPLE OF AN X-Y PLANE

Carefully study Appendix A Drawing 1, and the program coordinates shown. The absolute coordinates shown trace the upper half of the gauge. Each point is identified using its X or Y distance from the PRZ. The PRZ is the intersection of datums -A- and -B-. The following activity will allow you to see if you understand the Cartesian coordinate system and absolute value coordinates.

Problems 1-5 refer to Appendix A Drawing 1 and the table reproduced below.

Absolute Center Gauge Program Coordinates

Sequence Number	X Coordinate	Y Coordinate	
1	.650	0.0	Center of left vee groove
2	0.0	0.375	Upper tip of left vee
3	1.056	0.375	
4	1.20	0.125	Bottom of small vee
5	1.344	0.375	
6	1.55	0.375	
7	2.20	0.0	Point on center/line

1. How was the 1.344 arrived at in Sequence 5?

____

2. In Sequence 2, the X value is zero. It can be eliminated because it is a null entry.

   Is this statement ☐ True or ☐ False?

   If it is false, what will make it true?

____

3. The X value for the far tip was X2.20. How was this value arrived at?

____

4. Complete the set of program coordinates to trace the shape of Appendix A Drawing 1. Continue on from the right tip. Remember, any Y coordinates below the PRZ will be negative.

Sequence Number	X Coordinate	Y Coordinate
8		
9		
10		

5. Suppose that in Appendix A Drawing 1, the PRZ has been moved. The absolute coordinates for the ¼″ diameter hole on the left side are: X–1.2, Y0.0

   A) What location would have been chosen for the PRZ?

   ____

   B) What would the coordinates be for the upper point of the large vee groove?

   ____

6. Write a program set of absolute coordinates for Fig. 4-9.

   A) Choose the intersection of datums -A- and -B- for the PRZ.

   B) Mark the correct symbol on the drawing for the PRZ.

   C) Program only the significant position points.

   D) You may omit null entries or you may include them.

Sequence Number	X Coordinate	Y Coordinate	Point
1			A PRZ
2			B
3			C
4			D
5			E

Sequence Number	X Coordinate	Y Coordinate	Point
6			F
7			G
8			H
9			I
10			A Return to PRZ

7. Assume that in programming problem 6 above, you accidentally add an extra 1.0″ to the X coordinate for point E: X3.646, Y1.604. All other coordinates are correct. Sketch the resultant part shape.

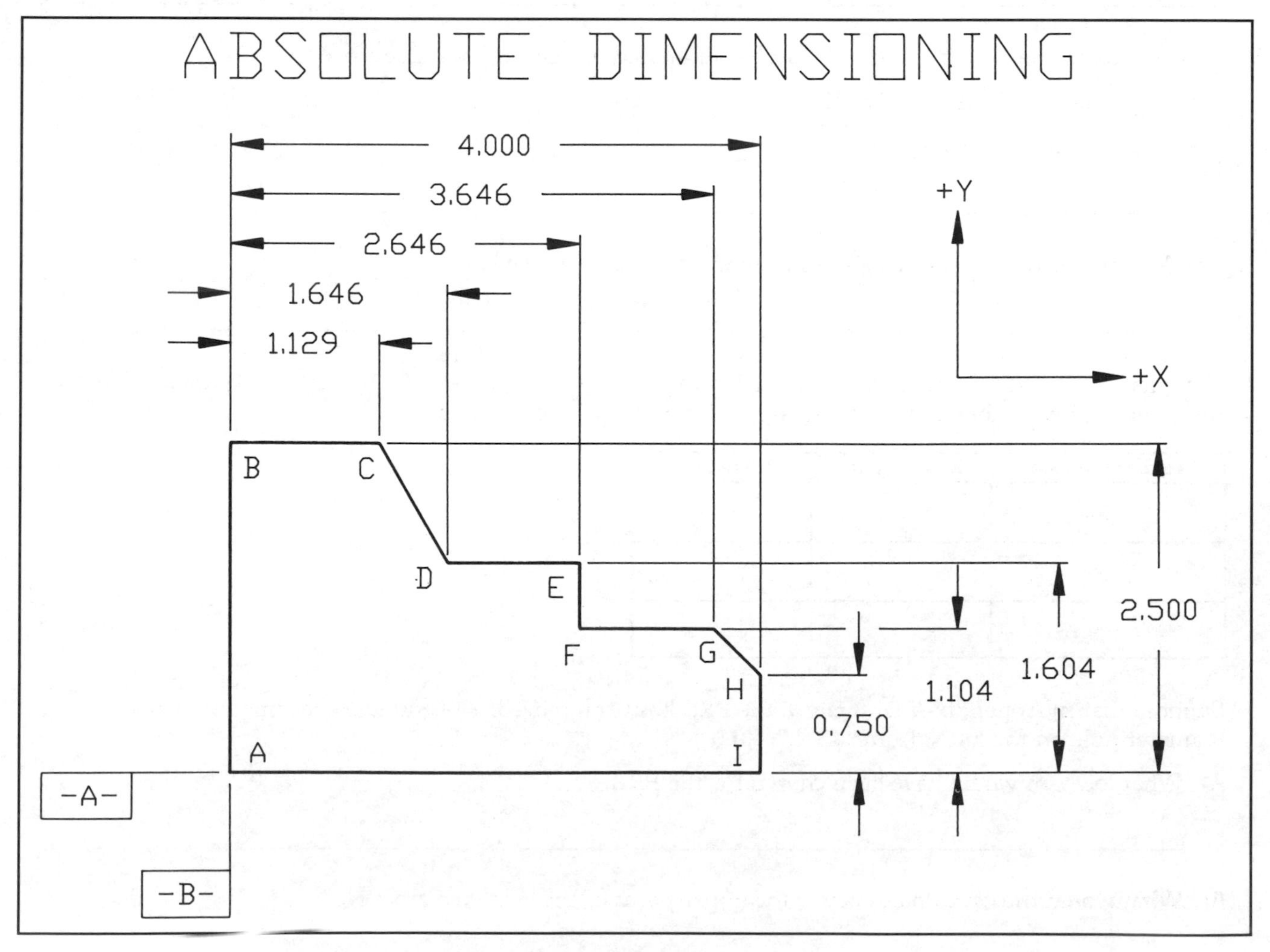

Fig. 4-9. Part drawing for Activity 4-2.

GLENCOE PUBLISHING COMPANY
15319 Chatsworth Street, Mission Hills, California 91345
(818) 898-1391

Dear Instructor,

Enclosed is the text you have been looking for. COMPUTER NUMERICAL CONTROL SIMPLIFIED, by Michael Fitzpatrick, is the first numerical control text written especially for the beginning student. Reflecting the latest developments in the C/NC technology, this easy-to-read text provides interactive experiences on real machinery to train students to become entry-level C/NC operators. The emphasis throughout is on learning by doing. Hands-on activities prepare the student for the job as well as reinforce and verify textbook learning.

Flexible and complete, COMPUTER NUMERICAL CONTROL SIMPLIFIED can be scaled up or scaled down and matched to any equipment. The text information can be tailored for individual class conditions through the use of twenty-two custom sheets. These sheets, which are bound into the instructors guide, are reproducible instructional aids that allow the teacher to relate core information in computer numerical control to specific items of equipment and specific machine operations.

Please take a few minutes now to discover the many advantages COMPUTER NUMERICAL CONTROL SIMPLIFIED provides student and instructor alike.

To order, simply phone 1-(800) 257-5755 (8:00 a.m. to 6:00 p.m. Eastern Time) or send purchase order to:

Glencoe Publishing Company
Order Department
Front and Brown Streets
Riverside, NJ 08075-1197

Sincerely,

George E. Provol

George E. Provol, Dir.
Post Secondary Sales

GP/mf
Encl.

4-5 Incremental Cartesian Values

Incremental Cartesian values always refer to the previous point identified. Each new point is identified by its distance from the last point. The last point, then becomes a temporary reference.

EXAMPLE

Refer to Appendix A Drawing 1. The set of coordinates shown identifies the same points shown in the absolute example. This time, however, they are incremental. Each point is identified as to how far in X and Y, it is from the previous point. Each point coordinate is relative to the last point coordinate.

Appendix A Drawing 1 is not practical for incremental values because it is dimensioned with absolute dimensions. The trick is to recognize when to use which value for the coordinates. To write these coordinates, much subtraction was needed to get the incremental distances.

Sequence Number	X Coordinate	Y Coordinate	
1	0.650	0.00	Center of left vee groove
2	–.650	0.375	Upper left point
3	1.056	0.0	Y = 0.0 indicates no movement. This could be omitted as a null entry to bottom of small vee
4	0.144	–0.250	
5	0.144	0.250	
6	0.156	0.0	
7	0.650	–0.375	Point of center gauge

Turn to Appendix A Drawing 2. This drawing is ideal for incremental programming.

Do you understand incremental values in the Cartesian system? The following activity will help you determine if you need further study.

1. A) Write a set of incremental program coordinates for Appendix A Drawing 2.
   B) Choose the intersection of datums -A- and -B- for PRZ.
   C) Start at the PRZ (Pt A) and move in a clockwise direction.
   D) You may omit nulls or include them.
   E) Your tool is positioned at point A to start.
   F) Indicate the PRZ with the correct symbol.

Sequence Number	X Coordinate	Y Coordinate	
1			B
2			C
3			D
4			E
5			F
6			G

Sequence Number	X Coordinate	Y Coordinate	
7			H
8			I
9			J
10			K
11			L
12			A RETURN TO PRZ

2. How many null entries are in the above set of coordinates?

3. Refer to Fig. 4-10.

   A) Write a set of incremental point coordinates.

   B) Indicate the PRZ at the intersection of datums -A- and -B-.

   C) The tool is initially positioned at the PRZ.

Sequence Number	X Coordinate	Y Coordinate	Point
			B
			C
			D
			E
			F

Sequence Number	X Coordinate	Y Coordinate	Point
			G
			H
			I
			J

4. In error, you add an extra 1.0″ to the move between points D and E (X2.0,Y0.0).

   A) What would happen to the shape of the part?

   B) What would happen to the line between points I and A?

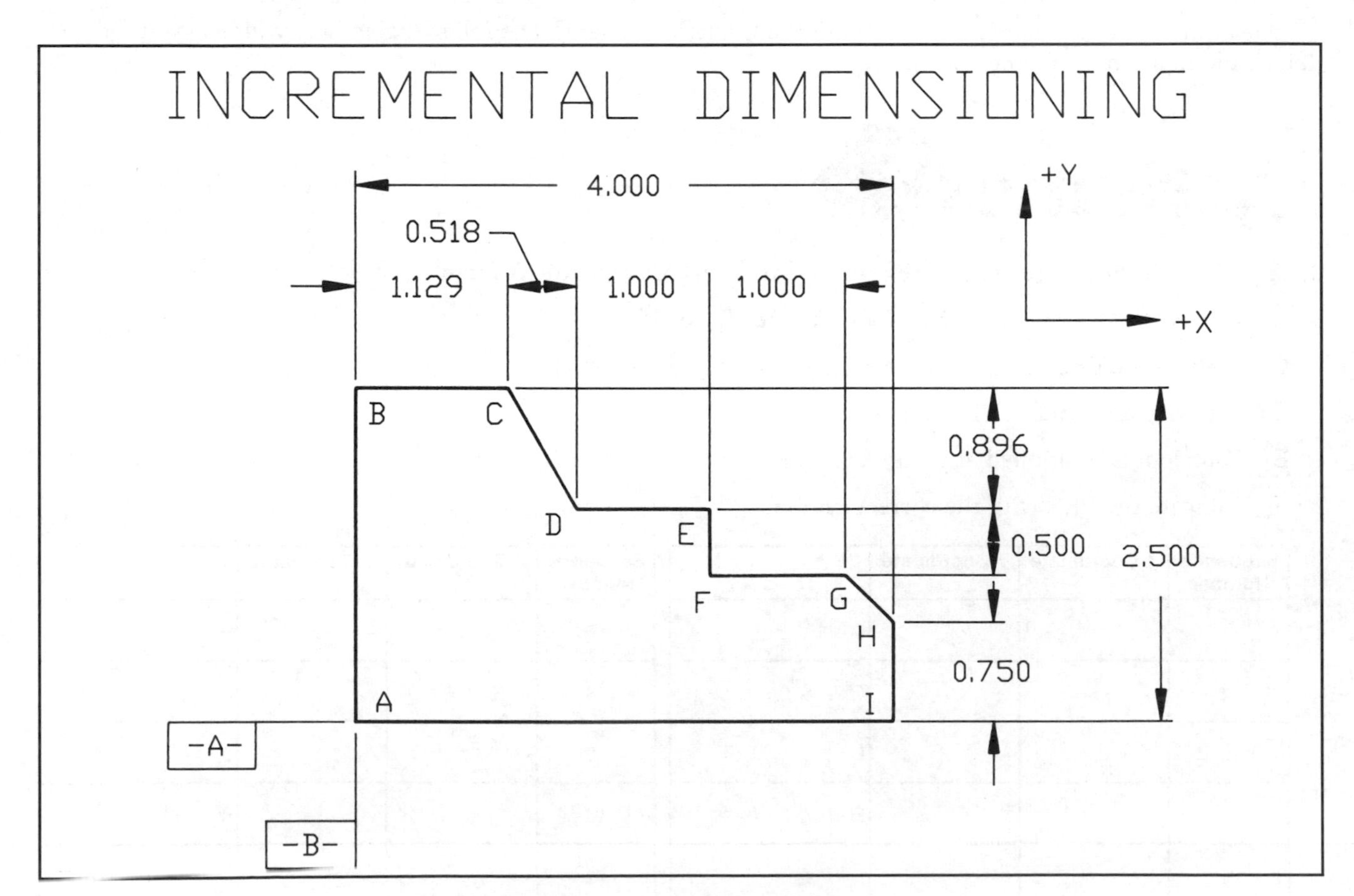

Fig. 4-10. Part drawing for Activity 4-3.

ABSOLUTE VALUES COMPARED WITH INCREMENTAL VALUES

Choose the value that saves the most time and math. There is no right value for coordinates. The drawing will dictate which values to select. If the drawing is drawn for absolute values, use them. If it is drawn for incremental value, then use them.

In selecting values, use the following guides:

1. If the drawing is dimensioned using *baseline dimensioning*, it is set to program using absolute values. See Fig. 4-9.
2. If the drawing is dimensioned using *incremental dimensioning*, then use incremental values. See Fig. 4-10.
3. If you know the absolute value and the incremental value, choose the absolute value. Properly used, either value produces exactly the same accuracy and results. Use of the absolute value, however, tends to eliminate errors.

Refer again to Figs. 4-9 and 4-10. Figure 4-9 uses absolute value, while Fig. 4-10 is dimensioned incrementally. What would be the problem in programming the first drawing in incremental values or the second in absolute values?

Choose the point coordinate values that fit the drawing. Do not convert the drawing into your favorite values. Use the value as it appears on the drawing.

Absolute and incremental values may be mixed in a program. However, the controller must be informed that the coordinates are either incremental or absolute. This will be discussed later.

Advantages of Incremental Values

- *Easily learned.* Incremental values are easy to learn because we tend to think in terms of the next task to be done. We think incrementally.
- *Simplified math.* Most math required will be easier using incremental values. The math is centered around the central problem and need not refer back to the PRZ for absolute calculations.
- *Easily checked.* A program written in incremental values is easy to check as long as the cutter returns to the PRZ. Adding all the plus moves, then adding all the negative moves on any single axis should result in the same number.

Disadvantages of Incremental Values

- *Errors affect all points.* A potential problem in using incremental values is that errors can be continued throughout the program. A second error will compound a first. Fig. 4-11. This problem is called accumulated error. Any error will be passed on to the next point when reference is made to the previous point. If the previous point is incorrect, the next point will be, too. If you make an error in a value in absolute, the part will be incorrectly made. However, error will not be passed on to other features of the part. It will affect only the specific point in question.

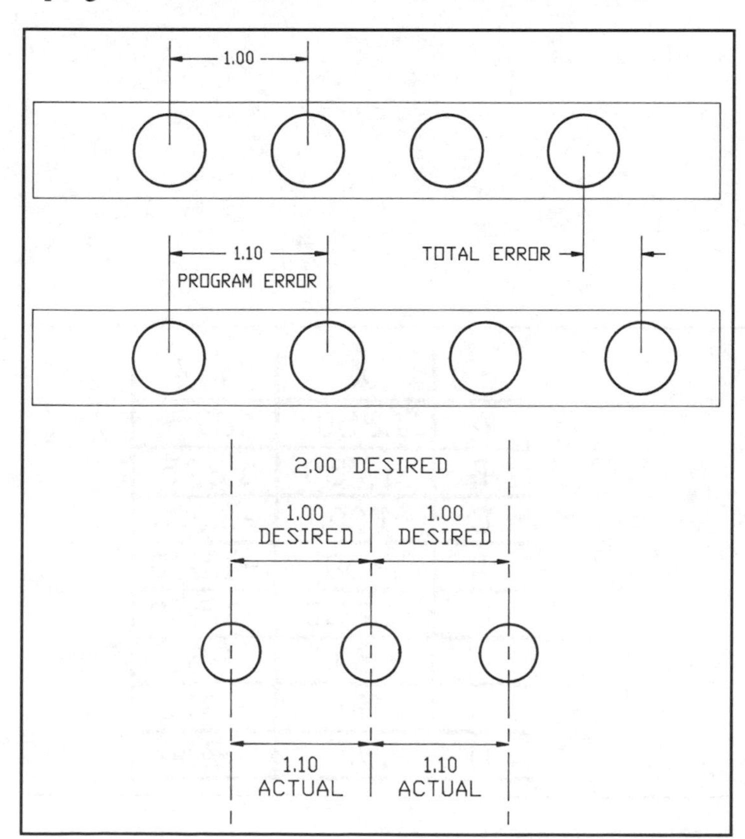

Fig. 4-11. Accumulated errors can occur using incremental valued coordinates. Each point beyond the error will "inherit" the problem.

4-6 Quadrants and Values

QUADRANTS IN THE ABSOLUTE GRID

The absolute grid is divided into four major sections called *quadrants.* The direction sign value of a coordinate (plus or minus) depends upon the quadrant in which the point lies. The PRZ is the center of the grid. In absolute values, one direction from PRZ is positive, and the opposite direction is negative.

For an example, refer to Fig. 4-12. Point A in the upper right corner, quadrant 1, would have a positive X value and positive Y. In quadrant 2, point B has a positive Y value and a negative X. In quadrant 3, point C has negative X and negative Y values. Point D then has positive X and negative Y values.

Absolute Value Example

No matter where the tool is located in Fig. 4-12, if the coordinates for point B are called out, X−1.75, Y1.875, the tool would position at point B because point B would be identified as unique. What are the absolute coordinates for point E?

Advantages of Absolute Values

- *Fewer errors.* Each point coordinate refers to the PRZ. Errors are not passed on to the next point coordinate. Each point coordinate is a new identification based only on the PRZ.
- *Slightly faster to program.* If dimensioned correctly (base line dimensioning), a program is slightly faster to write in absolute. For this reason, most drafters dimension in baseline if the part will be produced on C/NC equipment.

Disadvantages of Absolute Values

- *Abstract.* Absolute is an abstract concept. Always referring to the PRZ is somewhat foreign to new programmers. This is a small problem and easily overcome.

As a new programmer, it would be wise not to mix values at first. Get used to programming in absolute and incremental by writing individual programs in one unit, then the other. Once you are comfortable with programming, it is easy to mix values. Here are three additional guidelines:

- If both values are known, select absolute.
- If not, select the one you do know.
- Each type is useful and may be mixed in the program.

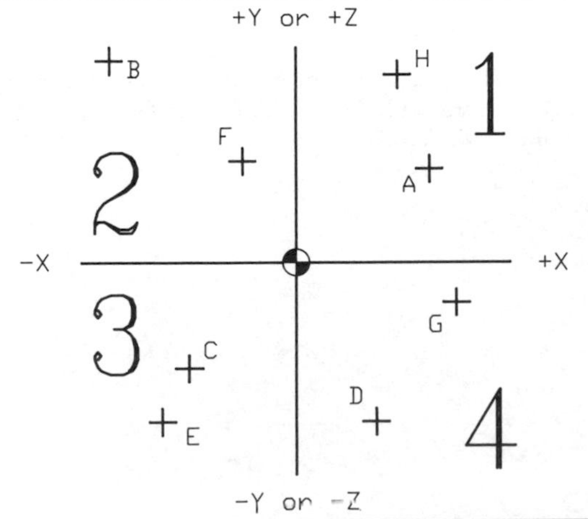

	X	Y,Z
A	1.250	.875
B	−1.750	1.875
C	−1.000	−1.000
D	.750	−1.500
E	−1.250	−1.500
F	−.500	.950
G	1.500	−.375
H	.95	1.750

Fig. 4-12. Coordinate values in the four quadrants in a Cartesian grid.

Incremental Value Example

The sign for an incremental value depends only on the direction, not on the quadrant in which the point lies. If you are positioned at point B, Fig. 4-12 and call out point A, the incremental coordinates would be X3.00, Y–1.00.

The direction in X is positive, while the Y movement is negative. Now, reverse the direction from point A to point B would be:

X–3.00, Y–1.00

In both directions, we crossed a PRZ-quadrant line, but the sign was dependent only on the direction of travel, not on the quadrant.

1. Do you understand the concept of quadrants?

2. Can you see how the absolute valued coordinates change value sign (plus or minus) as they pass the quadrant lines?

3. Do you understand why incremental values are positive in one direction and negative going the opposite direction?

4. Do you understand why the quadrant has no relationship to the sign value of the incremental value coordinate?

The following questions refer to Fig. 4-12.

1. In which quadrant would the tool locate if the absolute coordinates X1.45,Y–2.0 were called out?

2. What is the absolute value for the following points? Pay close attention to the sign of the coordinates.

Sequence Number	X Coordinate	Y Coordinate	Point
1			A
2			B
3			C
4			D

Sequence Number	X Coordinate	Y Coordinate	Point
5			E
6			F
7			G
8			H

3. Assume that you are located at the center PRZ. Write a set of incremental values for the points in their alphabetical order.

Sequence Number	X Coordinate	Y Coordinate	Point
1			A
2			B
3			C
4			D

Sequence Number	X Coordinate	Y Coordinate	Point
5			E
6			F
7			G
8			H

4-7 Quadrants and Selecting the PRZ

PRZ SELECTION FOR ABSOLUTE VALUES

The two primary concerns in selecting the PRZ are the primary datum intersection and the first quadrant.

The PRZ should be based upon the actual part dimensioning. The PRZ should coincide with the datum basis for the drawing. *Datums* are points, lines, and surfaces to which reference is made for distance on a blueprint. Thus, the PRZ should be chosen at the intersection of two primary datums on the drawing.

It is also considered good practice to select the PRZ in such a way that the part lies entirely inside the first quadrant if possible. By doing so, all absolute point coordinates have positive values.

In selecting the PRZ, however, the datum requirement takes precedence. It is more important to set the PRZ in reference to the datum basis than to have all positive absolute values.

The lower left corner (–X,–Y) of the part is chosen as the PRZ for most mill programs. By doing this, all absolute point values will be positive. They will be in the first quadrant as long as the part can be oriented so that this point is also the datum intersection point on the part.

To use all positive values on the lathe, the PRZ would need to be on the centerline and at the chuck face. This is dangerous and not often used. The PRZ is chosen on centerline, a certain distance away from or just touching the stock, in the Z axis. This standard practice is safer and quicker to coordinate. That means that all moves from the tip of the part toward the chuck would have negative absolute Z values.

On most lathes, it is not possible to machine in the first quadrant unless the PRZ is at the chuck face. Some turning centers have rearranged axis frames. Although contrary to the EIA recommendations and the "right hand rule," this allows first quadrant programming.

Safety Rule. *Before you operate any C/NC equipment, make sure that you completely understand the axis frame for that machine.*

INCREMENTAL VALUES

The quadrant selected is not as significant when programming in incremental values. In such programs, only the direction (left-right, up-down) determines the sign as being positive or negative.

For example, refer to Fig. 4-12. If the tool is located at A and we call out an incremental coordinate to B, the X value would be negative – movement to the left. However, if the tool is at B and we call out the coordinates for A, then the distance is the same along X, but the sign is positive. Only the relative direction of the coordinate determines the value. As you can see, we crossed a quadrant line, and it did not change the direction value. In addition, the PRZ could have been anywhere in the illustration, it would not change the incremental values from A to B or B to A.

1. Write a complete point identification list for Appendix A Drawing 4.

   Programming Instructions

   A) When programming a lathe part, it is most common to enter diameters, not radii. While some lathe controllers accept radii coordinates, the diameter can be more easily and quickly entered.
   B) You are turning only the right half of the part. Program from the .5625″ diameter to the 1.00″ diameter. The stock is finished cold-rolled steel that needs no turning on the 1.00″ diameter.
   C) Use all absolute values.
   D) PRZ is on centerline datum -A-, at datum -B-.
   E) No face cut will be taken on the end of the part. Start at .5625 diameter.
   F) No roughing passes. Finish cuts only.

4-8 Identifying Significant Points on the Part Geometry

The majority of programs are point coordinates. Some are for reference and most are for positioning. These points are mostly on the part geometry (shape). The rules and guidelines for selecting reference and position points are discussed here.

SELECTING POSITION POINTS

A point on the part geometry is significant if it has the following characteristics. Fig. 4-13.

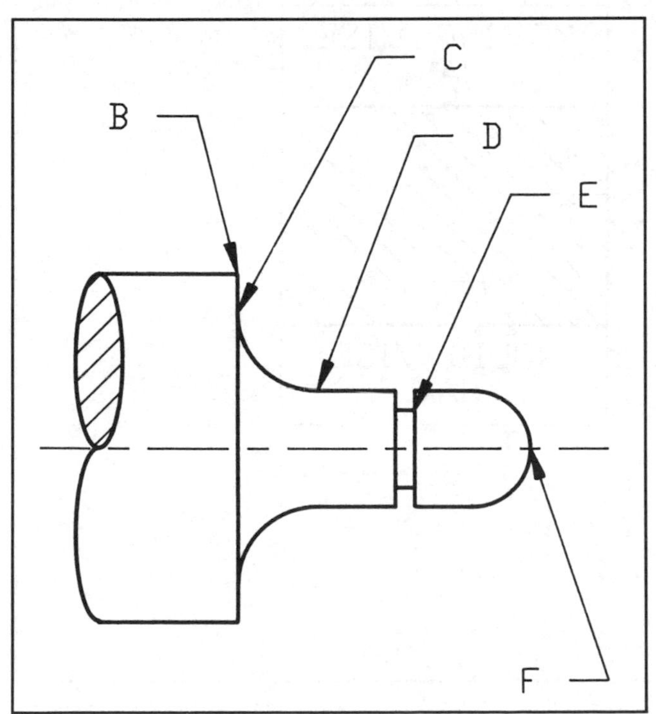

- It is a corner of the part. This is a point where one straight line intersects another. See point B.
- It is a tangent point. This is a point where a straight or curved line intersects with a curved line. Points C and D are tangent points.
- An intersection point. This is a place on the part where two surfaces or lines intersect. See point E.
- At quadrant lines. This is a point where the line crosses a quadrant line. Some machines will not produce a curve across a quadrant line. See point F.

In general, a *significant position point* is any place where the shape of the line changes. For example, the line may change from straight to straight, straight to curved, curved to straight, or curved to curved. Intersections are often significant points, too.

Fig. 4-13. The locations for significant points on this lathe part.

SELECTING REFERENCE POINTS

Program Reference Zero (PRZ)

The point chosen for PRZ is very important. Often the success of the program and the quality of the part can be traced to the choice of the PRZ. The PRZ may be located at any of the points listed above, but here are some additional guidelines.

- If possible the PRZ should be at the intersection of two datum references on the drawing. The tooling should then contact the part in the same relationship. The contact points should be against the same locating surfaces. If not, inaccuracy may be introduced before the part is even machined. It would be caused by the variation in size of the raw stock. Fig. 4-14.
- The PRZ should be in a location that is easily coordinated with the actual physical part.
- The PRZ should be chosen so that the part lies within the first quadrant, if possible.

Figure 4-15 shows a mill part that might have the center hole as the PRZ. It is an easily located central reference point. Before the central hole was made, the edges might have been used as a PRZ. For progressive setups, different features of the part might be the PRZ at different times.

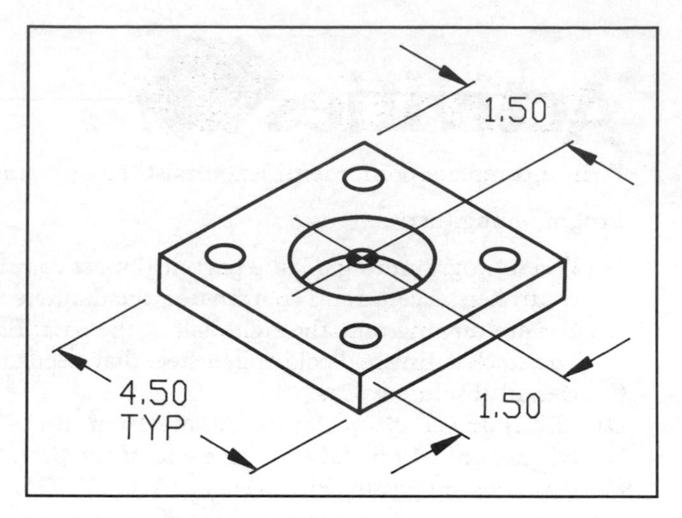

Fig. 4-15. A center hole might be a logical choice for PRZ when the part looks like this wheel adapter.

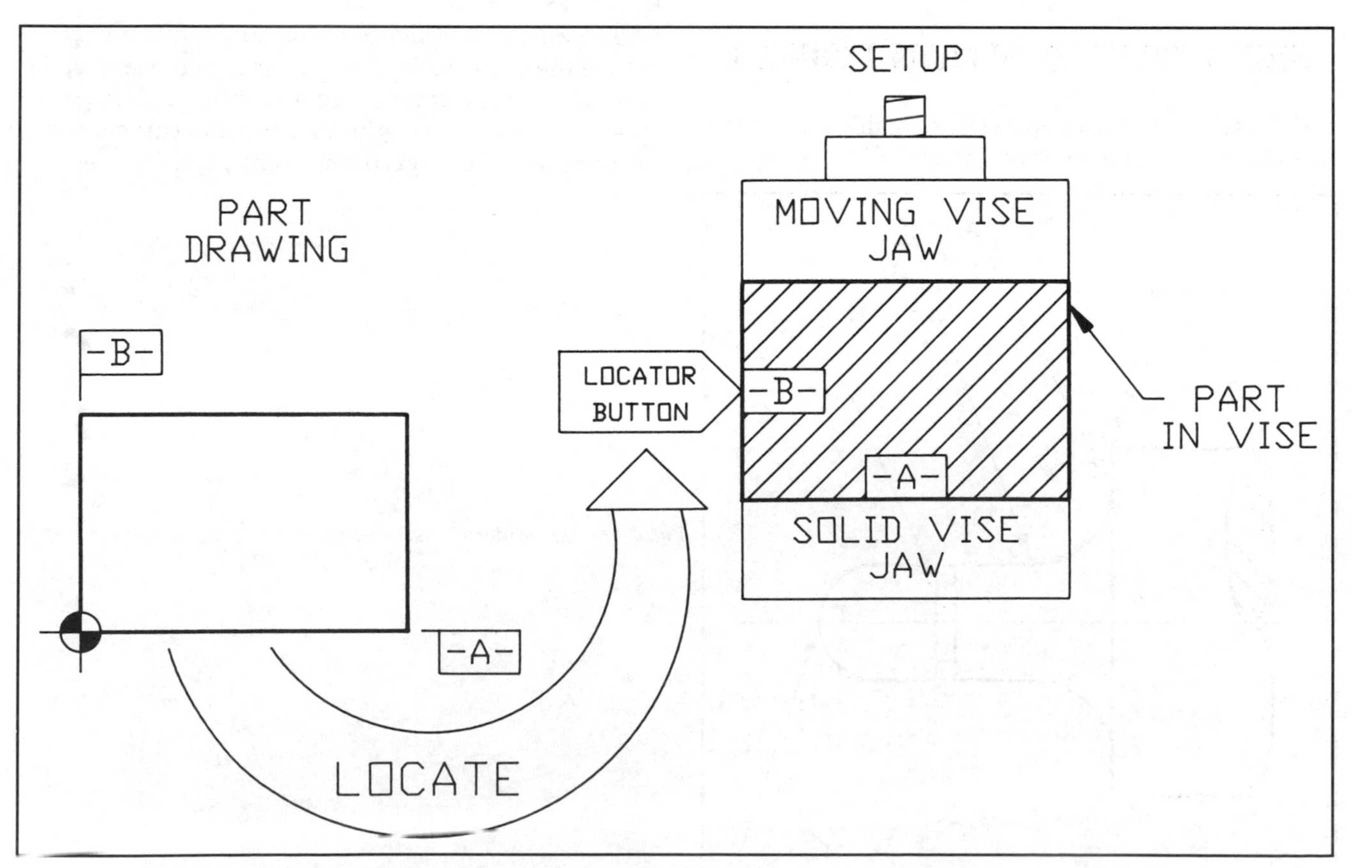

Fig. 4-14. The tooling, chuck, or holding fixture must always contact the part in the same relationship as the reference point on the part.

4-9 Coordinating the PRZ

There are three general methods used to coordinate a PRZ to the program location. Coordinating means to position the cutting tool at the PRZ or at a place that is a known distance from the PRZ. In other words, coordinating is parking the tool exactly where the program expects it to be, relative to the workpiece, when the program starts. One method is to use an electronic probe or sensor. If your machine has automatic tool sensing, read the manual. Two general methods of coordinating the PRZ are discussed here.

METHOD A

The method uses standard shop alignment tools and techniques directly on the part. A corner, an edge, or the center of a hole on the centerline are all places easily located. For this type of coordination, the part is usually held in standard tooling, such as a vise or a chuck. You will use Method A for all setups in this book. This method is for single or small batch part runs.

METHOD B

Sometimes—as on a rough casting—there is no reliable place to coordinate. The solution to this problem is to hold the part in a fixture, which provides a location (such as a small hole or a pad) that is a given height or distance from the chuck. This PRZ target is called a *tooling hole* or *tooling point*.

Using a dial indicator or feeler gage, position the machine spindle or tooling over the target on the holding fixture. The target provides a reliable location for the PRZ. This is usually done only on production runs where many parts will be run. It is most commonly used on mills.

Fixture Advantage

Use of tooling of this type shortens part-to-part cycles and holds each part more reliably. The time spent setting up the machine is usually shortened through use of a C/NC fixture. The time for setting up the machine is called *turnaround time*. A C/NC fixture also speeds up loading and unloading the stock to be machined. This is called *cycle time*. Cycle time is measured from one complete part to the next. Fig. 4-16

Tooling Points and Datum Targets

Toolmakers fabricate tooling points for production. They are fixtures. The points where the fixture contacts the part are set by the rules that govern datum reference planes in the geometric dimensioning and tolerancing system. Three contact points establish the primary contact surface, usually datum A. Two contact points establish the secondary datum B. One point establishes datum C. For more information on datums see ANSI document Y14.5M, *Dimensioning and Tolerancing.*

4-10 Types of Machine Movements

This section explains the different movements that C/NC machines can make. In Chapter 11, we will look at writing the actual programs that contain the various kinds of machine movements. A C/NC machine is capable of moving straight lines, angles, and circular arcs. The shapes produced using these movements can be classified in three categories.

- Straight line movement.
- Curved line movement.
- Nonregular shape movement.

Fig. 4-16. This industrial milling fixture holds the part in a rigid position to establish the PRZ and to speed up cycle time. The Boeing Company

STRAIGHT LINE MOVES

Point to Point

Refer to Fig. 4-17. Point to point is a movement in one or more axes, where each axis works independently of the others to arrive at a specific point. Each axis travels at a single speed, usually rapid travel. The axis travels independent of the other axes.

Point to point is best illustrated by a C/NC drill press locating over holes to be drilled. The path taken is immaterial. Only the location is important. This is the simplest of all machine moves. If two axes move at once, a 45° angle will be produced until the shortest travel axis is satisfied. Then the machine will move in a straight line parallel to the long move axis.

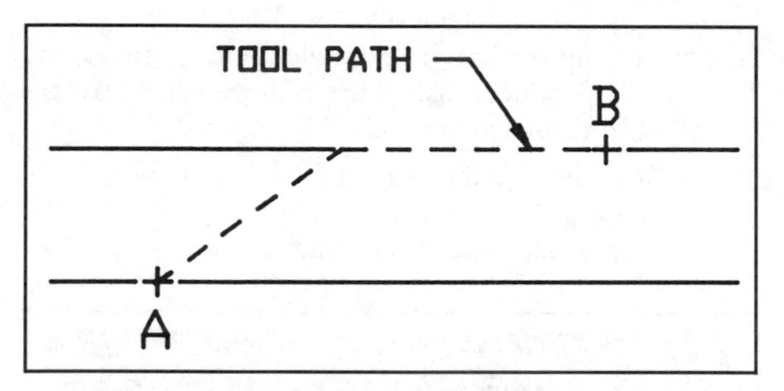

Fig. 4-17. Point-to-point movement is usually at rapid travel. Point-to-point is used only where the path taken is unimportant, yet the location is important.

Single-Axis Linear Movement

This movement produces a straight line parallel to a linear axis. Fig. 4-18. For example, milling a box or turning a straight shaft would require linear moves. It is one step higher than point-to-point movement because a specified feed rate is needed by the controller.

Some controllers may require the same code for this kind of movement as for angular linear interpolation. *Interpolation* means finding a value between two given values. A machine interpolation requires each axis to go at some intermediate speed to produce a tool speed. A single-axis straight-line movement is not strictly an interpolation, but some controls need to be put into the linear mode to move in a straight line at the feed rate rather than at rapid travel.

Multiple-Axis Angular Linear Interpolation

A straight line produced by the simultaneous movement of two or more axes is called linear interpolation. Fig. 4-19. The result is an angle produced by the axes moving at a specific ratio to each other.

On a lathe, the cutter would move at a 45° angle if both the X axis and the Z axis were to move at the same speed. If one axis moves more quickly than the other, the machine would produce a different angle. Some mill controllers are able to produce a straight line using the simultaneous movement of three axes.

NC machines can approximate curves using linear movement. However, true circular interpolation and the remaining machine movements beyond this point are possible only on a CNC machine. The output data to the axes require a microprocessor to do the calculation.

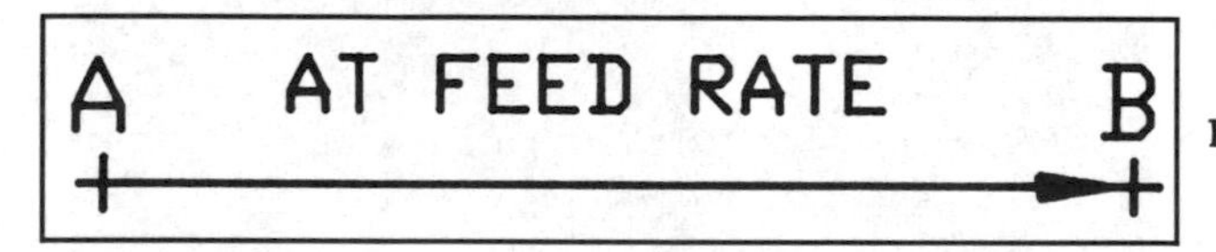

Fig. 4-18. Linear movement in one axis at the given feed rate.

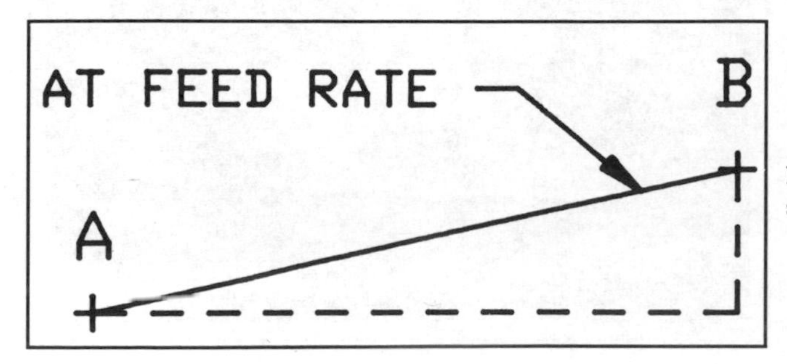

Fig. 4-19. Linear interpolation requires two or more axes moving simultaneously at the given feed rate.

CURVED MOVEMENTS

Circular Interpolation

In making a circular arc (part of a circle) or a complete circle, the axes involved must constantly change their speed ratio with respect to each other. However, the tool must move at the programmed feed rate. Fig. 4-20. Only a CNC machine is capable of this type of movement.

Circular interpolation is common on lathes and mills. It is usually in a single plane, XY, XZ, or YZ. A few advanced mill controllers are capable of doing circular interpolation in three axes.

Conic Section Interpolation

Not all regular geometric curves are parts of circles. There are other regular curved shapes called *conic section curves*. They are produced by slicing a simple cone at various angles. Examples of conic section curves include ellipses, parabolas (a flashlight reflector is a parabola), and hyperbolas. Advanced controls can produce conic section curves. Fig. 4-21.

NONREGULAR CURVES AND SURFACES

A crumpled and smoothed sheet of paper forms a nonregular surface. It cannot be described with plain mathematical formulas. C/NC machines cannot interpolate these shapes. The shape must be produced with a table of coordinate values for points very close together on the surface.

Computer software uses the *spline method* to help make nonregular shapes. This method mathematically slices the object into thin sections. Then the computer connects all sections with the smoothest curves possible. From this computer-smoothed model, a table of coordinates is generated. These coordinates are used to machine the surface.

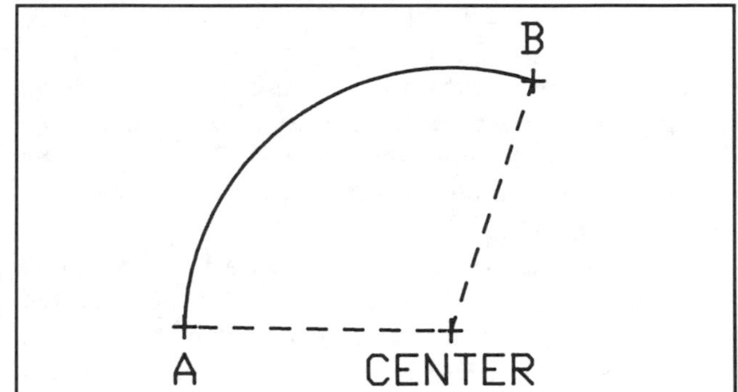

Fig. 4-20. Circular interpolation is a constantly changing ratio between two axis motors to produce a circular arc at the given feed rate.

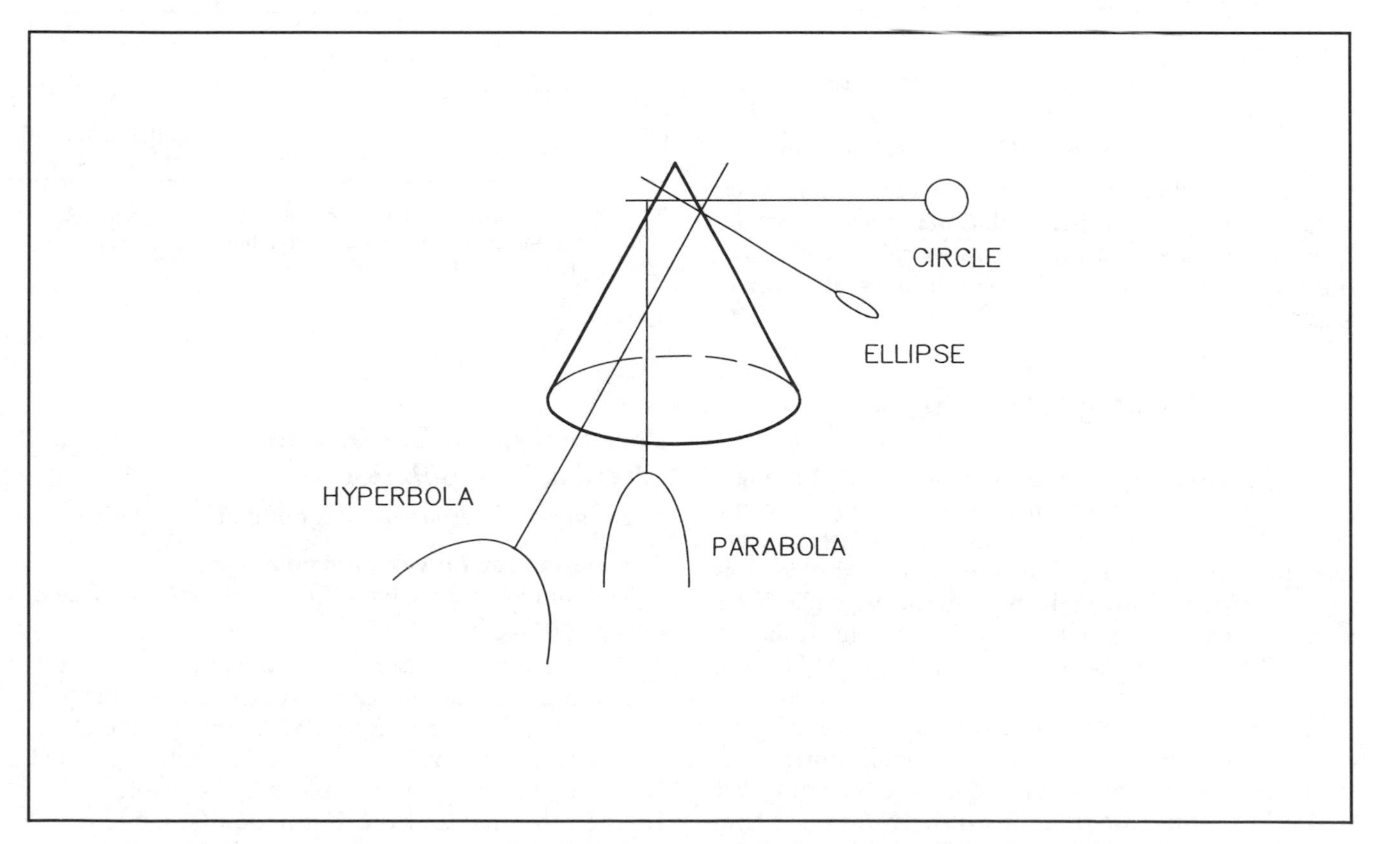

Fig. 4-21. Conic section curves are possible in CNC. As the cone is sliced at various angles, different curves other than a circle are produced.

4-11 Examples of Machine Movements

MACHINE MOVE EXAMPLE

To understand the concept of machine movements, think of yourself operating a hand-fed Bridgeport vertical milling machine. The X handle is your right hand and the Y in your left. Given some practice, you could crank the X axis at a different rate than the Y. If you kept the X and Y at different constant speeds, the result would be an angular move of the tool to the work. Fig. 4-22.

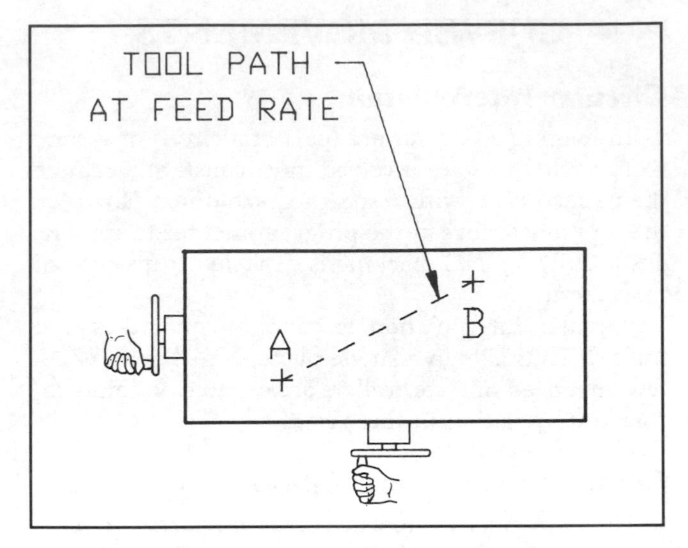

Fig. 4-22. With practice, you might be able to do linear interpolation but it is doubtful that you could maintain a constant feed rate. The computer does so with ease.

FEED RATE

If you wished a certain feed rate, you would have to crank both X and Y handles at calculated rpms so the overall result would be the given feed rate at the tool path. You would have to control the speed and ratio between each.

Example. Machine a 30° angle at a feed rate of 15 IPM. Referring back to the example of the hand-fed Bridgeport, you would have to calculate the correct inch per minute (IPM) rate for each axis. The Y axis feed rate is equal to the hypotenuse (15) times the sine ratio of 30°. Fig. 4-23.

Y axis feed rate
Sine 30 × 15 = .50 × 15 = 7.5″ per minute

X axis feed rate
Cosine 30 × 15 = .866 × 15 = 12.99″ per minute

The controller also makes this calculation when programming linear interpolation. If each axis responds simultaneously, at the rates calculated above, the tool would move at 15″ per minute at the correct angle.

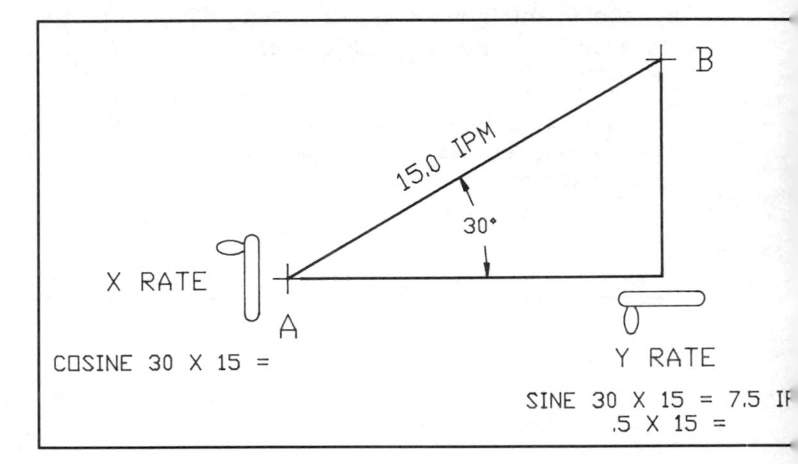

Fig. 4-23. A vector diagram of the calculations made by the MCU to initiate linear interpolation. Each axis requires an intermediate RPM to produce a specific tool feed rate.

INTERPOLATION

As discussed earlier, *interpolation* means finding a value between two given values. When a CNC machine interpolates, the controller calculates the intermediate feed rates for the individual axes so that the result is the programmed rate at the tool. In the example above, if you crank the Y axis at 7.5 IPM and the X axis at 12.99 IPM, you would produce an angle of 30° at a feed rate of 15 IPM.

If you could vary the X to Y ratio as you went along, a curve might result. You probably would not be able to create a smooth curve, although the CNC controller could do it with ease. It controls the rates and ratios of the axis drives to move the tool or table to make a circular path at a constant given feed rate.

Parameters for Linear and Circular Interpolation

Parameters are factors that define an operation.

Parameters for Linear Interpolation

The controller must have three parameters for linear interpolation.

1. A code or conversational command to cause the controller to initiate linear interpolation. This tells the controller that it is to move at the specified feed rate in a straight line to the coordinates specified and that more than one axis may be involved.
2. The coordinates of the destination (absolute or incremental).
3. The feed rate.

Parameters for Circular Interpolation

To draw a circular arc using a compass, you would need some parameters. Fig. 4-24. The controller needs the same parameters.

The programmer must identify the following five parameters:

1. *Start point for the curve.*
2. *End point for the curve.* This could be in terms of Cartesian coordinates or polar coordinates on many controls. For polar coordinates, the end point would be specified as the number of degrees to swing, rather than the X-Y or X-Z distance away.
3. *Center point location.* Once the controller knows the present location, and the location of the center of the arc, and the end of the arc, it knows the radius. It knows this because the center should be the same distance from the start as from the end point. Controllers using polar coordinates are able to accept the radius as one of the parameters. On some controls, the radius for the curve can be substituted for the center.
4. *A code or conversational command that defines circular interpolation.* Coded machines must be told if the curve is to be clockwise or counterclockwise.
5. *The feed rate.*

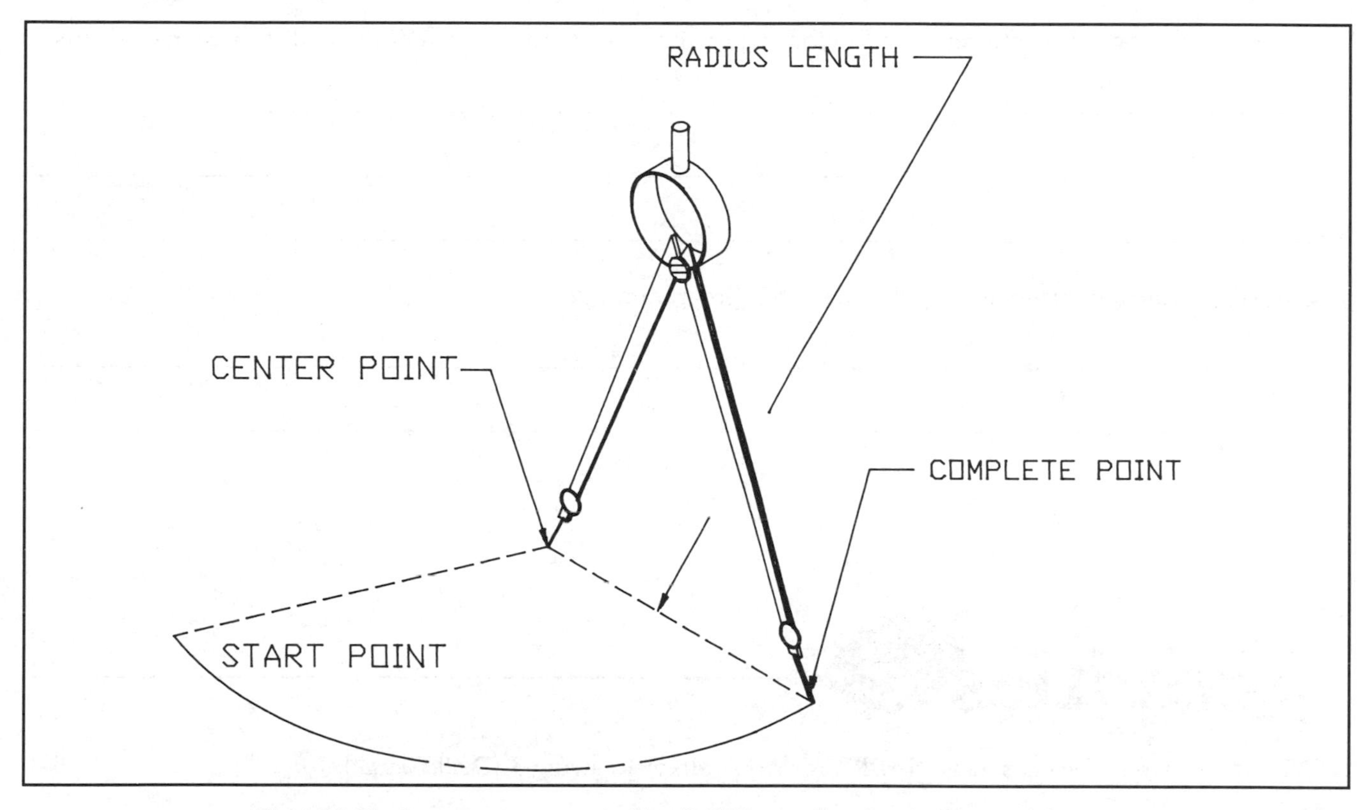

Fig. 4-24. The four parameters required by all CNC controllers for circular interpolation.

You have studied significant point identification and machine movements. Check yourself on these universal C/NC subjects by answering the following questions.

1. Can you select position points on a print?

____

2. Can you select reference points on a print?

____

3. Can you list the four rules for identifying a significant position point.

4. Do you understand why a PRZ should be chosen carefully?

5. Do you understand the difference in the four types of machine movements? Can you list three movements?

6. Can you define *parameter* as it applies to C/NC programming?

Activities

1. What point would be chosen as the PRZ for Appendix A Drawing 3 (Drill Gauge).

2. Why did you choose this point?

3. Linear interpolation moves at rapid travel until one of the axes is located. Then it moves at feed rate to the final location.

   Is this statement ☐ True or ☐ False?

   If it is false, what will make it true?

4. List the three parameters needed to initiate linear interpolation.

____

____

____

5. List four parameters needed to initiate circular interpolation.

____

____

____

____

6. Circle the movements that *only* a CNC machine can make.

A) Single-axis linear
B) Circular interpolation
C) Linear interpolation
D) Point to point
E) Parabolic interpolation
F) Elliptical interpolation
G) Nonregular curves and surfaces

7. Define the word *parameter* as it is used in C/NC programming.

____

____

8. The word *interpolate* means to find a value that equals two or more other values.

Is this statement ☐ True or ☐ False?

If it is false, what will make it true?

____

____

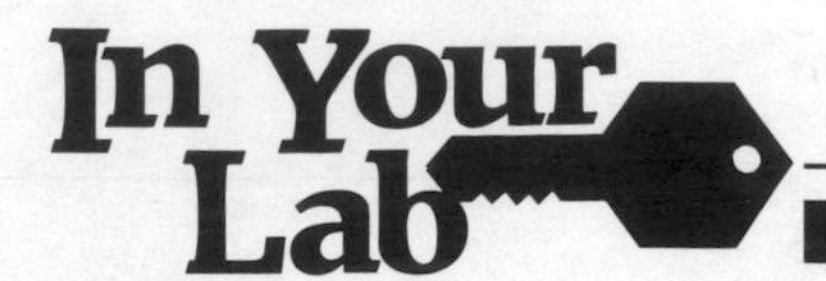

Your instructor will assign a machine or machines to answer the following questions. Questions 9 to 12 relate to a milling machine.

9. Is this machine capable of three axis linear interpolation?

____

10. Is the machine capable of three-axis circular interpolation? Some machines can do two-axis circular interpolation while moving along a third axis in a straight line. This is called a two and a half axis machine.

____

11. What is the largest radius curve this machine can produce? This may require a local center point reference beyond the work envelope.

____

12. Circular interpolation requires much computation within the controller. Many CNC machines must limit the feed rate while doing circular interpolation because the large amount of data output taxes the controller's ability to keep up with the feed rate. This limit is called a *default* feed rate. Does this machine require a default feed rate for circular interpolation?

____

Questions 13 to 16 relate to a lathe.

13. Can this lathe produce curves of less than 90°?

____

14. What is the radius of the largest curve the lathe will produce?

____

15. Circular interpolation requires much computation within the controller. Many CNC machines must limit the feed rate while doing circular interpolation because the large amount of data output taxes the controller's ability to keep up with the feed rate. This limit is called a *default* feed rate. Does this lathe require a default feed rate for circular interpolation?

____

16. What feed rate is possible when doing linear interpolation? (Does this machine have a default feed rate for interpolation?)

____

UNIT 3

Operating Programmed Machine Tools

Machine operation will probably be your first job experience in C/NC. A machine operator is responsible for the production run. C/NC operators load parts, monitor production, do routine machine maintenance, and perform secondary operations on completed parts. They may even have some authority over the program. Operators usually do not write the program nor do they do their own setups at first. Experience gained in machine operation can lead to other responsibilities and careers. With competency, operators are allowed to do setup work and may write some programs. This unit is designed to teach you the skills of machine operation.

CONTENTS

CHAPTER 5

THE COMPONENTS OF AN NC SYSTEM

This chapter presents an overview of numerical controlled (NC) equipment. Although this book is focused on the more modern C/NC, information on NC machines is essential. There are many in use in industry.

OBJECTIVES

After studying this chapter, you will be able to:

- Identify the components of an NC machine system.
- Describe the purpose of each component.
- Explain data storage media in C/NC (tape and computer disk).
- Understand how information is impressed on a punched tape.
- Be able to identify tape codes in Electronic Industries Association Binary Code Language System (EIA-BCL) and American Standard Code for Information Interchange (ASCII).

5-1 Numerical Control Components

FOUR COMPONENTS OF A NUMERICAL CONTROL SYSTEM

NC machines do not analyze data. They follow instructions in the way that a music box does. For example, a music box has a program built into the revolving cylinder. Each time the program is passed through the reader, music is produced. Modification of the program would require a new cylinder. NC machines are similar. Their program is usually on a flat tape with holes punched in it to transmit the program data to a reading device. There are four components in an NC system.

- The machine.
- The control, which includes the tape reader.
- The tape (program data).
- The tape punch.

The Machine

An NC machine is a machine designed or adapted to be operated by numerical data fed into the control by a hard media, such as tape or cards. NC machines come in many forms. Unit 1 discussed many different applications of NC in manufacturing. Any machine may be driven by NC. Some are complex, while others are simple. All follow the axis system and use coordinates to locate cutting tools. Other than the controls, NC machines appear to be similar to CNC machines. Fig. 5-1.

Many of the more complex, expensive NC machines are being upgraded to CNC. Basically, the machines are the same. Mechanically, there is little or no difference between an NC and CNC machine, with the exception of the control. The machine shown in Fig. 5-1 is now being upgraded to CNC.

An NC machine axis must have a drive system designed for programmable control. The control may or may not operate all of the functions of the machine. For example, an NC drill press might have an automatic positioning table but leave the Z axis drilling for manual or semiautomatic operation.

The Control and Tape Reader

NC controls perform two functions.

- They read and execute commands from the tape.
- They allow manual movement of the machine for setup or simple manual operation.

The tape reader is an integral part of the control for NC. Although other program storage methods have been used, punched tape is the most common media. Other methods include magnetic tape, decks of punched cards, and insert peg boards.

An *NC controller* is the device that interprets the program commands and translates them into axis moves and machining operations. Fig. 5-2. The controller receives the program *one command at a time* through the tape reader. A second function of the controller is to allow the operator manual control of the machine for setup work and adjustment during the machining operation.

Fig. 5-1. This huge C/NC Spheromill (turning mill) will be upgraded to C/NC. The Boeing Company

Fig. 5-2. An industrial NC machine control. Note the tape reader. Eldec Corporation

Tape Cycles

The tape reader senses the data optically. As the tape is fed past the sensor, a light source is shown through the tape. Sensors are triggered when holes pass by. These pulses are translated into the "words" that commence machine operation.

A *cycle* is one full run of the tape or one complete run of the program on a part. Each time the machine is cycled (a part is made), the NC tape must pass through the reader.

The NC Tape

An NC tape is 1" wide paper laminated to mylar plastic. Fig. 5-3. This makes a nearly indestructible media. Each line has eight tracks in which a hole may be punched. The hole indicates a signal to the sensor. No hole is a null (zero) for that position. The use of "on-off" data is called a *binary system*. *Binary* means a number system with a base of 2. The digits available in the binary system are 1 and 0. We will briefly study binary number arithmetic and tape codes in Section 5-2.

An *NC tape* is the total storage media for the program data. Data is coded onto the tape by means of holes punched in eight tracks on the tape. To change or edit the program, a new tape must be made.

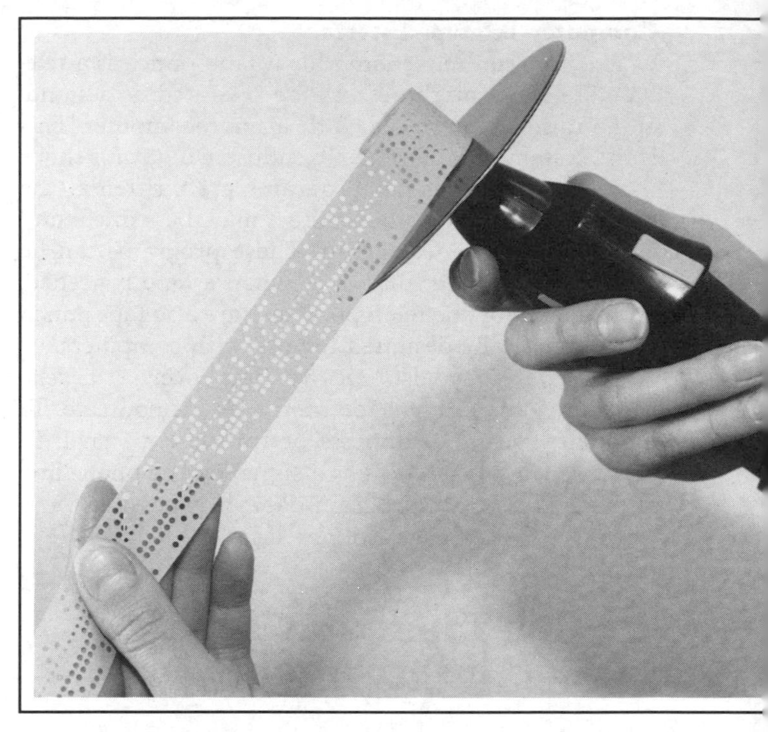

Fig. 5-3. The industrial standard 1" NC tape. This tape is Mylar plastic laminated to paper. Eldec Corporation

The Tape Punch

Although the tape punch is not part of the actual machining operation, it is necessary in the NC process. Fig. 5-4. C/NC tapes are made on a device that looks very much like a large typewriter. Due to the complexity of the tape punch-writer, it is usually not kept in the shop environment.

Tape punches have two different reels. Each performs a different function. The first reel is used to hold the tape as it is punched from the keyboard. The second reel can play back a tape program and print out the program on paper. Using this function, a tape may be duplicated or read onto paper for editing. These units are able to punch tape and type the program on paper, too. This is useful for two reasons.

- Existing tapes are easier to edit by playing back the correct parts and modifying only the bad parts. The new tape is made on reel one, while the old tape is run through reel two. A *hard copy* is a printed sheet that you can read. A hard copy can be printed directly off the tape by running the tape through the unit and having it type a sheet of paper with the program.
- The punch machine is also useful as a programmable typewriter. Now, however, these machines are seldom used in this capacity. Word processing can easily be done on microcomputers. In earlier NC shops, the tape punch was the only means of producing the tape. Many shops still use this system exclusively.

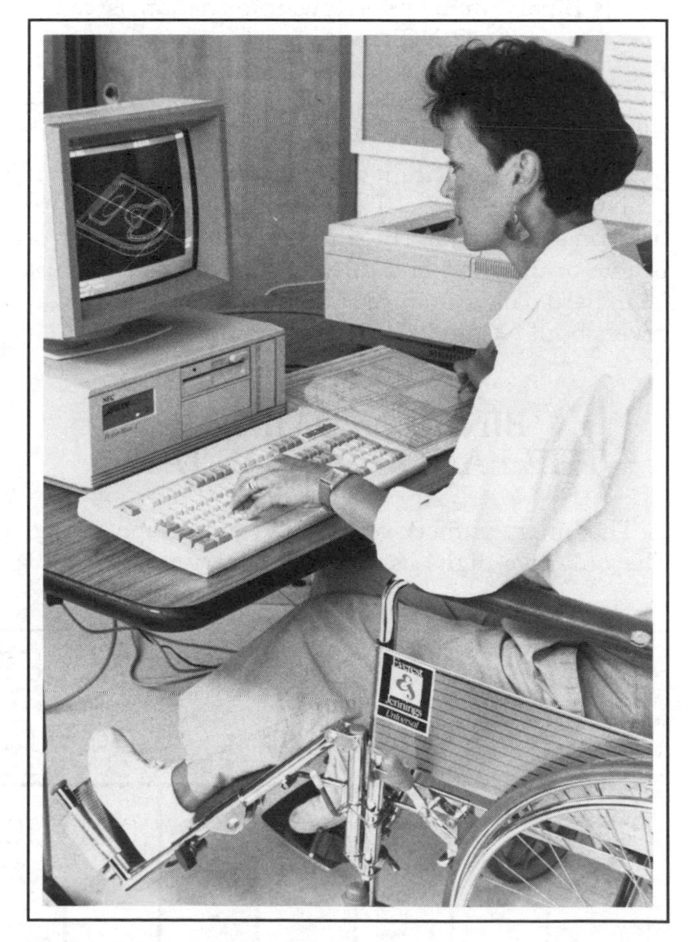

Fig. 5-4. A tape punch or teletype is a standalone program station. It may also be interfaced with a computer or phone line modem.

Computer-Teletype Interface

Another common name for a tape punch is a teletype. Teletype machines can be operated as a stand-alone unit or interfaced with a microcomputer. This is the system of preference in industry today for shops that use punched tape. Programs are written at the computer console, where they may be edited and viewed on the video screen. These programs can be stored on computer disk and, when a tape is needed, punched out by the teletype. There are also tape punch units specifically designed for use with computers.

Teletypes may also receive data over a *modem*. Modem is an acronym for Modulate-Demodulate. To modulate means to impress or translate a signal. In this case, data is received as a signal on the phone line and translated into program data. It is possible to receive C/NC data over a phone line.

5-2 Tapes and Tape Information

This section discusses the following three features of C/NC tape.

- Binary numbers—How information is coded onto the C/NC tape.
- Formats—How data is arranged on the tape.
- Code structures—The two types of code sets—EIA-BCD and ASCII-ISO.

There is no significant difference in the information on NC and CNC tape. The information may be more brief on a CNC tape. The smarter CNC master control is able to interpret and calculate much of what must be in hard data form for NC information. This section covers both NC and CNC tape coding.

BINARY NUMBERS: THE LANGUAGE OF TAPES

The binary number system is ideal for tape coding. The only two digits used in the binary system are 0 and 1. A zero is represented by no hole in the tape. A one is represented by a hole. This on/off situation is well-suited to digital electronics. Zero is off and one is on.

To understand the binary number system, let us first review base ten numbers. Binary numbers have similarities to base ten numbers. The only difference lies in the value placed on each column right and left of the decimal point.

Base Ten Values for Columns

Each position (column) left and right of the decimal has a different value, expressed as a power of ten. Fig. 5-5.

In base ten, a number placed in the first column to the left of the decimal point represents the digit shown multiplied by ten to the *zero* power.

$$10^0 = 1$$

A digit placed in the second column is the digit times ten to the *first* power.

$$10^1 = 10$$

A digit placed in the third column is the digit times ten to the *second* power.

$$10^2 = 100$$

Numbers to the right of the decimal point become *ten times smaller* with each column further to the right. The first column's value to the right equals

$$10^{-1} = .1$$

The second column's value to the right equals

$$10^{-2} = .01$$

The third column's value to the right equals

$$10^{-3} = .001$$

Example: The number 27.35 in base ten is equal to 2×10 plus 7×1 plus $3 \times .1$ plus $5 \times .01$

Starting from the leftmost column, we added the value of the digits times the value of the column.

Each column has an ascending value of ten. The system is based on multiples of ten.

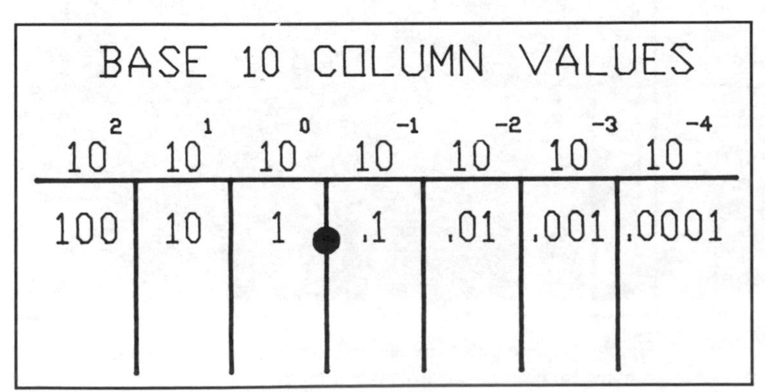

Fig. 5-5. Base ten columns. Each column right and left of the decimal point represents a power of ten.

In a binary system, the columns are ascending values of two. Fig. 5-6. Each column has a value of a power of two. The digit one (1) in the column represents the value of the column. A zero in the column is a place holder, indicating no value.

Here are two examples of numbers expressed in binary.

Example: The number 27 expressed in binary is 11011.

Starting from the fifth column to the left.

Base 2 Value $1\times 2^4 + 1\times 2^3 + 0\times 2^2 + 1\times 2^1 + 1\times 2^0 = 27$

Base 10 Value $16 + 8 + 0 + 2 + 1 = 27$

Example: The number 5.75 expressed in binary is 101.11.

Base 2 Value $1\times 2^2 + 0\times 2^1 + 1\times 2^0 + 1\times 2^{-1} + 1\times 2^{-2} = 5.75$

Base 10 Value $4 + 0 + 1 + .5 + .25 = 5.75$

5-3 Tape Standards

The standard C/NC tape is 1″ wide and has an eight-channel format. This tape size is standardized by the Electronic Industries Association (EIA). There are EIA specifications for all dimensions and features of the tape.

TERMS

The following terms are used to describe tapes and the data they hold.

Channels. There are eight information channels in EIA tape. If the tape is held vertically, the channels go from right to left. There is a sprocket drive channel between channels 3 and 4. The drive channel is off center, so that the tape cannot be put in upside down. Fig. 5-7. Channel 5 is used for error detection and not for program data. Fig. 5-7 on page 72.

Channel 8 is used to indicate the end of a block only. A punched hole in channel 8 indicates a complete command to be carried out by the machine.

Row. A *row* is horizontal across the tape. One row represents one *character* in the program.

Words. Program *words* are made up of letters and numbers. Words occupy rows. A complete word may require from two to several rows of punched information.

Commands. Words are combined into a group to prompt a machine movement. These word groups are called either *commands* or *blocks*.

5-4 Punched-Tape Coded Systems

CODED SYSTEMS FOR TAPE-PUNCHED INFORMATION

Information punched on tape includes numbers, letters for axis identification, and symbols such as minus and block end symbols. To simplify the coding of program data and binary numbers on the tape, two separate sets of standards are used—EIA/BCL and ASCII.

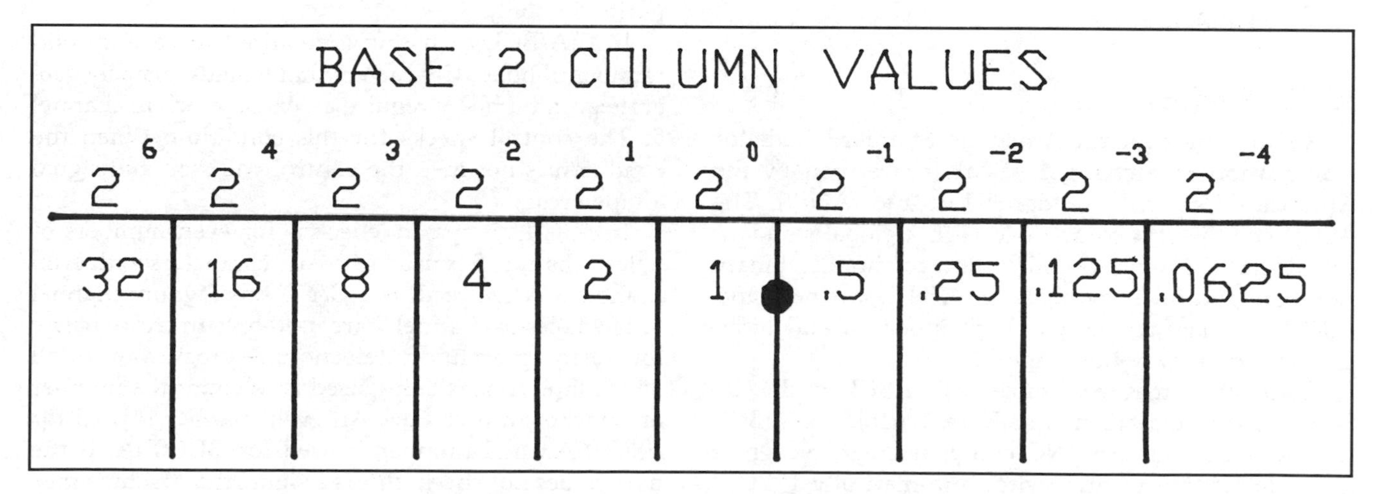

Fig. 5-6. Base two columns BINARY. Each column in a binary system represents a power of two.

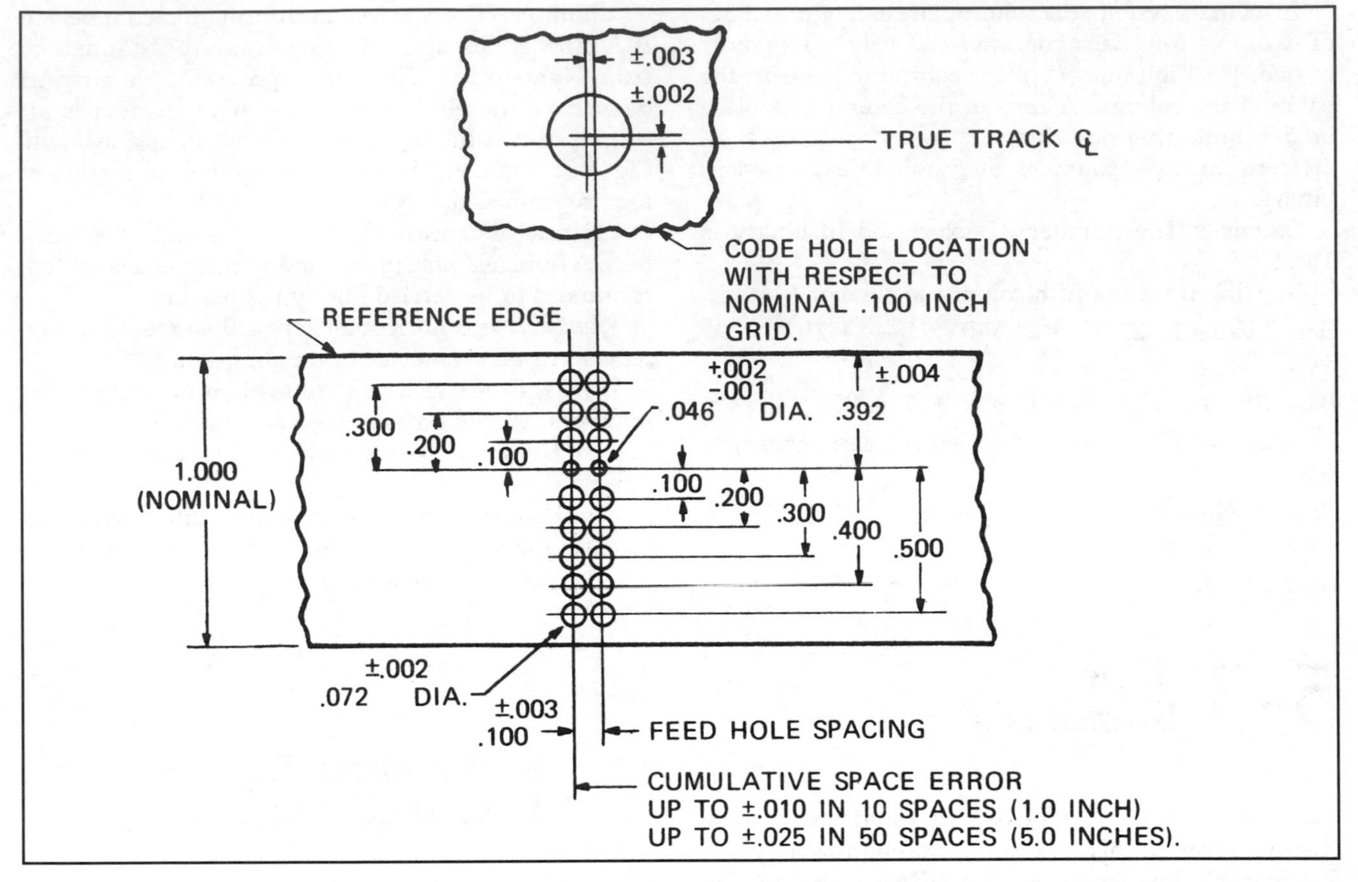

Fig. 5-7. A standard eight-channel tape per the EIA recommendations. Note the sprocket drive row. Reprinted with permission from EIA-227-A

EIA System

A system called the *binary coded language* (BCL) is also standardized by the EIA. This is simply a method of listing what each character or number will look like when punched into the tape. From the brief development on binary numbers, you can see that even short decimal numbers are complicated. BCL simplifies that problem. By using the placement of the punch on the tape, the numbers and symbols are shortened and standardized. Fig. 5-8.

ASCII System

ASCII stands for the American Standard Code for Information Interchange. ASCII is overseen by the American National Standards Institute (ANSI). The ASCII code is also compatible with a global standard called ISO. This is a coded system to simplify binary numbers and other symbols for C/NC punched tape. ASCII code has provision for both uppercase and lowercase letters. Fig. 5-9 on page 74.

Controllers may recognize both ASCII and BCL. Some controllers are manually switchable from BCL to ASCII. Advanced C/NC controllers can recognize the code being used and switch automatically. If BCL/ASCII switching or recognition is not part of the control, then the correct code must be used by the programmer. Otherwise, the program will not work.

Parity Checking

Another difference between ASCII and BCL is in parity check. *Parity check* is a system used to detect some kinds of errors. In parity check, each row is required to have either an even or odd number of punched holes. Channel 5 in the tape is used for a parity line only.

In EIA/BCL each row is required to have an odd number of holes. Characters that would normally generate even numbers require an extra punch in channel 5. The control checks for this condition. When the condition is not met, the control will stop and signal a tape error.

In ASCII, the parity check is for even numbers of holes. Channel 5 is used to create characters with even numbers. When reading a C/NC tape, ignore channel 5. The holes in channel 5 are used only to create parity for—error prevention/detection. A quick way to tell which tape code is being used is to count the number of punches in each row. An even number is used for ASCII. An odd number is used for BCL/EIA. If the parity does not check, this is a signal that the tape may be mispunched.

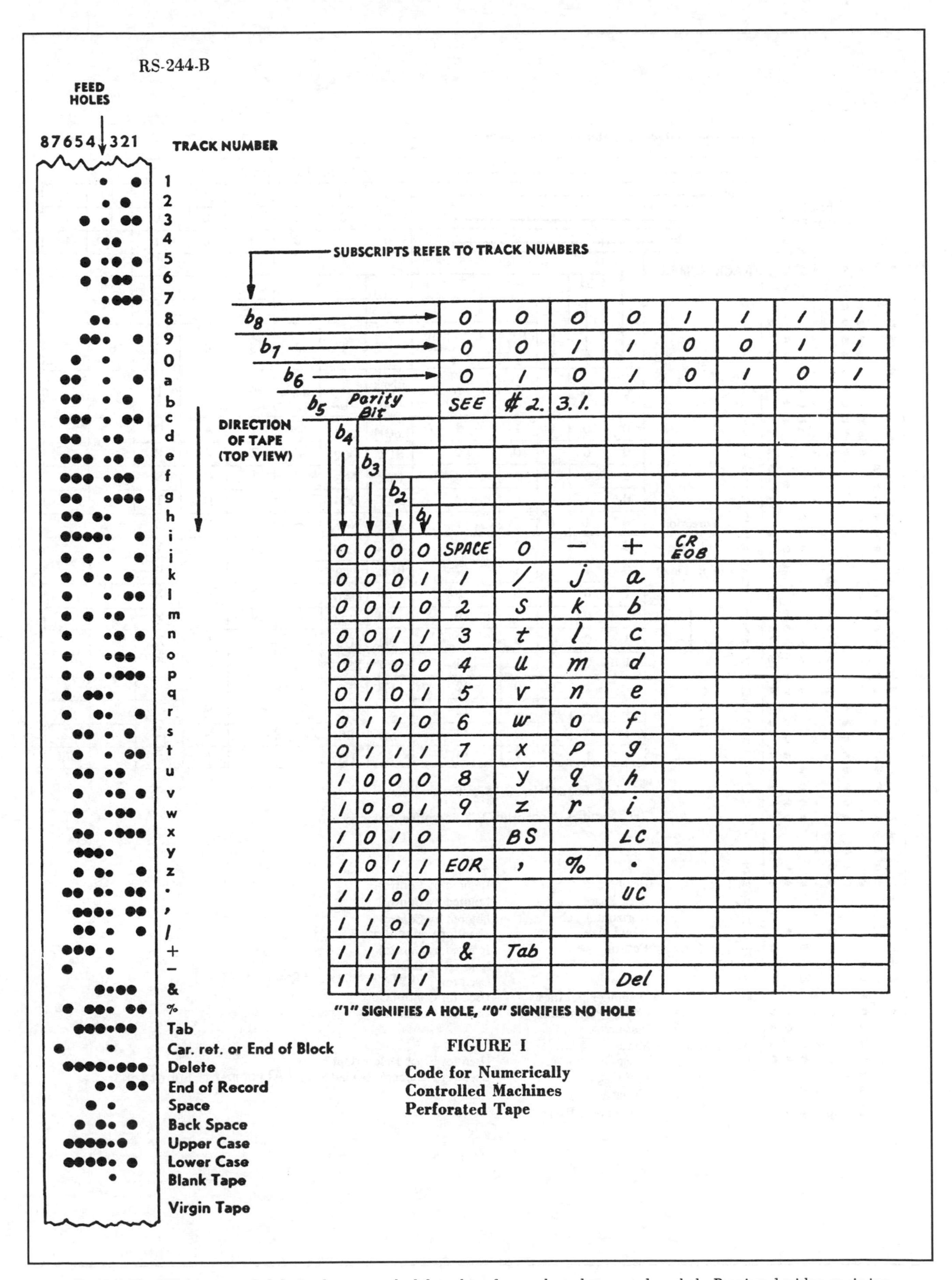

b_4	b_3	b_2	b_1	b_8 = 0 b_7 = 0 b_6 = 0	b_8 = 0 b_7 = 0 b_6 = 1	b_8 = 0 b_7 = 1 b_6 = 0	b_8 = 0 b_7 = 1 b_6 = 1	b_8 = 1 b_7 = 0 b_6 = 0	b_8 = 1 b_7 = 0 b_6 = 1	b_8 = 1 b_7 = 1 b_6 = 0	b_8 = 1 b_7 = 1 b_6 = 1
b_5 Parity Bit				SEE	# 2.	3. 1.					
0	0	0	0	SPACE	0	—	+	CR EOB			
0	0	0	1	1	/	j	a				
0	0	1	0	2	s	k	b				
0	0	1	1	3	t	l	c				
0	1	0	0	4	u	m	d				
0	1	0	1	5	v	n	e				
0	1	1	0	6	w	o	f				
0	1	1	1	7	x	p	g				
1	0	0	0	8	y	q	h				
1	0	0	1	9	z	r	i				
1	0	1	0		BS		LC				
1	0	1	1	EOR	,	%	.				
1	1	0	0				UC				
1	1	0	1								
1	1	1	0	&	Tab						
1	1	1	1				Del				

"1" SIGNIFIES A HOLE, "0" SIGNIFIES NO HOLE

FIGURE I

Code for Numerically Controlled Machines Perforated Tape

Fig. 5-8. The EIA binary coded decimal tape-punched data chart for numbers, letters and symbols. Reprinted with permission from EIA 244-B

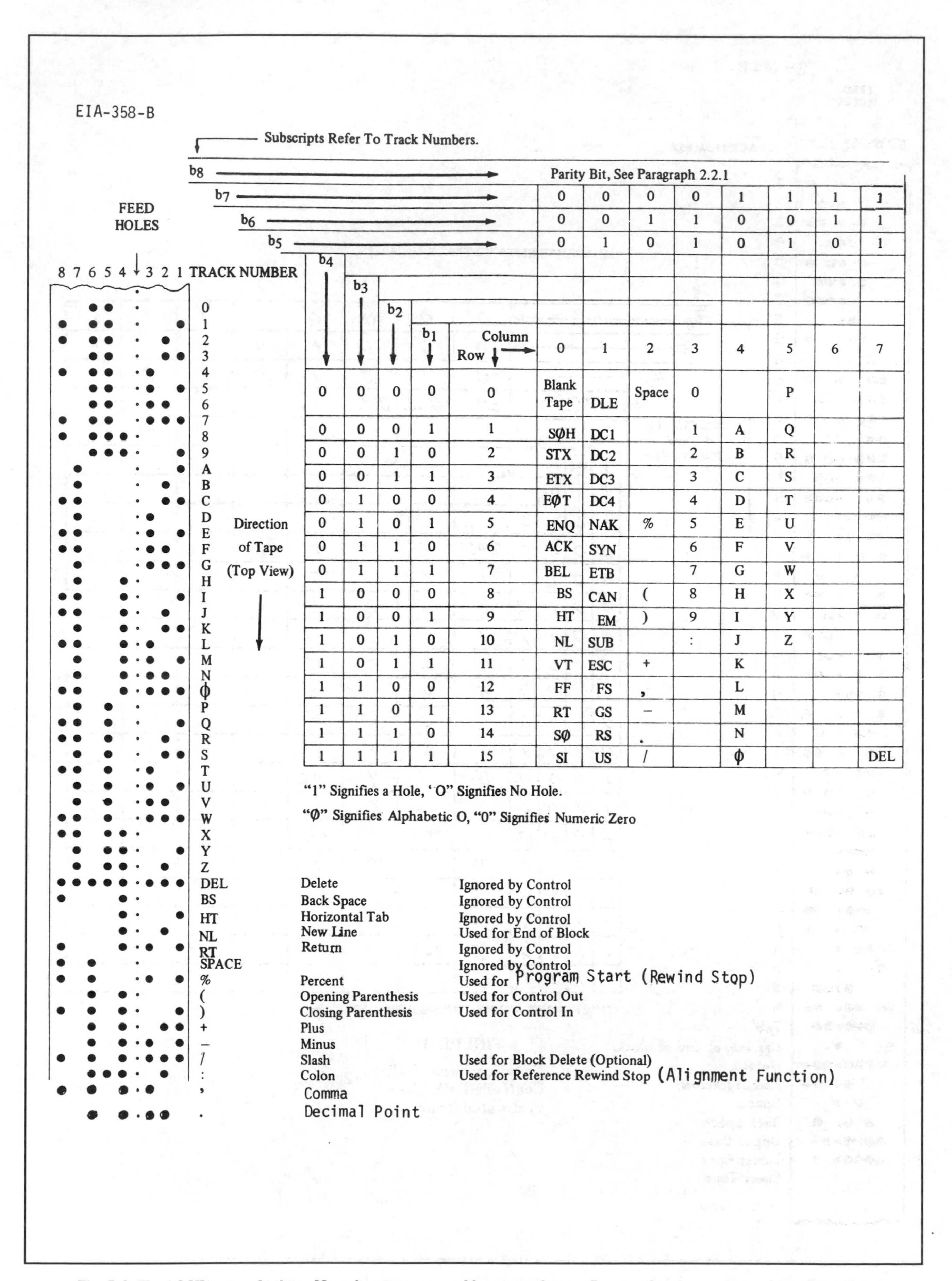

EIA-358-B

Subscripts Refer To Track Numbers.

b_8 → Parity Bit, See Paragraph 2.2.1

				b_7	0	0	0	0	1	1	1	1
				b_6	0	0	1	1	0	0	1	1
				b_5	0	1	0	1	0	1	0	1
b_4	b_3	b_2	b_1	Row ↓ / Column →	0	1	2	3	4	5	6	7
0	0	0	0	0	Blank Tape	DLE	Space	0		P		
0	0	0	1	1	SØH	DC1		1	A	Q		
0	0	1	0	2	STX	DC2		2	B	R		
0	0	1	1	3	ETX	DC3		3	C	S		
0	1	0	0	4	EØT	DC4		4	D	T		
0	1	0	1	5	ENQ	NAK	%	5	E	U		
0	1	1	0	6	ACK	SYN		6	F	V		
0	1	1	1	7	BEL	ETB		7	G	W		
1	0	0	0	8	BS	CAN	(	8	H	X		
1	0	0	1	9	HT	EM	)	9	I	Y		
1	0	1	0	10	NL	SUB		:	J	Z		
1	0	1	1	11	VT	ESC	+		K			
1	1	0	0	12	FF	FS	,		L			
1	1	0	1	13	RT	GS	–		M			
1	1	1	0	14	SØ	RS	.		N			
1	1	1	1	15	SI	US	/		Ø			DEL

"1" Signifies a Hole, "O" Signifies No Hole.

"Ø" Signifies Alphabetic O, "0" Signifies Numeric Zero

Code	Name	Use
DEL	Delete	Ignored by Control
BS	Back Space	Ignored by Control
HT	Horizontal Tab	Ignored by Control
NL	New Line	Used for End of Block
RT	Return	Ignored by Control
SPACE		Ignored by Control
%	Percent	Used for Program Start (Rewind Stop)
(	Opening Parenthesis	Used for Control Out
)	Closing Parenthesis	Used for Control In
+	Plus	
–	Minus	
/	Slash	Used for Block Delete (Optional)
:	Colon	Used for Reference Rewind Stop (Alignment Function)
,	Comma	
.	Decimal Point	

Fig. 5-9. The ASCII tape code chart. Note the uppercase and lowercase figures. Reprinted with permission from EIA 358-B

5-5 Tape Formats

The arrangement of information within a complete block or command has undergone evolution in C/NC tapes. The various ways of arranging the program are called *formats*. Here are two typical blocks of C/NC information:

N001 G00 X4.75 Y1.5625 Z1.5
N002 G01 X0.50

This simple program rapid-travels in three axes. It then feeds in a straight line using the X axis only. The words in the program have the following functions:

N001 and N002 are the sequence numbers.
G00 means "go at rapid travel."
G01 causes linear interpolation and feed rate.

You already recognize the X, Y, and Z words. Now, we will see how this program information looks in each format. We will start with earlier formats. We then will look at the improved tape formats in use today.

EARLY TAPE FORMATS

Early tape formats included fixed sequence, tab ignore, and tab sequential. The same program is shown below in each of these formats.

Fixed Sequence Format
00100047500015625015000
00201005000000000000000

Notice that the blocks that had no information within had to be filled in. Also notice that the X information has to fall in the right place, with the decimal inferred by the placement of the words. Everything must be in the right place. This format is very difficult to read.

Tab Ignore Format
001 00 04750 015625 01500
002 01 00500 000000 00000

Now there are spaces caused by the tabs written into the program. They do not show on a printout. They are ignored. Every word must still be in the right place. Blank words must be filled in. There is no easy way to know what the words are other than by knowing the location in which they are shown. The tab ignore format is, however, a little easier to read.

Tab Sequential Format
001 00 04.750 01.5625 01.500
002 01 00.500

Notice that words that do not need to be repeated are ignored on a printout. This helps in reading and programming.

The early tape formats required rigid placement of the data. If the command was an X word, it had to fall in the right place in the line. The command was identified only by the location in which it fell. These formats were difficult to read. They required that every line be filled, even though a part of the line was not being used. Much programming was repeated.

INDUSTRIAL TAPE FORMATS TODAY

The two tape formats in most common use today are the *word address* and *interchangeable word address* formats. In the word address format each line contains only the needed information. The lines need be only as long as required. For example, if only an X movement is needed, a Y word command could be omitted. Commands that are repeated from one block to the next are omitted. A command to move in linear interpolation is needed only once. Then for all blocks beyond, linear interpolation is automatic until changed. This saves time and simplifies the program.

Modal Commands

The repeating of information is called *modal repetition.* A code or command that continues until changed or cancelled is called a *modal command* because the control is put into the mode by the command and stays there until told to change or cancel.

Word Address
N001G00X04750Y001.5625Z01.5000
N002G01X00.5000

Each word in the command is addressed with a letter. This aids in reading the program. Note the lack of spacing.

Interchangeable Format
N001 G00 X047500 Y015625 Z015000
N002 G01 X005000

The interchangeable format is the same as the word address format, but adds two extra features.

- Tabs are allowed between words to increase readability.
- Words may be interchanged within the line. This allows flexibility in programming.

All coded programs in this book will be in the interchangeable word address format, but will add another feature, decimal point representation. Most programmers would identify the interchangeable commands above as being word address. Word address is the common term used for both. From this point on, *word address* will be the term used to refer to this type of format.

LEADING AND TRAILING ZEROS

Earlier controllers required a certain number of digits in each number block. If a digit was not needed, the space was filled with a place-holding zero. Zeros were placed before and after significant numbers to fill the full block. Today, data on most offline computers and CNC controllers does not require leading and trailing zeros. For example, in sequence number 001 in the example above, the X coordinate was:

Old Style

Leading zero   Trailing zeros

X04.7500

New Style

X4.75

The zeros at the front and the back had to be in place so the number occupied the correct number of columns. This places the decimal point in the correct position. Computers eliminated the need for this practice. For ease of reading and general math consistency, all programs are presented in the interchangeable word address format without consideration for leading or trailing zeros.

You may encounter an older system that does require zero management. If you do, it is a simple matter to adapt it. You may place leading or trailing zeros on most C/NC controls. As long as the decimal is properly placed, the computer will understand.

N001 G00 X4.75 Y1.5625 Z1.5
N002 G01 X0.5

Programs then will appear in this form, which is easy to read. A zero will be placed in front of the decimal point for numbers less than one such as 0.50 above. With computer data, leading or trailing zeros may be added as long as the decimal point is in the correct place.

CONVERSATIONAL LANGUAGE

Some CNC controllers do not use coded information. These controllers use a system called *user friendly*, or *conservational language*. We will look at this system in later units. Conversational language can be put on tape by first translating it into coded words. Usually, though, it is stored on disk or in internal memory. Conversational programming is sometimes called "shop language programming."

In conversational language, some codes are replaced with English words or their abbreviations. The general arrangement of blocks, the programming logic, and the overall process remain the same.

Most conversational control units are *interactive*. This means that they question the programmer with prompts, analyze data as it is input, signal when something is incorrect, and work intelligently with the pro grammer to write programs. There is no conversational NC.

You do not have to be able to read C/NC tapes to be a good operator. You should understand how the information is encoded on the C/NC tape. After reading Chapter 5, you should be able to:

1. List and describe the four components in an NC machining system.
2. Name the industrial standards for:
   A) tape size and information tracks.
   B) the number system used in tape coding.
   C) common tape formats today.
3. Describe parity.

Have you reached these goals? The following student activity should help you test yourself.

1. In an NC machine, a program is contained
   A) On the tape.
   B) In the onboard memory.
   C) In binary code in the MCU.

2. If a double-spindle NC milling machine makes two parts per cycle, how many times must the tape be run through the MCU to produce 126 parts?

____

3. The tape reader is part of the NC controller.

☐ True ☐ False

If this statement is false, what will make it true?

____

4. In terms of RAM, what is the difference between NC and C/NC? (You may need to refer to Unit 1 for the definition of RAM.)

____

5. List the four components in an NC system.

____

____

____

____

6. Which component is not connected to the machine?

____

7. Write the binary equivalent of the base ten number 4.25.

____

8. The binary number 110.01 is equal to what in base ten?

____

9. How could you quickly tell if an NC tape was EIA/BCL or ASCII/ISO?

____

10. Parity is a system used to ______________________________.

11. Track eight is used to indicate the letter A on C/NC tape.

☐ True ☐ False

If this statement is false, what will make it true?

____

12. A complete group of words that cause a machine movement are called a ______________________

or a ______________________.

13. Refer to Fig. 5-8. Darken the correct punch points for a BCL/EIA punched tape for:

    Number 4

    Letter X

    Symbol – (minus)

					XX			

					XX			

					XX			

14. Refer to Fig. 5-9. Darken the correct points to represent an ASCII punched word for:

    Number Character 4

    Letter X

    Symbol – (minus)

					XX			

					XX			

					XX			

In Your Lab

Does your lab have a machine that uses punched tape? If so, answer the next five questions.

1. What tape format does your machine use?

    ____

2. Can your machine use both EIA/BCL and ASCII/ISO? If not, which does it use?

    ____

3. Take a sample of the tape that your instructor assigns. Does it have odd or even parity? How can you tell?

    ____

4. Your instructor will supply you with a simple program on punched tape from a machine in your shop. Interpret a single block that your instructor assigns.

____

____

5. Does your controller require leading and trailing zeros? Write the following numbers in a way your controller will understand.

   A) .56 ___________________________________

   B) 1.01 __________________________________

   C) 15. ___________________________________

   D) 2.0022 ________________________________

THE COMPONENTS OF A CNC SYSTEM

The microprocessor greatly expanded the capabilities of programmed machine tools. This chapter will familiarize you with the major components of a CNC system.

OBJECTIVES

After studying this chapter, you will be able to:

- List and describe the major components in a CNC system.
- Describe the types of data storage available.
- Compare the various methods for programming a CNC control.
- Describe and compare open-loop and closed-loop systems.
- List the special features of CNC.

6-1 The Four CNC Components

There are four components in a CNC system.

- The machine.
- The control.
- Data storage devices.
- Programming stations.

THE MACHINE

As pointed out in Chapter 5, when comparing the actual machine, there is no significant difference between an NC and a CNC machine.

NC Machines May Be Upgraded

Many CNC machines have been upgraded from NC equipment. An upgrade might require different drive motors and feedback senders to make the machine compatible with the new controller. A better set of ball screw drives is usually added. Any NC machine may be converted to CNC by adding a microprocessor-controller.

In Unit 1, we looked at the concept of *feedback*. This was the signal from the machine axes to the control that told the control the movement command was being carried out correctly. Feedback is accomplished by either direct-reading scale type devices or indirect-reading rotary devices fastened to the machine axes. Machines that have feedback are called *closed loop*. The machine must have a controller that has the ability to analyze feedback signals.

Open Loop

Drive motors can be stalled by dull cutters, fast feed rates, or cutting too much material. If stalling occurs on a machine without feedback, the controller has no way of knowing that something is wrong. If the cutter lags behind the controlled commands, the shape of the part will be changed. The cutter will move less distance than commanded. It is even possible to stop an axis completely. The controller will never know unless it has feedback.

Closed Loop

Closed loop controls can compare the actual move command to the position of the axis. The machine knows where its axes are at all times and can tell if there is a problem with the synchronization between command and position. If an axis stalls through too much feed or a dull cutter, the controller can take corrective action—within limits. The controller can bring the cutter back into coordination if the error is not beyond a set tolerance for deviation. Otherwise, it must stop all motion and signal the operator that an "out of synchronization" state exists. Dealing with changing loads is the main advantage of closed loop technology.

As might be expected, open-loop machines are far less costly. Not only are the feedback senders not present, but the control needs less capability. It does not need to compare a returning signal.

Advanced Features

All the advantages of CNC machines are due to the controller's ability to analyze data. CNC machines can sense dull tools by either pressure, sound, vibration, or using optics. CNC machines can change to sharper tools when needed. They can optically sense tool positions. They can use electronic feeler gauges to set the PRZ. CNC machines can detect their own mechanical backlash and compensate for it without interference from the operator.

THE CNC CONTROLLER

There are major differences in NC and CNC controllers. Because it has a microprocessor, a CNC machine has valuable capabilities.

It can *run programs from internal memory.* The program is held in the active **random access memory (RAM)** during machine operation. If the machine does use tape, it only needs to be run through one time to dump the program into RAM.

It can *save programs to permanent internal memory.* Some CNC controls have a memory storage area for programs that is separate from active RAM. They allow the control to store programs and recall them to active RAM when needed.

It uses *manual data insert programming.* A CNC controller acts as a programming unit as well as a machine control. *Manual data insert (MDI)* is possible. This allows new programs to be written at the control. Also, previous programs may be edited or modified. The ability to modify programs at the control can be very useful. For example, a shop that makes different bushings on a CNC lathe might have a basic bushing program saved in the control's permanent memory. When needed, the program is uploaded to active RAM and modified to fit the particular part print.

A CNC *can operate without a program.* This ability is used in the following three ways.

- *Manual data operation.* A CNC machine may be operated with no program at all. The operator supplies the commands from the console key pad. This is called manual data insert (MDI) machine operation. It is slower than program execution. However, it has limited use in setup and in making one-time-only parts. This technique can also be used to develop a program as you go along, using progressive programming.
- *Progressive programming with first part run.* During setup, it is possible to develop the program using the method above and save each step to memory as it proves successful. Each program command is written in from the keypad. Then the command is executed. If correct, the command is stored in memory. As the part is developed, so is the program—one block at a time. A skilled setup person can write a program and produce a good part at the same time. This practice is called manual data insert (MDI) programming.
- *Digitizing.* Some advanced CNC controllers allow the operator to "lead the machine through the shape." The operator physically moves the machine to each significant point. The machine will remember the path and reproduce it. The tolerances will be only as accurate as the operator's "eyeball" placement of the spindle. In certain situations, this technique is very useful.

  Digitizing is also used elsewhere in manufacturing. It is common in CAD drawing where the draftsperson uses a digitizing probe to "lead" the drawing lines around on the screen. Another common application is in the programming of robots. Here the setup person leads the robot through its movements. Meanwhile the computer remembers the positions and the path taken to get there.

  CNC digitizing is not done by the programmer unless he or she is also the setup person. Because digitizing lacks accuracy, it is useful only where exact precision is not important. A good example would be in endmilling a pocket in a part. For accuracy, the outside edge dimensions would have to be programmed in the usual way. The interior of the pocket could be digitized. If the pocket shape was complicated, the digitizing moves could be programmed more quickly than they could be calculated and written out.

It can employ graphic part and program representation. This feature is available on some CNC controllers. Graphics can be used for programming by visualizing each step on a screen as it is written. The video display on the control can simulate the move as it is written. Graphics can also be used to dry test a program after it is written. The machine moves are shown on the screen. Each move can be tested for acceptability without actually running the machine.

It is difficult to verify dimensional accuracy with this method, but many program errors are obvious—such as omitting a minus sign on an axis move. It will still be necessary to run a trial "first part" after graphic testing the program. The graphic image shows the part being produced in real time. Latest versions show the part in one color and the tool in another color. Rapid moves are shown in phantom lines. The tool path, tool shape, and progressive passes of the tool are shown. This feature saves time and helps ensure safer programs. Graphics also enable the operator to monitor the program during actual machining. The screen shows the operator exactly what is going on in the machine, though the actual cutting may be blocked out by cutting fluid, chips, and safety guards.

Part plotters are also available. These draw a simple straight line representation of the program. Without graphics or a plotter, it is a common practice to place a pen in a milling machine spindle and draw the program on paper.

It can use interactive graphic programming. Interactive graphics is a less common feature because it requires more intelligent (and expensive) internal software. Interactive graphics will be explored more in Unit 7. New CNC controllers such as that shown in Fig. 6-1 can develop their own programs from visual descriptions of the part. These visual descriptions are supplied by the setup person, who enters the part shape on the MDI keypad. This image is created on the screen much the same as a drawing is created through CAD. Once the part shape image is complete, the controller then asks some important questions. These relate to the material type, cutter selection, holding methods, and clamp locations. From these parameters, a program is written *by the controller.* This feature is not available on all controllers. There are also software packages that allow CAD units to create programs and then translate them into specific C/NC machine language. This is called *postprocessing.*

Fig. 6-1. For safety and convenience, this operator can monitor the part run from the control graphics.

Other Features Found on Modern CNC Controllers

Tool Compensation

CNC controls can compensate for cutters of a different size and shape. If a cutter is changed, the new cutter need not be the same size to produce a part of the correct shape and size.

Part Scaling

Some controls permit increasing the size of a part without changing the program. A percentage of growth or shrinkage is written into the program. The control automatically scales up or down. Some advanced controls allow scaling on one axis at a time, thus changing the shape of a part. For example, a circle in the X-Y plane could be scaled in one axis only and made into an ellipse. A square could be scaled into a rectangle. Fig. 6-2.

Mirroring

This feature is also found on advanced controls. A symmetrical part needs only be programmed for half of its shape. The control can be commanded into the mirror mode, and the program run through a second time. In Fig. 6-3, the right half of the program produces the part in solid lines. After commanding the machine to mirror the X axis, the program is run again from the same start point. This time each positive X command actually causes a negative X movement. The machine produces a mirror image or left-hand version of the program (the dashed line in the example.)

There is a problem in using mirroring. Note that the cutter would be climb milling on the right half and conventional milling on the mirrored left. This difference in cutting action can cause problems. This problem is often solved by using a reverse-rotating (left-hand) cutter for the mirrored portion of the program.

DATA STORAGE DEVICES IN C/NC

There are five common ways that programs are stored for future use in C/NC.

- Punched tape.
- Magnetic tape.
- Computer disk—floppy.
- Computer disk—hard.
- Bubble memory.

Punched Tape

Advantages

Punched tape is useful for hard storage of programs. Though it can be physically damaged, punched tape is very stable. Once it has been downloaded into the CNC RAM, the tape may be stored. The tape is not needed to run the machine. As mentioned, punched tape can be used in combination with computer disk memory. The computer is used to write and store the program. Then the tape is used to program the controller.

If the shop has no **direct numerical control** interfacing (DNC), the programming department works on computer disk. When the program is complete, a punched tape is made from the computer data. This tape is then taken to the shop.

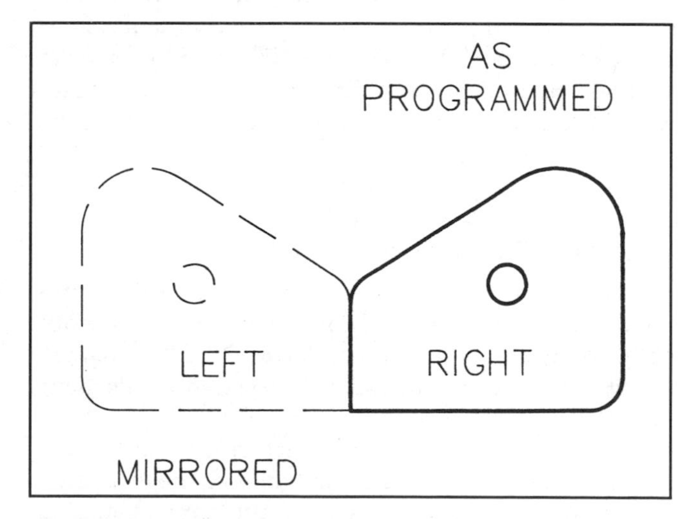

Fig. 6-3. The solid line represents the base program. The dotted line is the second half after mirroring. Note the reversed cutter direction.

Fig. 6-2. This circle could be changed into an ellipse by upscaling the X axis. The square would then become a rectangle.

Disadvantages

Punched tape takes up space. A second problem relates to the mechanical reader. The reader must be in the shop. It requires frequent maintenance. A tape punch must also be maintained. This is a complicated device that also requires maintenance. For these reasons, tape is becoming less popular. Compared with tape, computer memory requires less space, is more compatible with CNC, and can be quickly edited.

Magnetic Tape

Advantages

Magnetic tape is the least expensive of the options available. A cassette recorder and interfacing are usually all that are required for a setup.

Disadvantages

The major disadvantage of magnetic tape is that it is slow. Because tape is linear, if several programs are on a single tape, the user must look through each program in turn until the correct one is found. It is also prone to accidental erasure if it contacts magnetic or electrical sources. The readout from the tape is much slower than from disk and bubble memory. Magnetic tape is not popular. It is chosen usually only to save cost.

Computer Disk (Floppy)

Advantages

This option is common in industry today. Most shops now have their programming done on microcomputers that use floppy disk or hard disk memory.

Direct Numerical Control

The most efficient way to download from computer disk to controller RAM is with a direct electronic connection from the microcomputer to the C/NC controller. This is part of a shop computer management system called direct numerical control (DNC). Floppy disks are inexpensive, can contain many programs, and require little storage space. Information on floppy disks can be quickly accessed.

Direct Connection from Micro to Controller

DNC is a direct connection from the offline computer to the controller. DNC is a common way of loading a program into the controller RAM. The program is stored in computer memory then sent (downloaded) directly to the controller.

If the shop has no DNC interfacing, a disk drive may be incorporated in the controller or the disk drive may be an option connected to the controller. The stored program will be put on a utility disk and taken to the controller in the shop. There it will be loaded into active memory for a part run.

DNC also allows a host computer to control the CNC machine through the controller. This allows much larger programs than could be stored in the control RAM.

Hard Disk

Hard disk storage holds many more programs, and retrieval time is less. Hard disk storage may be in the controller or in an offline microcomputer.

Bubble Memory CNC

There are new data storage devices on the market. Bubble memory cassettes have already been used on several machine controls.

PROGRAMMING STATIONS FOR CNC

A CNC control may be programmed in three general ways: tape punches, direct connection, and manually at the controller console.

Tape Punches

The difference between NC and CNC tape programming stations is that nearly all CNC punched tape is generated by a microcomputer interfaced with a tape punch or teletype machine. However, a tape punch can be used to program a C/NC machine.

Direct Connection—Download from a Microcomputer

A computer may be used as a programming station. The program can then be sent down to the C/NC controller RAM in the following ways:

- From a program already stored on disk.
- From a program generated by a CAD/CAM unit and post-processed into a C/NC program by the computer.
- From a program generated by a special programming software package. This software generates the program from geometric images the programmer creates. This is much the same as CAD/CAM.
- Manually by the programmer. Programming at the computer console is the most common today in industry. Once this program is written and saved on disk, it is downloaded to the controller on the C/NC machine.

CAD/CAM stands for computer-aided drafting/computer-assisted machining (or manufacturing). The term CAM is used for a CAD unit that generates a C/NC program from a graphic image of the part using some additional software. CAM is also applied to dedicated programming software that writes C/NC programs from interactive graphics at a computer rather than a control. Either the CAD or CAM generated program is then post-processed to fit a particular C/NC controls format.

Manual Programming at the Controller Console

The most common programming station is the CNC keypad. All C/NC machines have this capability. Programming can be accomplished by the following three methods:

- Simple manual programming (MDI).
- Graphic programming at the control.
- Digitizing.

The objective of this unit is to describe the components of a C/NC system so you will be a better machine operator. In Chapter 7, you will be able to practice simple program changes at the machine keyboard.

Much of this information nearly duplicates the previous chapter. That is because most of the difference between NC and CNC is in the control. As a CNC machine operator, you must be able to:

1. List the function of each component in a CNC system.
2. Compare the various ways you will be able to load programs into your control in your shop.
3. List the ways CNC machines can be programmed.
4. Describe the ways CNC data is stored and retrieved.
5. Describe and list the above for your lab.

1. Circle only the ways in which punched tape is used in CNC work.
   A) As a program storage media.
   B) To produce parts.
   C) To edit programs in RAM.
   D) To assist MDI programming.
2. What would be a positive factor in choosing an open loop CNC machine tool?

   ____
3. A closed loop machine tool uses feedback signals to compare the commanded moves to the actual move accomplished.

   Is this statement ☐ True or ☐ False?

   If it is false, what will make it true?

   ____
4. The feature of CNC machines that allows the controller to duplicate half of a symmetrical part program by reversing one axis is called

   ______________________________.
5. List two safety features gained by the use of graphic program representation on a video screen.

   ____

   ____

6. Briefly describe the difference between graphics and interactive graphics for CNC.

____

____

7. List four ways a CNC machine may be programmed using an offline computer.

____

____

____

____

8. List the ways in which a CNC control can be programmed. Count all uses of an offline computer as a single answer.

____

____

____

____

____

____

____

____

9. Of the methods you listed in Question 8, which is the inaccurate way of programming a CNC.

____

10. CNC part scaling is used to measure machine movements.

    Is this statement ☐ True or ☐ False?

    If it is false, what will make it true?

____

11. Circle the letter of the uses of non-interactive graphics on a CNC machine.

    A) To visualize the program as it is being written.

    B) To monitor the program without actually seeing the work being done.

    C) To dry-test the program.

    D) To ensure dimensional accuracy.

12. In Question 8, select four programming stations that cannot be used on an NC machine.

____

____

____

____

13. Why could the programming methods listed in Question 12 not be used on an NC machine?

____

____

In Your Lab

If you have CNC equipment, answer the following questions. If you have more than one kind of controller, you may wish to confine some answers to two or three.

14. List the media your shop uses to store program data.

____

____

____

15. Does your shop have a control capable of digitizing?

____

16. List the closed-loop controls in your shop.

____

17. Does your shop have graphics. If it does, circle the letter of the type of graphics.
    A) Offline CAD.
    B) Offline CAM.
    C) Controller graphics.
    D) Interactive graphics.
    E) Plotting either from computer or on the machine.

18. Have you ever used computer disks? If not, have your instructor show you the correct way to use them. They can be damaged.

19. Do you have a direct computer link to your controller?

____

CHAPTER 7

SAFETY IN C/NC MACHINE OPERATION

In C/NC, safety takes on a new meaning. Some C/NC machines will move at 700-plus inches per minute and interpolate at 300 inches per minute. The operator has some heavy responsibility.

As you can see from the objectives below, C/NC safety requires new skills. In manual machining you, the operator, read the print and set up the machine. You then decide what cuts to make, as well as how and when to make them. As a C/NC operator, you will not make all the decisions. The programmer will make many of them. The programmer chooses the cuts, but you are responsible for making them. To avoid program failures (called "crashes") there must be *good communication between the programmer and the machine operator*. This communication must be *written*. It is called the *setup document*. A properly written setup document will help ensure a safe program.

This chapter is designed to teach you how to read a C/NC setup document. In Chapter 8 we will actually run a program on one of the machines in your shop. Before that, however, you must know how to be a safe operator and what to do if something goes wrong.

OBJECTIVES

After studying this chapter, you will be able to:

- Use the C/NC document for safety and efficiency.
- Test a C/NC program for safety.
- Test the setup.
- Stay "on top of the run."
- Plan an emergency crash procedure.

7-1 C/NC Safety Skills

In this unit, you are being trained as an operator/setup person. We will discuss what must be done to get the setup running from "ground zero." In this case, there is neither a program nor a setup. Nothing is ready. In Chapter 8, we will look at the complete operation. For now, however, we are focusing the discussion on safety. These are the basic steps.

1. Read and follow the setup document.
2. Test the program.
3. Do the production run *safely*. This includes:
   A) Preplanning for a crash.
   B) Staying alert.
   C) Monitoring the safety.

7-2 The Setup Document

A setup document is an organizational tool, as well as a safety tool. It speeds the setup, helps to ensure correct parts, and shows the correct setup. The setup document shows:

- The cutter tooling required.
- The holding tooling.
- The sequence of tool changes.
- The location of the program reference zero.
- The tool path. This can be a drawing or a hard copy of the program. Graphic capability on the controller can replace this information with an image of the cut sequence.
- The sequence of clamp or chuck changes.
- Any special requirements for the job: *coolant requirements, fixture problems, quality checks, material type,* and *surface finish requirements,* plus anything that previous operators may have learned the last time the job was run.
- The machine and part number for which the program was written.

Each of the items in the above list must be considered when setting up a C/NC machine. If any item is overlooked there could be a dangerous accident. For example, an accident could result from the wrong tool for the job, the right tool at the wrong time in the program, clamps in the wrong place, or cutters coordinated to the wrong location.

Even if you have made the setup yourself, read it. Know the program. Figure 7-1 shows an actual setup document from a C/NC shop. Note that all the information on the preceding list is shown on two pages.

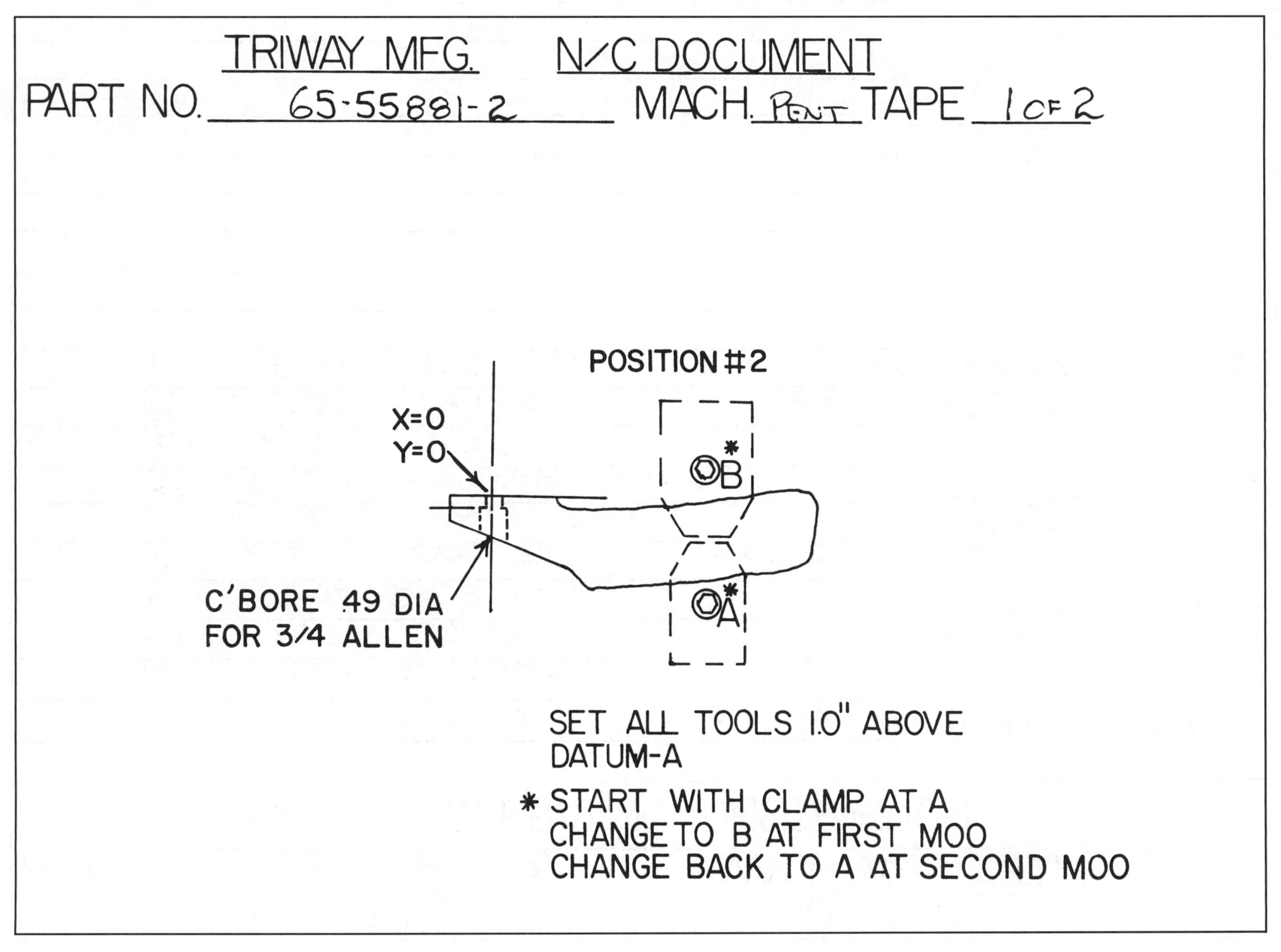

Fig. 7-1. A typical C/NC setup document from industry. Triway Manufacturing

TRIWAY MFG. N C DOCUMENT

PART NO. 65-55881-2 MACH. PENT TAPE 2

PROGRAM OPERATING DESCRIPTION

Step	Operation	Position
①	LOAD TOOL 1	POS D01
②	ROUGH 23.7 RAD	
③	STOP - CHANGE CLAMPS	
④	ROUGH 24.9 RAD	
⑤	LOAD TOOL 2	POS D02
⑥	FINISH 24.9 RAD	
⑦	STOP - CHANGE CLAMPS	
⑧	FINISH 23.7 RAD	
⑨	LOAD TOOL 3	POS D03
10	CUT .25 RAD	

MATERIAL = TITANIUM

TOOL 1 (ROUGH)
4 FL × 1" LONG S.C. 2½ DIA

TOOL 2 (FINISH)
4 FL × 1" DIA × S.C.

TOOL 3
4 FL × .5 DIA × 2" LONG S.C.

In very simple setups and when a software setup document is being used, a written setup document *might* not be required. Some CNC controls allow comments to be added to the program. These can be in the form of a software setup document.

VERY SIMPLE SETUPS

For simple parts the setups do not vary. For example, a machine may be used to make only bushings. These are held the same way every time. Thus, the tooling is loaded into the turret in the same position permanently. The PRZ is always on centerline at the far tip of the part.

SOFTWARE SETUP DOCUMENT

With a software setup document, vital information is written into the program. This information may be displayed on the video screen at the machine. In effect, this display becomes the setup document. However, the information is in the program data rather than in a separate document.

7-3 Testing a C/NC Program

The program has been loaded into RAM. You have bolted everything down and loaded the tools called out. Are you ready to load a part and start the machine? "Not in a million years" should be your answer. There are far too many "grinches" that could be waiting to cause an accident.

To prevent accidents, *any* program must be carefully tested. Everyone makes mistakes. The use of computer-generated programs, graphic representations of the program, and intelligent controllers will help reduce—but not eliminate—accidents. *Always test the program before running a full cycle.* There are seven levels of program testing. Not all are required. Often, one test is enough to verify a good or bad program.

TESTS

Test 1

Read the hard copy or scan the program lines on the video screen. This test is the least useful. Because it can be tedious, you may not find errors. Also, long programs are hard to read and interpret. However, this may be the only way to test a program for an NC machine.

Test 2

Run a graphic check on the video screen. While this is the most desirable method, not all controls have this ability. No NC machine has this feature and only some of the newer CNC controls can image the program.

Test 3

Dry-run the program. This is the most common testing procedure. There are two ways to dry-run a program.

A) Run the program full cycle with no cutters or with the cutters pulled back from the part. This is used mostly on lathes.

B) Many CNC controls have dry-run cycles. These disable the axes that would carry the tool to the part, but allow the movement that describes the part. This cycle is often used to test programs on mills, where the X and Y axes are left to operate while the Z axis is disabled. With this test, you do not know for sure what the Z axis is going to do. You will, however, be able to verify that the tool path is correct.

Both of the dry-run techniques can be run at a feed rate higher than that used for the actual part run. Many are run at rapid feed to save time.

Test 4

Single-step the cycle. In this method, the program is run with all axes operating. The cutters are in place and a part is loaded. The control is set to execute only a single block of information when the operator touches the start button. In this mode, the control runs one block at a time. Test material, instead of the actual metal, might be used.

Analyze the test results. The objective here is to look on the readout or program hard copy and determine what the machine is going to do next—for each block. Analyze the data. Ask yourself, "Is this reasonable?" If it is, then touch the START button.

The problem with this method is that new people forget to look at the readout. They do not know what is going to happen if they touch START. So while intently watching the cutter and part with their face right in the way, they touch START. You must analyze each step before it happens—not after.

Test 5

Optional stop testing. This method must be built into the program or inserted at the first part run. Optional stop symbols or codes are put in the program at the completion of each feature of the part. As the program is run, the machine will stop when this command is read. The operator verifies that the program is acceptable to that point. Then the operator continues.

Optional stop testing also helps measure and correct the dimensions as the first part is run. If the part run is safe and the part is acceptable, the optional stop switch is turned off. This means that the controller will ignore the command and run the program straight through.

Test 6

Plot the program. This simple program test is common on mills, but can also be used on lathes. A pen in the spindle of the mill and a pad on the table will work. A better plotting attachment can be made easily. The lathe needs a pad on the bed and an attachment that holds the pen vertically. Some shops have special plotters.

Test 7

Use soft test material. Before putting the cutter to the metal, it is common for some shops to run a part from some soft inexpensive material. Special machinable wax or foam are used for this purpose. Although soft, these materials retain the shape produced. They can be inspected for dimensional accuracy. They are usually used when the material to be machined is very expensive. Wood can also be used. Caution must be exercised with wood and foam because they can clog relays and actually pass by seals intended to keep metal chips out of machinery. Wax is clean and can be recycled. Use a small vacuum to reclaim the wax chips.

7-4 Planning A Safe Production Run

Once the program is up and running successfully, you must stay on top of the run. Here are three very good guidelines:

- Plan an emergency crash procedure.
- Plan a way to stay alert all the time.
- Plan a regular schedule of monitoring the setup.

PLAN AN EMERGENCY CRASH PROCEDURE

Have you had a serious accident on a machine? Do you remember how you reacted? Did you turn off the feed or spindle first? Did you back the cutter away from the part? Did you freeze, or do something foolish? In a crash, you must react immediately. If you have not planned ahead, you may not do the right thing. Protecting yourself and the equipment in an emergency requires planning and practice.

Even experienced machinists will sometimes make the wrong move in an emergency. By pulling the wrong lever or pushing the wrong button, they may compound the problem. To avoid doing this, plan a crash procedure. Follow these basic steps.

A) Determine a two-level procedure.
   - *For level one,* there is no immediate danger but the run must be stopped. If not, a dangerous situation could develop.
   - *For level two,* all operations must be stopped immediately. Some level-two situations can be dangerous.

B) Practice this procedure until it is second nature. Review it every day. Here are the options. Use any or all of these in your plan.
   - *Turn off the feed.* Most machines have a method of overriding the feed. If so, you might be able to stop the cutting. This is useful only in a level-one problem. Further action would be required, but stopping the feed might buy you some time. On some machines, the feed override might be called *slide hold.*
   - *Hit the cycle stop button.* Most C/NC machines have a button that will interrupt the automatic cycle. The cutter will continue in contact with the work. The cutter may leave a mark where it dwelled in one spot. This procedure is not immediate. The machine will stop only at the end of the current cycle. This procedure could be used only in a level-one problem. In a long cycle, it might not be in time to prevent a more serious problem.
   - *Hit the emergency stop button.* All machines have an emergency stop button, which stops everything. This button is used in level-two problems. Its use may have serious consequences for the setup or the program. It might erase the program if the control does not have battery backup. On some controls, you might loose the setup coordination and your position.
   - *Back the cutter away from the work.* On some machines, such as a vertical mill with a manual knee, this is possible.
   - *Reduce the feed or spindle speed.* Chatter and heat are caused by excessive speed or feed. Accidents can sometimes be prevented by alert overriding of the feed or speed. Not all machines have this capability. On those that do, however, the feed and speed should be retarded (overridden) for the first part run.
   - *Dive under the bench.* If you sense that you may be injured, dive under the bench. Obviously, such a decision depends on quick analysis.

STAYING ALERT AND MONITORING THE SETUP AND RUN

The biggest challenge to a machine operator is to stay "on task" after the run becomes routine. Often, a perfect setup will deteriorate as the forces of machining begin to take their toll on the equipment. When you do not pay strict attention to the machine, you are asking for trouble. Heat, cutting forces, turning forces, and vibration can cause changes. In operating a machine, watch for the following:

- Dull cutters.
- Loose cutters.
- Loose chucks or fixtures.
- Loose setup bolts.
- Flawed work materials.
- Lost pressure in air or hydraulic chucks.
- Incorrectly loaded parts.
- Malfunctioning controllers.

Other Duties

Plan a way to constantly monitor the above list. This will help you stay alert. You may also be required to check the size and finish of the parts. Four other duties you must do during the production run are:

- Do secondary operations such as burring parts during the part cycle only if there is time. Remove the completed part and set it aside. Load a second part and start the machine. Then burr the first part.
- Make sure the coolant is reaching the cutters. Heat and chip buildup can destroy the cutter and the part. Neglect of this detail can ruin parts, dull cutters, and cause needless heat.
- Clear chips from the setup during the run. Chips must be blown or brushed out of the way. If a lathe is producing long, stringy chips, a chip hook might be used. A better way to solve this problem is to create a chip-breaking cutter. Most carbide insert tools have special geometries for this. However, if the cut is light, the chip might not be breakable. This is a dangerous situation. The program might need modification to interrupt the cut to break the chip.
- Do routine maintenance. Watch lubrication and coolant levels. Watch the air or hydraulic pressure if it is part of the machine. Clean and adjust guards. Check general machine adjustments.

Use your senses to monitor the machine operation. Listen carefully to the machine. All journey-level machinists will tell you that you will often hear a problem before you see or feel it. Listen for unusual vibrations, chatter, bumps, and bangs. These might be early warnings of an impending accident.

HELPFUL HINTS

In working, be sure to follow these four general safety guidelines.

- *Do not have music playing in the workplace.* Music is distracting and interferes with your hearing. Often an unusual sound is the first indication of a problem.
- *Do not leave the immediate area of your machine during the run.*
- *Check the setup at each shift change.* Make this check even if the setup is running perfectly.
- *Try to make the operation more efficient.* Constantly seeking better or faster ways to make the run is an effective way of remaining attentive. Good housekeeping comes into this area, too.

Check the setup on a regular schedule. You decide how often to check the setup, but stick to the schedule. Do not check or clean moving machinery except to blow or brush chips away.

Safety is easy to forget, especially when the job is under control and the C/NC machine is working perfectly. To be a safe professional C/NC operator, practice the following:

- Always read and understand the setup document. Know what is going to happen.
- Have a two-level "emergency crash plan" all set and practiced.
- Test your programs.
- Stay alert.
- Always monitor your machine. Use all your senses to make sure things are going right.

1. List the items that should be in the setup document.

2. What is a hard copy of a program? How is it used to make a C/NC job more safe?

3. List the six possible ways to test a program before running it.

4. If a single-step test is to be used safely, an operator must ______________ each program step.

5. List the three ways to plan a safe production run.

6. There are five possible actions you may take in an emergency. Which of these actions is not immediate and has no consequences to the run?

____

7. Without a ________________ ________________, hitting the Emergency Stop Button might erase the program or the machine coordination.

8. Why might the use of Cycle Stop not be a good level-two procedure in an emergency?

____

____

In Your Lab

Your instructor may wish to assign one or two specific machines for the following questions.

9. Do the machines in your shop stop immediately when Cycle Stop is depressed?

____

10. Is there any consequence to hitting Emergency Stop on the machines in your shop?

____

____

11. Choose a typical emergency from the list below. Then write a level-one crash procedure to deal with the emergency. Select more than one option. (Assume that the first will not work.) Possible emergencies include:

    1. A lathe part is chattering beyond safe limits.
    2. A mill tool is chattering beyond safe limits.
    3. A part has slipped slightly in a mill vise but shows no further signs of moving.
    4. A lathe cutter has a broken tip from machine wear.
    5. A mill cutter has loaded up with aluminum chips and is not cutting.

    Write the number of the possible emergency you have selected: ________________

    Outline your crash procedures.

____

____

____

____

____

____

____

12. Choose two typical emergencies from the list below. Write a level-two procedure to solve the problems. Since a level-two problem must be solved immediately, there might not be a second option other than to dive beneath the bench.

    Possible emergencies include:

    1. A part is slipping out of a mill fixture.
    2. A wheel hub is coming out of a lathe chuck.
    3. A large cutter is coming out of a vertical mill spindle.
    4. A control or a bad program has just driven a cutter into the chuck of a large C/NC lathe.
    5. The bolts are loosening on a mill vise, which is moving across the table.

    Possible emergency 1 ____________________

    Outline your crash procedure.

    ____

    ____

    ____

    ____

    ____

    ____

    ____

    Possible emergency 2 ____________________

    Outline your crash procedure.

    ____

    ____

    ____

    ____

    ____

    ____

    ____

13. List the methods used to test programs in your lab.

    ____

    ____

    ____

    ____

    ____

14. Does your lab equipment have graphic capability? Is it interactive?

____

15. Which of your physical senses is best for monitoring a C/NC run?

____

CHAPTER 8

MACHINE SETUP FOR A PROGRAM RUN

Chapter 8 presents the setup process from beginning to end. This is the point for which you have been waiting. After reading this chapter, completing the activities, and passing any examinations your instructor may require, you will be allowed to operate a C/NC machine. You will not write the program. It will be supplied by your instructor. This is typical of the progression of responsibility in industry from beginner to journey-level C/NC machinist. On your first job in C/NC, you are restricted to the simple operation of the machine. Later, with experience, you will be allowed to do more of the setup work. As your competency increases, so will your responsibilities.

OBJECTIVES

After studying this chapter, you will be able to:

- Select the correct cutter tooling for a C/NC job.
- Select and use holding tools such as fixtures and chucks.
- Perform routine maintenance.
- Download the program, or tape load on a NC machine.
- Load the raw material.
- Align and coordinate the machine and the tools.
- Enter tool offsets and cutter geometry.
- Test and run the program.

8-1 Selecting the Cutter for the Job

CUTTER GEOMETRY

Tools Required

Before selecting the correct cutting tools for a C/NC job, you must read the setup document. The programmer had specific cutting tools in mind when the program was written. The correct cutting tools must be used. During the setup, you have only limited control over the size and shape of the cutter used. These limits are imposed by the program type and the shape of the part.

Program Type

Is the program *computer compensated* (CComp) or *manually compensated* (MComp)? If the program was planned and written for CComp, there is some flexibility for cutter shape. If the program is MComp, there is no choice. The exact cutter for which the program is written is the only one that will produce the correct part.

Manually Compensated (MComp)

With a manually compensated (MComp) program, a cutter of the exact required size and shape must be used. An MComp program has a built-in allowance for the tool shape (cutter path). A CComp program contains the actual part shape and will generate the correct part only when the control is told the shape of the tool being used. MComp programs have cutter path data, while CComp programs use part path data and the computer compensates for the cutter shape. Fig. 8-1.

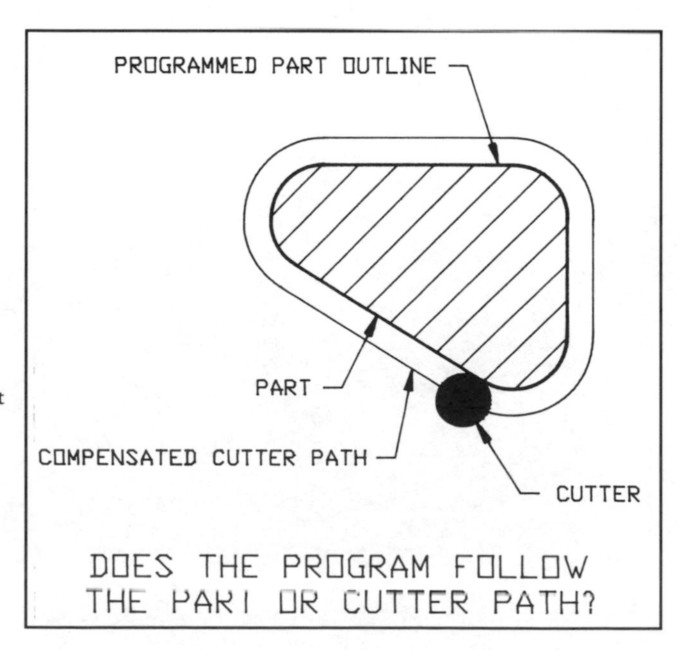

Fig. 8-1. Does this program follow the part or cutter path? Computer cutter compensated (CComp) programs use a part definition path, but follow the cutter path.

NC machines are able to perform only cutter path programs (MComp). Therefore, a cutter of the specified size is required to form the correct part. CNC programs may or may not have cutter compensation. Without CComp, you must use an exact cutter specified by the setup document.

Computer Compensated (CComp)

To be set up correctly, the CComp program must have the correct commands to initiate CComp and have part path as the program body. The setup person then has some choice of cutter size and shape. The control must be told the shape and size of the tool being used. *That is the responsibility of the setup person.* This tool data is entered in the control at setup.

Tool Geometry Information Required for CNC Compensation

To initiate tool compensation, the following parameters must be stored in the controller tool compensation memory.

Mills: Cutter diameter or cutter radius (Fig. 8-2)
Cutter length offsets
Tool number and/or turret position
X and Y adjustment (offsets)

Lathes: Cutter nose radius (Fig. 8-3)
Cutter shape
X and Z axis adjustment for size control (offset)
Tool number (location in the turret)

PART SHAPE LIMITS ON CUTTER GEOMETRY

If the program was written using CComp, the part path is programmed and the correct program commands are given. Thus, the CNC control can compensate this path for the cutter size and shape. This will give you some selection over cutter geometry—up to certain limits. In size and shape, the cutter should be close to the cutter for which the program was written. It does not, though, need to be exactly the same size. The limits of this flexibility are set by the shape of the part to be machined.

Some cutters are too big for the machine. Others are too small for practicality. Two other situations limit cutter selection:

- Internal features size—part shape (slots, grooves, etc).
- Internal corner radii.

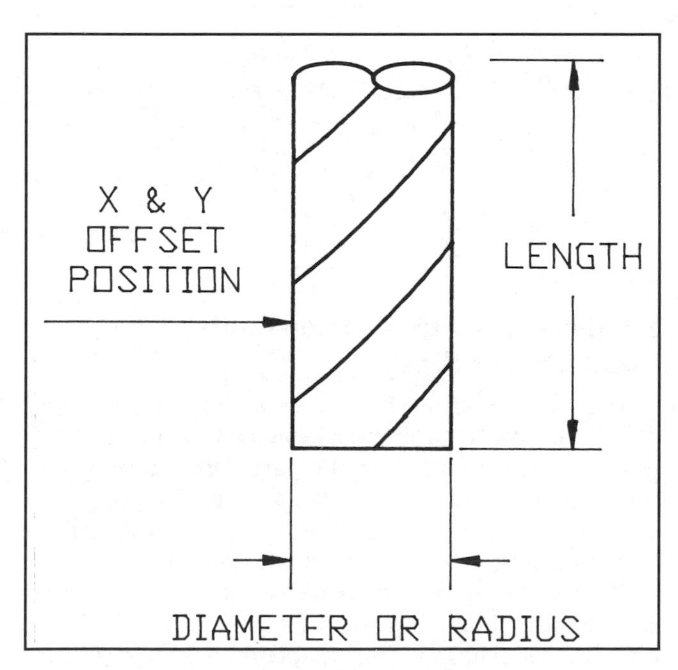

Fig. 8-2. Mill cutter compensation parameters.

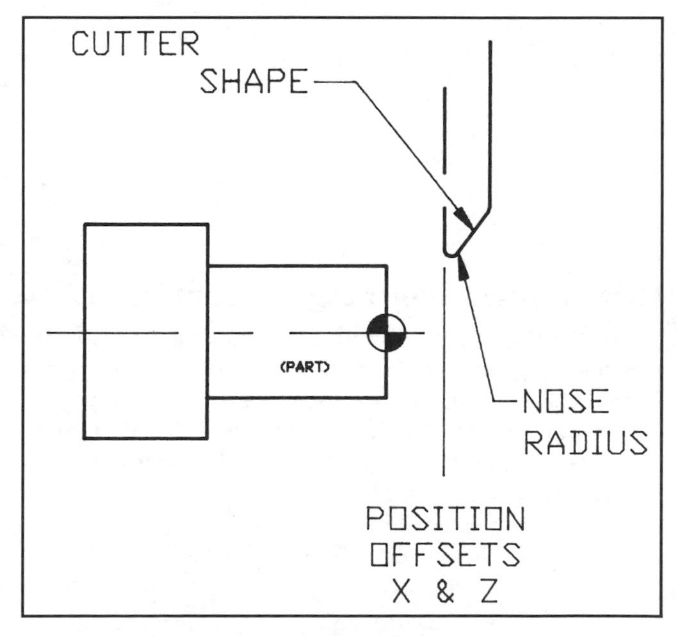

Fig. 8-3. Lathe cutter compensation parameters.

Look at Appendix A Drawing 3. This part is to be machined by peripheral milling. Notice that there are two features that could limit the end mill size used. The following exercise should help you find them. In this exercise, we are profile milling the shape with an end mill. Answer the following questions.

1. The tee slot cleaning head on the top of the drill gauge has a slot on either side. How wide is this slot?

2. What is the biggest end mill that could machine the slot? In other words, what is the largest cutter that could machine into the slot?

3. Would this end mill produce a part of the correct shape? Why not?

4. What are the two radii at the back of each slot?

5. What cutter diameter would be the upper limit to produce these radii?

6. Could the cutter have a smaller diameter?

Generating or Forming Internal Radii

Machining a curved path is called *generating*. In an internal corner with a radius, a programmer has a choice of whether to let the cutter radius form the corner radius or to machine a curved path with a smaller cutter (generating). Many factors will influence this choice. The size of the radii, machine finish, cutter strength, possible corner chatter or undercut, and time used to execute are just a few. Fig. 8-4.

Smallest Internal Corner Radii Limit Cutter Size

If only one cutter is used in a program, the tool radius for lathe or mill must be as small as the smallest internal corner radius on the part. Often a programmer will use the tool radius corners for roughing. Generating the internal radii for finish cuts would then be done using a cutter with a smaller radius.

If the cutter radius matches the part's internal corner radii, then the program forms these radii. However, if the tool has a radius smaller than internal corner features, the program must generate the radii.

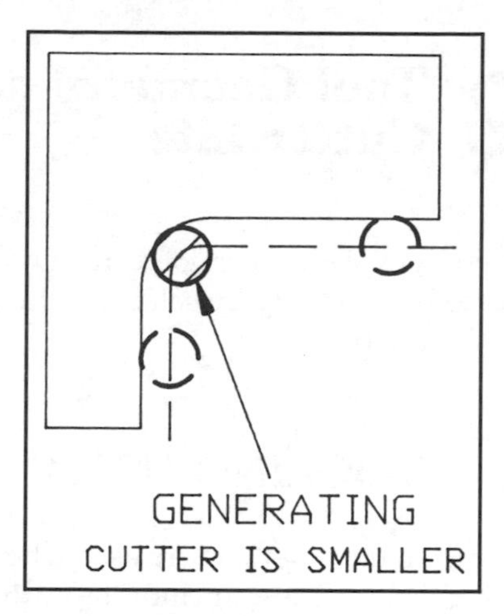

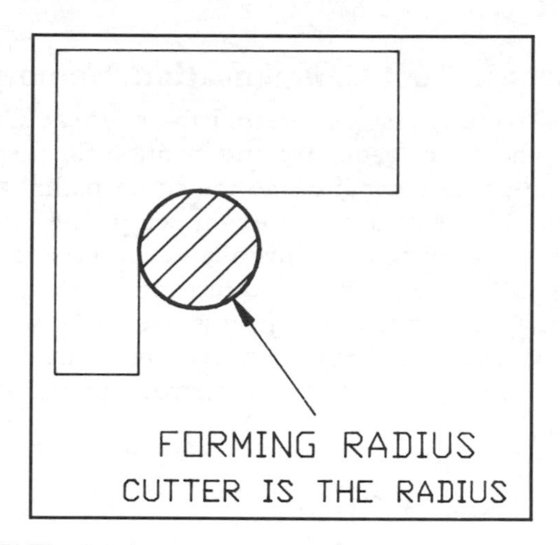

Fig. 8-4. (A) A cutter follows a curved path to GENERATE a radius.

(B) The actual cutter corner radius will FORM a radius.

Because computer-compensated programs are complex, your instructor might choose for you to limit your first experience to MComp programs. Using an MComp program will not limit the value of the setup experience.

You must know what type of program you are running. Manually compensated programs do not require that information on tool size and shape be entered into the control. Ask your instructor if your program will be MComp or CComp. Then continue your reading.

Lathe Tool Shape Considerations

Tool compensation is more complicated for lathes than for mill. The lathe must compensate for the tool nose radius (TNR) just as a mill control must compensate for the cutter radius. However, the lathe must also compensate for the shape of the tool. The overall shape of the tool must be considered when moving or cutting because the cutting edges away from the nose and the flanks of the tool can touch material or chucks.

Standard Tool Library

Most CNC lathe controllers have an internal set of standard tool shapes that can be entered in the compensation memory for each tool used. These give the nose radius and the actual shape and size of the tool. These types of tools must be mounted in an exact position. The shank of the holder must be parallel with the X or Z axis (true to the machine) for the part shape compensation to be carried out correctly.

The control requires the tool to be in a standard position. Using standard lathe tooling and compensated programs, the tools must not be rotated. They must be parallel with the axes of the machine. Fig. 8-5.

OFFSETS

In CNC work, an *offset* is a variable parameter to which the controller refers to make adjustments to the part path for the specific cutter used. Offsets are variable numbers stored in the control but outside the program. Offsets are used for reference in the program for cutter compensation.

Fig. 8-5. These tools are correctly mounted. They are true to the machine and against the qualifying post in the center.

Offsets in Tool Compensation Memory

Tool parameters are referred to as *offsets*. Offsets define the cutter geometry and position for the computer to use in compensating the cutter path. Offsets are put into the control during the setup. They are not in the actual program. This allows the offsets to be changed. Thus, a variety of cutters can be used. When the cutter is changed, the offset is re-entered with the new parameters. The computer reads these new parameters and adjusts the cutter path accordingly.

Three Uses of Offsets

Offsets have three uses.

1. By changing the offsets, an operator can make a slightly smaller or bigger part. The program is adjusted to fit the cutter available.
2. Offsets can be used to compensate for tool wear. As the tool dulls, the offset can be compensated to take slightly more material. This practice has limits, and should not be used in place of good sharp tools. Some advanced controls will change tool offsets on a schedule to automatically compensate for tool wear.
3. A roughing and finish pass may be taken using the same program and cutting tool, but using two different sets of tool offsets. By informing the control that the cutter is bigger or closer to the work than it actually is, it is possible to have the control back the cutter away from the part. This leaves a small amount for finish. Then the true cutter information is used for a second set of offsets. The control then produces a finished part.

Offsets do not reside in the actual program. They are loaded into the tool compensation memory. This is a separate part of the control from the program RAM. Later in this chapter, we will look at ways of entering the offset values. Entering offsets is one of the later steps in the setup. They are entered just before the program is tested.

Zero Diameter Cutter

As an example, if a vertical mill program is plotted, a CComp program will draw an oversize picture of the part if a cutter diameter is given. To test the program, enter a cutter diameter offset of zero diameter. The machine will move the cutter path to the part path and plot the actual part. In effect, the control will then be compensating for a cutter of zero diameter. The center of the spindle will trace the actual part shape.

Remember, this unit is training you to set up a machine. You will learn how to enter tool information in the controller later in this chapter. For now, you must understand the limits regarding tool selection.

8-2 Tool Geometry and Cutter Life

This section explains how you can help reduce costs and increase productivity by selecting the correct cutting tools for a job.

TOOLING USED

Standardized cutting tools are used whenever possible in C/NC work. These include taps, drills, boring tools, spade drills, reamers, center drills, parting tools, end mills, mill cutters, countersinks, counterbores, and carbide-insert tools. Good cutting tools and good cutting practices remain the same from manual to programmed machining. Coolant should always be used even on "dry materials," such as cast iron. Tool life, heat buildup, and safety are improved with the application of coolant. Note these standard guidelines.

- Climb milling usually prolongs tool life and improves machine finish.
- Coolant prolongs tool life and improves finishes.
- Rigid fixturing avoids chipping tools.
- Keep tools sharp. Stop the machine and change tools if they are dull.

IMPROVING PRODUCTIVITY THROUGH TOOL CHOICES

C/NC machines are costly. Therefore to justify their expense they must work more. The objective in any C/NC shop is to keep the machine working more time. The time the machine is working is "up time." The time it is not working is "down time." Ideally, the machine should be working 100% of the time. This is not possible. Most C/NC shops agree that 75-85% is a good figure for up time. Maintenance, setups, and part changing require nonproductive time. Another reason for down time is tool failure. Setting up the best tool for the job can reduce tool failure.

Carbide-Insert Tools Improve Up Time

Most C/NC shops use carbide-insert tools. These tools produce better parts. They also save time in setups and part runs. Carbide-insert tools are available in a large selection of shapes and grades. Fig. 8-6. Though carbide-insert tools cost more, they help the machine deliver more up time during the production cycle.

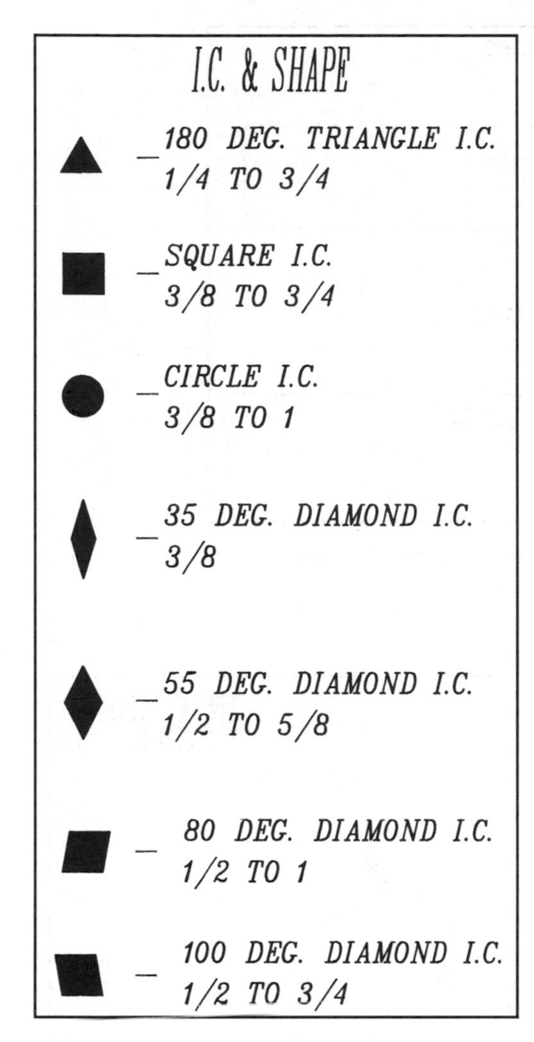

Fig. 8-6. A selection of available carbide insert shapes.

Selecting the Correct Tool Increases Up Time

You can increase up time percentage by selecting the best shape and type of tool for the job. As an operator, you should also change tools before they break from the pressure caused by dull cutting edges.

Nearly 100% of all shops have turned to indexable-throwaway carbide tools for production. The inserts are manufactured to a close tolerance. The toolholders hold the insert in an exact position. These carbide tools are usually *qualified*. This means that the tool nose will be in an exact position every time when pushed against qualifying pins in the tool turret. This remains true from insert to insert and holder to holder. Thus, the programmer can count on this from tool to tool.

The larger the tool, the greater its strength. The cutter shape also affects tool strength. This factor is most evident in lathe tooling, though it also applies to mills. The following guidelines apply to mill cutters.

- The larger the cutter, the stronger it is.
- Cutters with sharp corners tend to wear or break before radius cutters but radius corner cutters may chatter more.
- The number of flutes or teeth is governed by the type of material. Soft material requires fewer flutes to allow chip ejection. Cutters that must plunge into the material similar to drilling, must be specially ground for this purpose. Most two flute cutters will plunge.

INSERT SHAPE/STRENGTH RATIOS

Blunt Tool Noses Are Stronger

Figure 8-7 shows the relative strength of each insert. The pointed diamond insert is the weakest and the round insert is the strongest. This does not mean that you could select a round insert each time. The part shape is the first consideration. What is the smallest

Fig. 8-7. A gradient of tool strength to shapes. Blunter tools are stronger but have fewer applications. Kennametal, Inc.

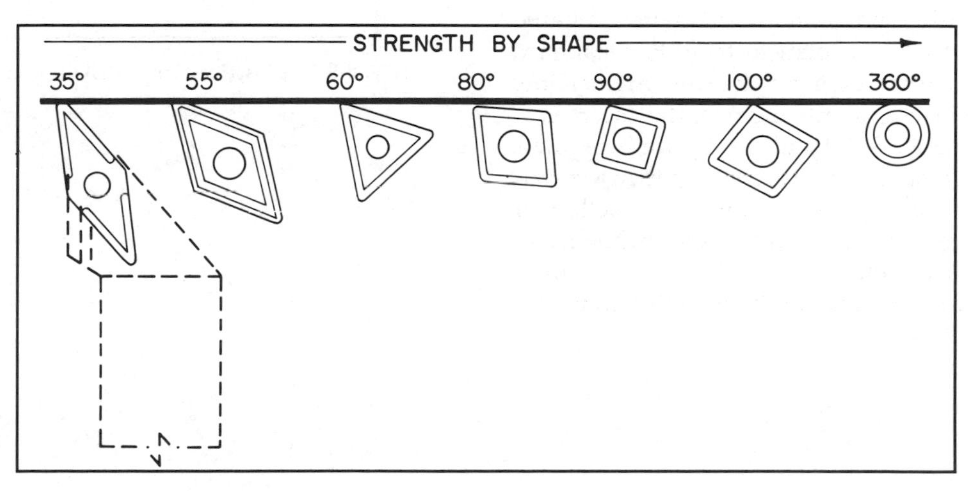

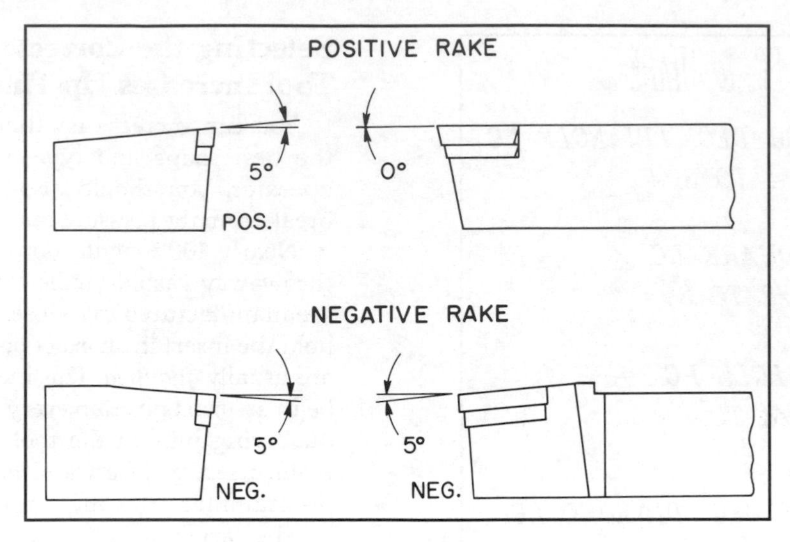

Fig. 8-8. A comparison of negative rake and positive rake tools. The negative rake is stronger. Kennametal, Inc.

internal radius that must be machined? Choose the largest point angle and radii that can machine the smallest internal feature on the part. Once this selection is made, the shape and nose radius must be entered into the control tool compensation memory. *Remember that this is true only if the program is CComp.* If not, you have no choice of cutter size or shape on mills or lathes.

Positive and Negative Rake

Another tool geometry feature that adds strength is the top rake surface of the insert. Fig. 8-8. Negative rake insert tools require more power and more rigid holding methods. However, they will last longer and take bigger cuts. They usually have more cutting edges per insert. The cutting edge is more blunt for a negative rake tool. Thus, it avoids chipping, dissipates heat, and withstands shock better.

Size Related to Cost

The cost of a tool insert increases with its size. The industrial standard for maximum efficiency is based upon the deepest cut you plan to take. For standard carbide inserts, if the cut depth is over one-half the size of the total insert inscribed circle, the insert is at maximum usage. Fig. 8-9. Until this figure is reached, it is more cost effective *not* to use a bigger insert size. Choosing an oversized insert means the toolholder must also be larger. Insert cost increases but the result is not that much better. Use carbide inserts to their maximum capacity for maximum cost effectiveness.

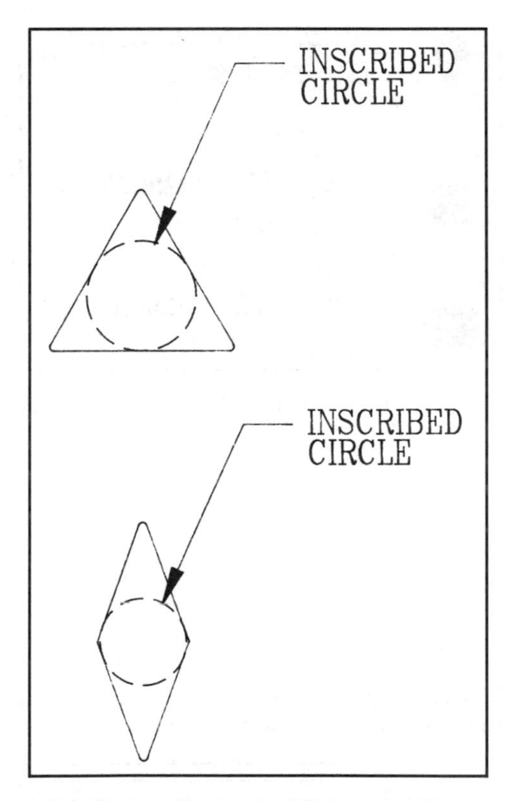

Fig. 8-9. The inscribed circle (IC) for a carbide insert.

8-3 Selecting Carbide Grades for Tool Life

SINTERING

Carbide tools are not made in the same way as standard high-speed steel (HSS) and cast alloy tools. The material in carbide is too hard and melts at too high a temperature to allow casting the correct shapes. The material in the carbide insert starts in a powdered form. It must be pressed like a charcoal briquette and then fired to fuse the materials. This process, called *sintering*, allows for different mixtures of the materials. Thus, there are different grades of hardness.

Standard carbide is made of two general components—tungsten carbide and cobalt. The tungsten carbide is often referred to as simply carbide. This is the actual cutting material. It is hard, but also very brittle. Cobalt is the glue that holds the powder together. It also adds toughness. Cobalt has little hardness, but it is not brittle.

Heat and Pressure

When the carbide-cobalt powder is pressed and then heated, the cobalt melts and cements the carbide grains together. The more carbide in the mixture, the harder the tool. However, this also makes the tool more brittle and more prone to chipping on interrupted cuts. If the cobalt percentage is increased, the tool becomes more shock-resistant. However, it also becomes less hard and more subject to wear. In choosing a carbide grade, follow these guidelines.

- What is the material hardness?
- What kind of cutting action will the tool be subjected to?
- Are the setup and machine rigid? Will there be chatter?

A harder tool will machine more parts before wearing out. However, if the tool is too hard, the edge will chip from the machining action.

Grades

Industry standards grade carbide from C-1 (toughest grade) to C-8 (hardest grade). Each manufacturer has charts showing the application of the grades and expected tool life.

Special Alloys and Coatings

Special cutter alloys and materials are available. New cutting tool materials are being introduced constantly. While these have advantages over carbide, they cost more and they must be correctly applied to gain any advantage. Many inserts are electro-coated (plated) with even harder materials to increase their resistance to wear. Inserts made from ceramic materials rather than carbide are also available. Coated inserts perform as though they were harder. They can cut amazing amounts of material with less wear. A complete discussion of these special cutter types is beyond the scope of this book. You may, however find manufacturer data and descriptions very interesting. Your objective will always be to find a cutter that will be cost effective and work quickly and efficiently.

8-4 Selecting Tool Holders

INDUSTRIAL STANDARDS FOR TOOLHOLDERS

Milling Machine Standards

There are standards for mill tool holders. The holding devices are usually tapered and drawn into the spindle by a drawbar or draw ring. There are a large variety of tapers. Several methods are used to draw the holders in for automatic tool changing. Fig. 8-10. Morse and American Standard are two of the tapers used. You should be familiar with the draw method and the holder type (shape and size) of any milling machine you operate.

Safety Rule. *Be sure you fully understand the draw process and use the right taper. If you do not, you could seriously damage the machine and tooling or cause personal injury.*

Fig. 8-10. A variety of standard machine tapers.

Lathe Holder Standards

The toolholders used for carbide insert work have definite standards. These standards determine the size, shape, and direction of cut for each holder. This direction is called the "hand" of the tool.

Right Hand and Left Hand

On a standard engine lathe today, the tool post is on the operator side of the bed ways. When correctly using a right-hand toolholder with the tool post in front of the bed ways, the tool actually moves to the left. Why? On early lathes, the tool post was on the back side away from the operator. A right-hand tool moved to the right and a left hand moved to the left. When modern engine lathes moved the tool post to the front, this reversed the actual movement of the tool. This can be confusing to new students using standard lathes. Today, many C/NC lathes have again mounted the tool turret behind the bed ways. If the tool turret is mounted behind the bed ways, *the right-hand tool moves to the right and the left-hand goes left.*

To keep this straight, *always view the toolholder from the insert end,* as though it was on the back side of the lathe. If the tool mounting is on the front of the lathe, remember that the right-hand tool will move to the left toward the chuck. Fig. 8-11.

Fig. 8-11. Right, neutral, and left-hand tools as viewed from the insert side.

NUMBERING SYSTEM FOR TOOL IDENTIFICATION

Standard Insert Numbering System

There are standard numbers for the identification of carbide insert shapes. Each part of the identification number tells a fact about the insert's shape and intended use. Fig. 8-12.

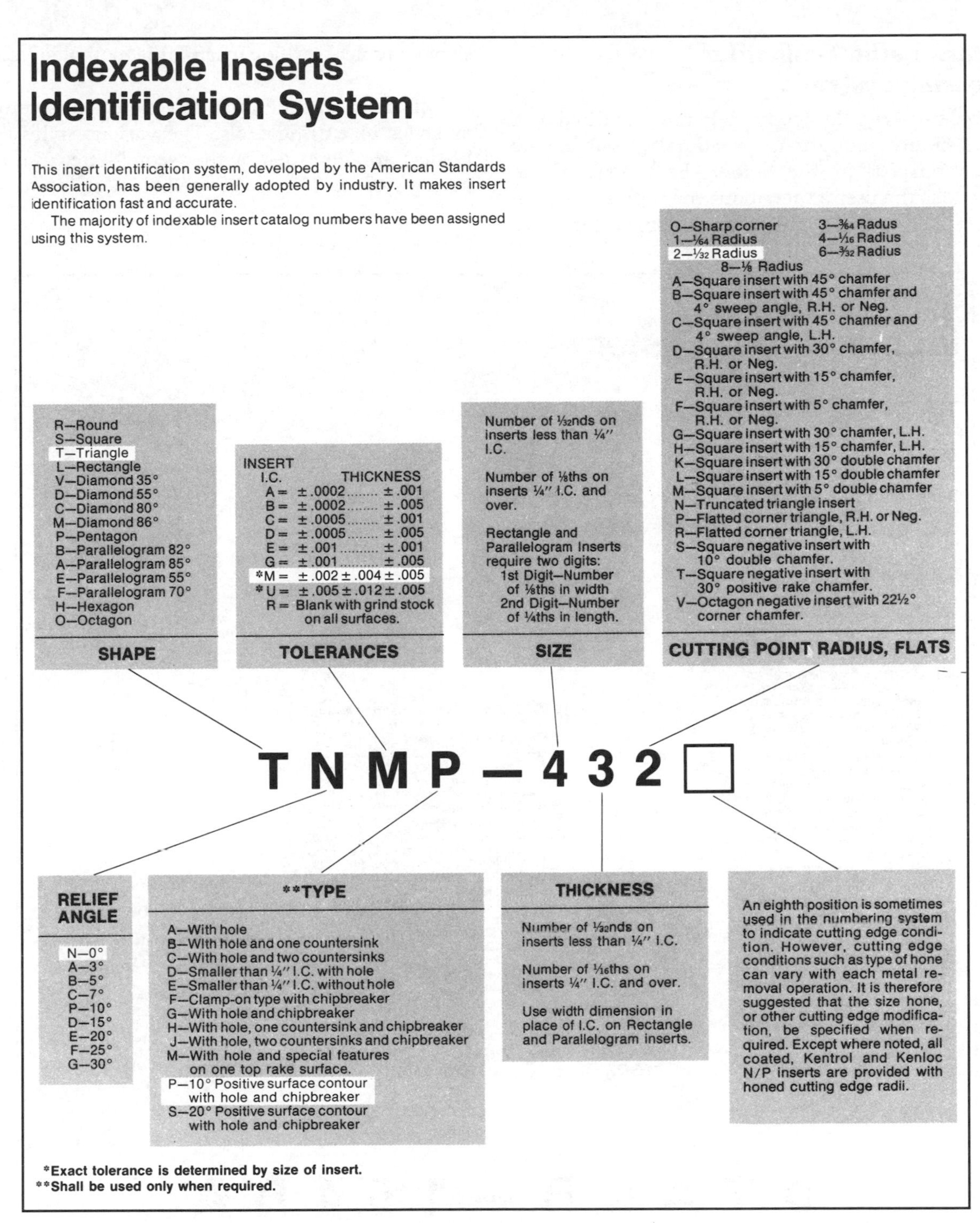

Fig. 8-12. Standard insert identification numbers. Kennametal, Inc.

Standard Lathe Toolholder Numbering System

Carbide insert lathe toolholders are identified by a standard numbering system. This number tells you the intended purpose of the holder. Each part of the number tells the user a fact about the tool. Each manufacturer also adds special information. Figure 8-13 shows a typical carbide insert lathe toolholder selection chart.

Different cuts and materials require different types or styles of cutting tools. The various styles of toolholders are identified by the second letter in the numbering system. The major toolholder styles are shown in Fig. 8-14.

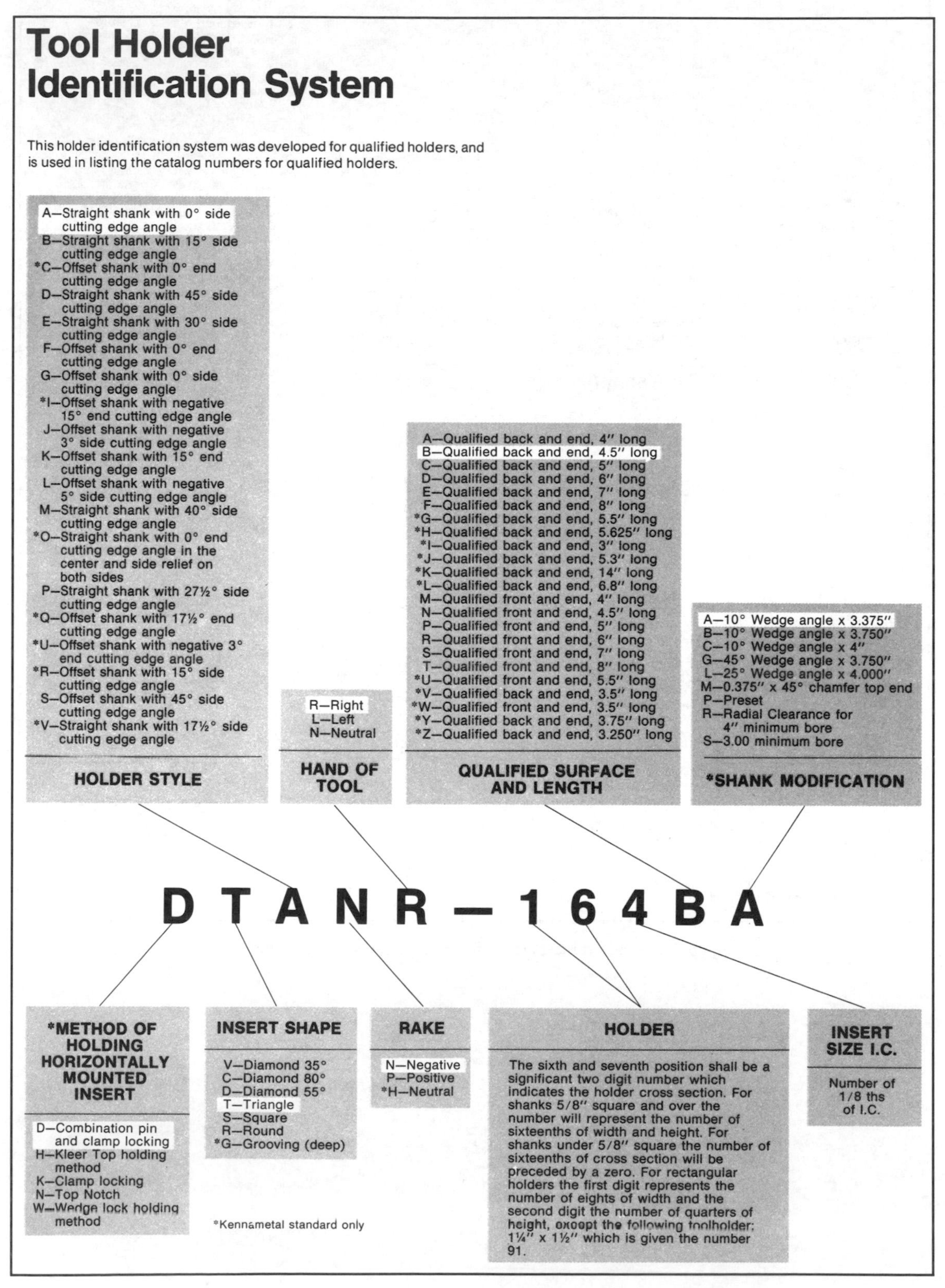

Fig. 8-13. Standard lathe tool holder selection identification numbers. Kennametal, Inc.

Standard Tool Styles

Kennametal tools conform to industry standards and are made in styles and sizes for all general types of machining operations:

Style A
for turning, facing or boring to a square shoulder.

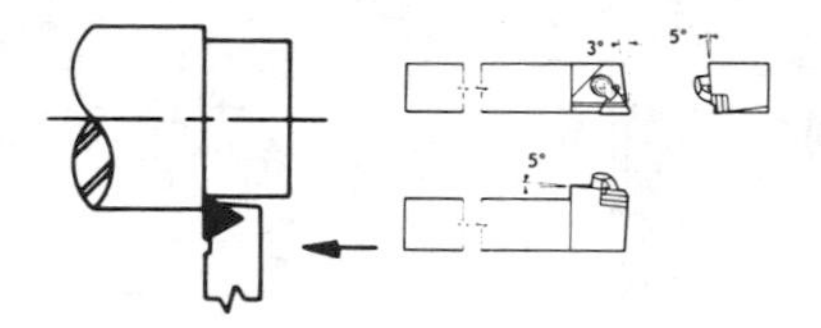

Style G
for turning close to chuck or shoulder, or facing to a corner.

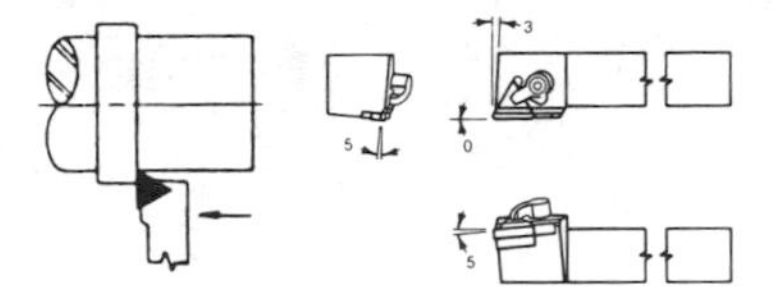

Style R
for rough turning, facing or boring where a square shoulder is not required.

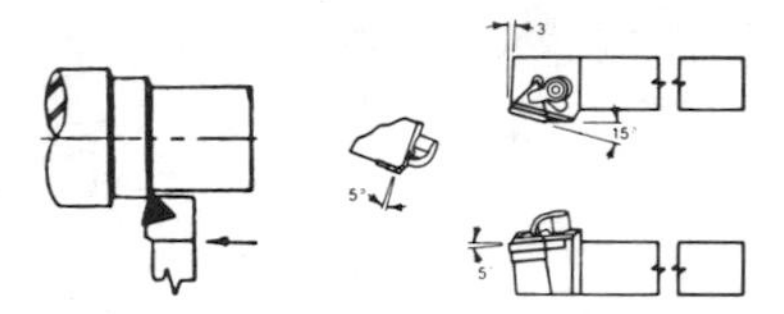

Style GCH
for deep grooving.

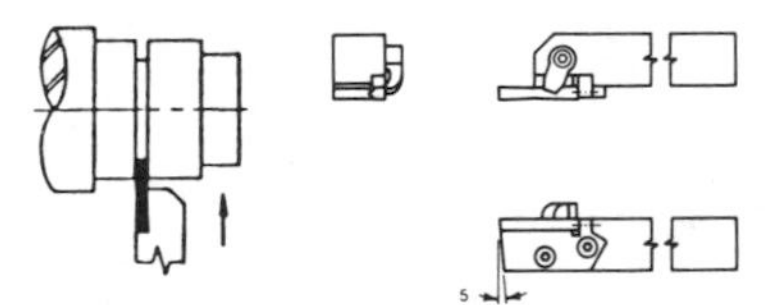

Style F
for facing, straddle facing or turning with shank parallel to work axis.

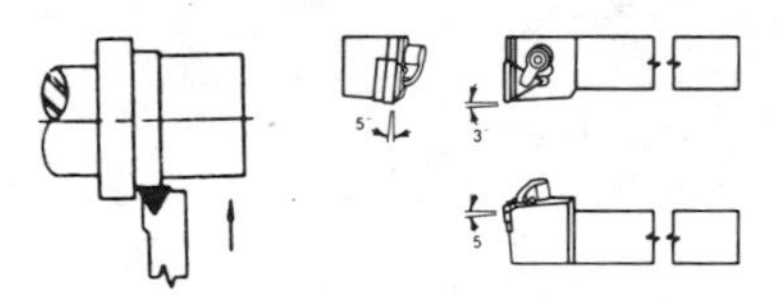

Style J
for profiling and finish turning.

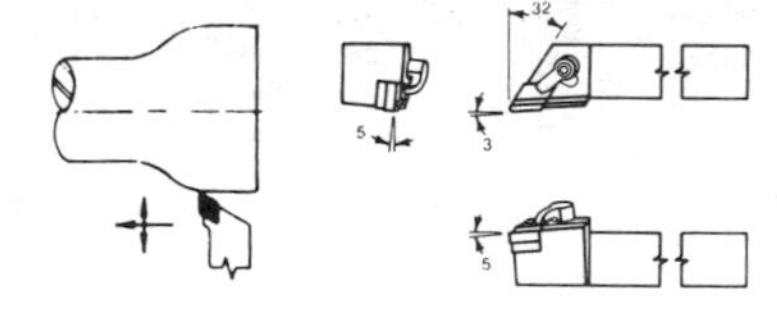

Fig. 8-14. Toolholder styles are identified by an industry standard system. (continued on next page)

Standard Tool Styles

Style K
for lead angle facing or turning with shank parallel to work axis.

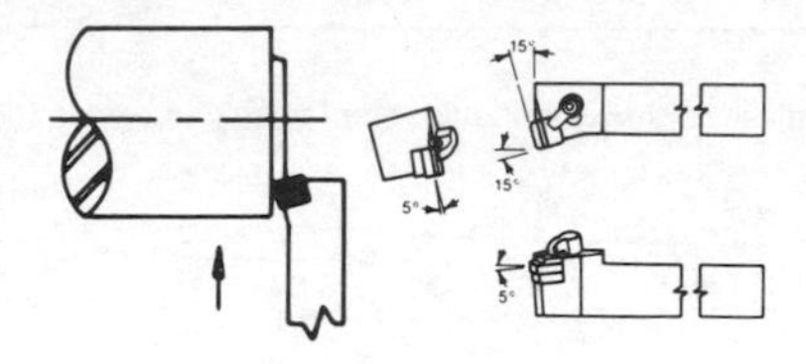

Style NE and Style NS
for threading and grooving operations. Inserts available for 60° V, Acme, Buttress and API threads.

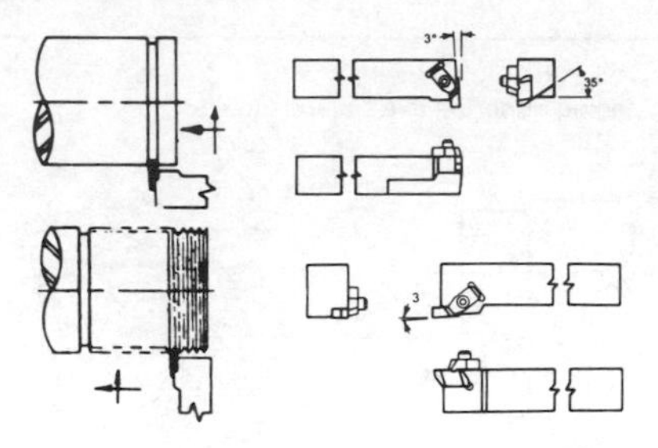

Style L
for both turning and facing with same tool. 80° diamond insert.

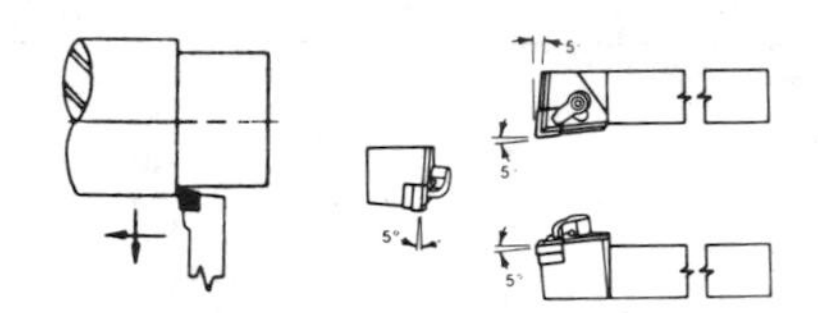

Style S
for chamfering and facing 45° lead angle.

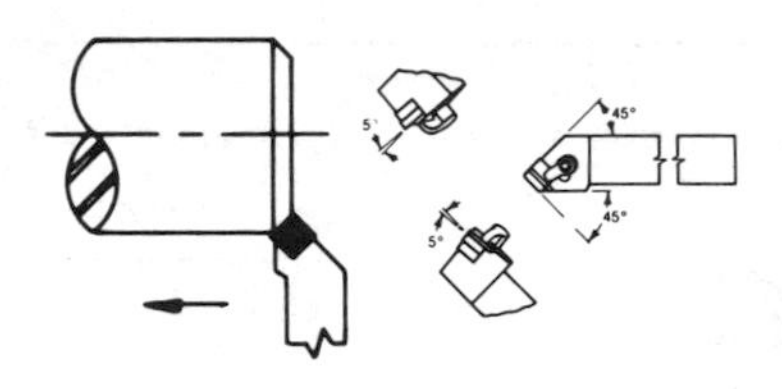

Style P
for profile machining. Insert centrally located.

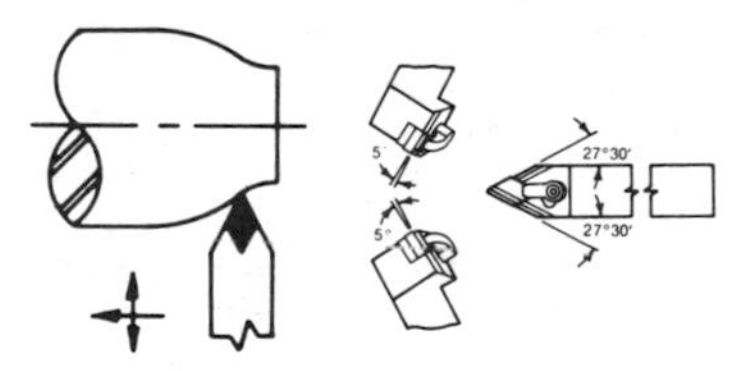

Style V
for profile machining. 35° insert.

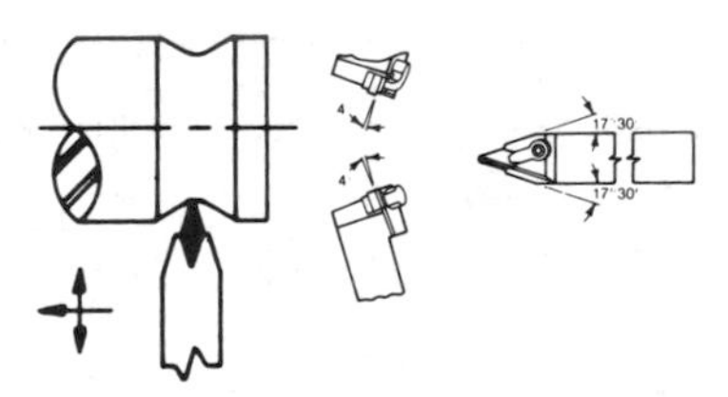

Fig. 8-14 (2 of 2) Toolholder styles are identified by an industry standard system.

The following questions refer to tool selection for NC and C/NC setup.

1. At initial setup, what is the first fact that must be known about the program to begin the tool selection?

___

2. Circle the letter of the two factors that limit tool size and shape selection in setting up a program already written.

   A) Available cutters, type of machine.

   B) Program type (MComp or CComp), shape of part.

   C) Cutter grade, material hardness.

   D) Machine capacity, available cutters.

3. If you were setting up an MComp program, what would be the four limits of cutter selection?

___

___

___

___

4. List the four tool geometry offset parameters for mills.

___

___

___

___

5. List the tool geometry offset parameters for lathes.

___

___

6. Circle the letter of the term or terms that describes qualified toolholders.

   A) Made of quality materials.

   B) Made to exact standard sizes and shapes.

   C) Made exactly to fit a particular custom application.

   D) All of the above.

7. If a program is MComp, you have limited choices about cutter size and shape.

   Is this statement ☐ True or ☐ False?

   If it is false, what will make it true?

8. Define the term *offset* in your own words. Be sure to include the terms *variable* and *compensation*. Tell where the offset resides (is stored) in the controller.

9. A pen is put in a milling machine spindle. A plot of a simple rectangular part is made. The part was supposed to be 4.00″ long, but the pen trace produced a length of 4.750″. Fig. 8-15. For what cutter diameter was this program compensated?

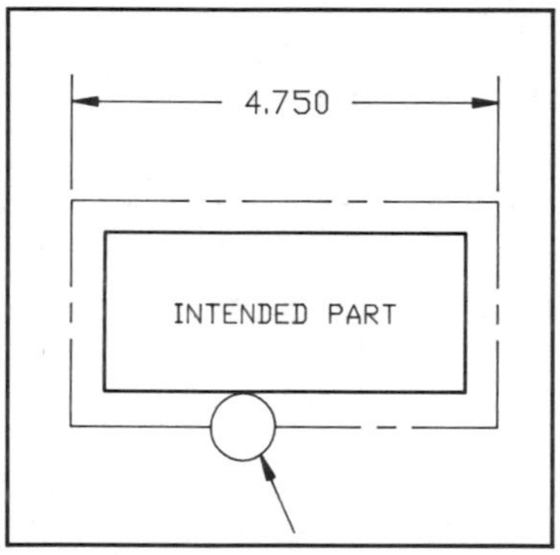

Fig. 8-15. What diameter cutter was planned into this program?

10. If the program in Question 9 was a CComp program, could the pen be made to trace the exact size of the part? How?

11. Refer to Fig. 8-6. Which would be the best carbide insert tool, in terms of wear and strength, for producing the part shown in Fig. 8-16.

    How many tool shapes would be needed for the job?

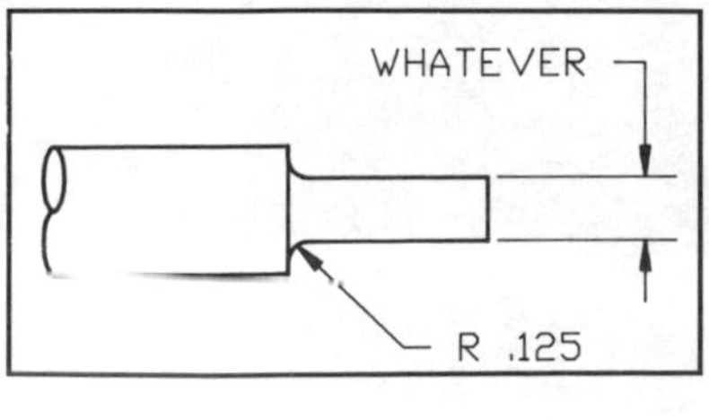

Fig. 8-16. What insert shape, nose radius, and IC would be best?

12. Which of the insert shapes shown in Fig. 8-6 would be the strongest selection that could turn the part shown in Fig. 8-17?

____

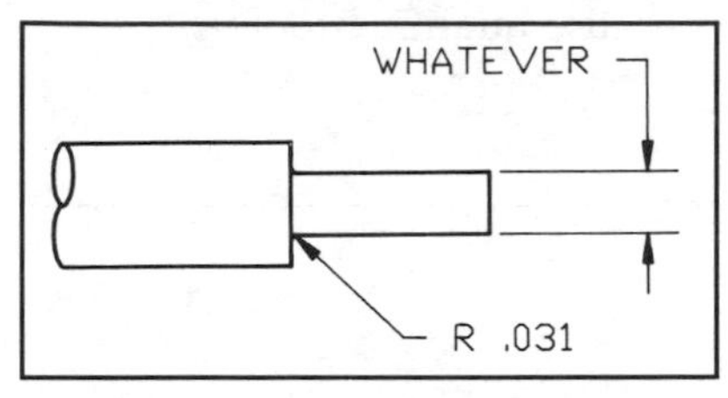

Fig. 8-17. Select the strongest carbide insert shape that can machine this part?

For Questions 13, 14 and 15, refer to Fig. 8-13, and this lathe toolholder identification number: DDAPL - 12 4 P.

13. What is the shape of the insert in this holder?

____

14. What is the size of the insert (inscribed circle)?

____

15. Is the tool right hand or left hand?

____

16. A standard carbide insert is numbered RPMP-432. In answering the following questions, refer to Fig. 8-12.

   A) What is the insert shape?

   ____

   B) What is the cutting point radius?

   ____

   C) Is this a positive or negative rake insert?

   ____

   D) What is the size of the insert (inscribed circle)?

   ____

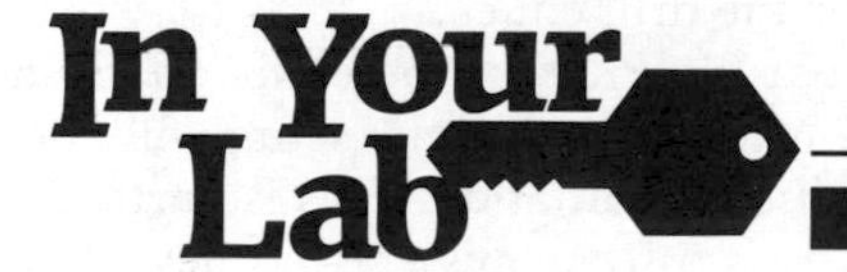

17. Does the control you are going to operate have computer tool compensation?

____

18. Do you use carbide insert tooling in your lab for C/NC work?

____

19. Does your lab use qualified tooling? (The probability is low in a school lab because qualified tooling becomes cost effective only in a production situation).

____

8-5 Work-Holding Tools and Fixtures

Work-holding tools are basic to the machine setup. When a program is written, the programmer must consider the whole process. He or she must have all the facts before writing a single line of program. He or she must know how the raw material is to be held in the machine. This is the first decision the programmer must make.

When setting up a production run, you must make sure the part is held exactly as specified in the setup document. Failing to do this can cause a crash.

GENERAL STANDARD TOOLING AND SPECIFIC FIXTURES

There are two broad groups of work-holding tools for C/NC work: general basic tooling and special fixtures. General basic tooling is universal, while special tooling is made for a specific job or part.

General Basic Tooling

The Setup Person's Responsibility

The choice of general tooling components is usually left up to the setup person. Standard vises, clamps, pins, and holding tools are used. A set of specially machined soft vise jaws might also be used. On lathes this means using standard three- and four-jaw chucks. These might have soft machined jaws or have collets and sometimes faceplates. The position of the part and the contact points will be defined in the setup document.

To master setup work requires years of experience. Every job has a new twist. The following procedures are standard.

1. Always clean every component that is to be used in the setup. Chips will destroy the accuracy of the setup. When squeezed down between two precision surfaces, even an aluminum chip will dent a hard steel plate.
2. Make sure that the part is held or gauged against solid surfaces, such as the solid jaw of a vise.
3. Make sure the part is held or located against the surfaces that create the program reference zero points. This is very important for part quality and accuracy. The theory behind this concept is the same as the theory behind the concept of tooling points in Geometric Dimensioning and Tolerancing. The primary reference surface must rest against three support points. For further information on geometrics see *ANSI Y14-M*.
4. Make sure you are mounting the fixture in a position on the machine that the spindle can reach. Do not, for example, bolt a vise so close to the end of a machine that the machine hits a travel limit before machining all of the part.
5. Make sure that you will be able to easily measure the part without destroying the accuracy of the setup. For example, if possible, you would not want to unchuck a lathe part to measure it or take a part out of the clamps on a mill.
6. Make sure that the parts can be changed easily and accurately.
7. If parts can be loaded incorrectly in your setup, make sure that all involved understand the right way to load them. A note in your work area is a good idea here.
8. It is usually better to use too many clamps than too few. Over-clamping is better, but be careful not to dent or mar parts. Remember that using more clamps requires more time when changing parts. This can affect cycle time.
9. Make sure that your setup will not interfere with the spindle. For example, the clamp should not be too tall for the drill chuck.
10. Make sure that the PRZ can be easily established in the setup. Make sure that the setup will not interfere with the coordination. This is particularly important. It is often missed by new C/NC machinists. They may hold the part well but fail to allow for the establishment of the PRZ. Think before you start. Read the setup document. Find out what part features are used to establish the PRZ. Do not obscure the PRZ with the setup.

Fig. 8-18. A typical lathe fixture in industry. The Boeing Company

General Basic Tooling Concepts

The use of universal setup tooling is basic to your machining education. Vise alignment, part indication, and the use of clamps and chucks are part of a general course in machining.

If you have difficulty with any tools or processes, ask your instructor for review material. You will need to develop certain skills and understand certain concepts to be competent in C/NC work. We will discuss only the holding and fixturing work that pertains to C/NC setups. Aligning and indication of tooling on the machine require a knowledge of manual movement of the machine. This will be covered in Section 8-9.

Special Fixtures

The tooling required for holding the work may be a special fixture designed and built for this job. A part number will be stamped on this fixture. The part number must be matched to the job, the program, and the part print. This tooling is usually not built by the setup person. A toolmaker will make these fixtures ahead of time if they are complicated. Fixtures have applications on lathes as well as mills. Fig. 8-18. The setup document will contain the information necessary to use the fixture.

Sometimes a simple fixture can be built for a job during the setup if the time can be justified. Shop-made fixtures are then stamped with the part number and job number. A sketch is made and recorded for future setup documents.

It is your job to make sure that the fixture is treated with care, and used and stored correctly. You must make sure that the job number and the fixture number match. You must make sure that the fixture is mounted correctly on the machine.

Fixture Advantages

Fixtures have several advantages. Even though a fixture costs more, it is cost effective in shops where repeated setups are needed or where a very special part is nearly impossible to hold for machining by standard methods. Fixtures also reduce down time. Fixtures have the following advantages.

- Shorten machine setup time.
- Shorten part-to-part turnaround time.
- Make it impossible to load the part incorrectly.
- Give an accurate way to locate difficult shapes.
- Increase safety by firmly holding parts.
- Make the PRZ easy to establish and coordinate to the machine. Some parts are difficult to coordinate because the PRZ is derived from a feature that is difficult to indicate such as a rough casting. In this case, the fixture itself will have a pad or a hole that the program refers to for the PRZ. You will need to indicate this point for coordination instead of the part.
- Make it possible to produce parts more accurately.

Handling Special Fixtures

Special fixtures are costly. They are made to precise standards. Tolerances for production fixtures are less than one-third of the production tolerance for the part to be held. *Treat these fixtures with care*. Follow these guidelines.

- Never modify a fixture without permission from your supervisor.
- Make sure that you keep all the clamps, pins, and other components with the fixture when it is stored. Nothing is more frustrating than to get most of the way through a setup and then find that part of the fixture was lost the last time it was used.
- Before storage, apply a light coat of oil to precision parts.
- Always clean the fixture before mounting and storing.
- Never hit the fixture with a hard hammer when aligning it to the machine. Fixture components could be damaged.
- Always double-check to be sure you are using the correct fixture.

8-6 Performing Setup Maintenance and Ongoing Maintenance

SETUP MAINTENANCE

C/NC machines left in your care need attention. The cost of a new C/NC machine will seldom be less than $50,000. Most cost far more. Maintenance must be performed by the setup person and operator at setup time and on a set schedule at other times.

The following steps should be performed during machine set up and on a set schedule during operation.

- Clean the working surfaces. Remove any chips that would interfere with the setup or be unsafe. When cleaning chips, follow your lab's policy on the use of pressurized air. Pressurized air works well in cleaning up but must be used with caution. Some labs ban the use of pressurized air on machinery. Never blow chips directly at the seals on the machines slides and bearings nor at the electronic panels.
- Check all fluid levels, including lubrication reservoirs. Fill low reservoirs. Check dipsticks. Add only the correct type of oil to lubrication reservoirs. If there is not a tag on the filler, find out the correct oil type.
- Check coolant level. Fill if needed. If the coolant is to be mixed with water, make sure you use the correct water/coolant ratio. Using too much coolant is costly—the concentrate is expensive. Too little concentrate in the water will rust machine parts and cause the coolant to be ineffective.
- Check air filters and hydraulic filters.
- Make sure the air pressure for fixtures, chucks, and air-activated brakes and lubrication systems is correctly adjusted.
- Check that all bolts are tight and safety covers closed. Make sure that covers seal out chips.
- Check all adjustable features such as belts and clutches.

Custom Sheet 1

At this point, you will need instruction on the particular maintenance features of the machine you are about to operate. Your instructor has Custom Sheet 1 prepared for you. He or she will present information on its lube points, air and hydraulic needs, and general maintenance. Your instructor also will cover safety or maintenance points that are special for that machine. A professional machinist not only makes parts, he or she also cares for the tools and equipment.

8-7 Downloading the Program into the Control

LOADING A PROGRAM INTO ACTIVE RAM

The transfer of program data from a computer to a CNC control is known as downloading. *Downloading* is the transfer of pre-written program data into the CNC control RAM from an external device. Downloading infers the transfer of information from a superior source (computer) to an inferior one (CNC control). However, in CNC work several processes that program the control are often called downloading. These are not strictly downloading but are often referred to as such.

The setup has been made. You are ready to begin working with the control and the program. The program must be loaded into the active RAM of the controller. First, you need to know how to turn on the control. The procedure for this is found on Custom Sheet 2.

There are four possible situations.

1. The program is stored in permanent internal memory inside the control. It will need to be transferred to active RAM. This process is called *transferring*, not downloading.
2. The program is on punched tape.
3. The program will be downloaded from an external device such as a computer or tape recorder.
4. The program must be hand-entered because the controller you are using has no offline data management system. Hand entry or MDI program entry is not called downloading.

Custom Sheet 2

In each case, you will need instructions as to how to turn on the control and download or transfer a program into memory. This information is presented on Custom Sheet 2. Your instructor has Custom Sheet 2. Read it now. Then read the following general procedural hints.

1. Be cautious that you do not destroy a program that is already in RAM. If there is a program in the control, find out if it is important before you erase it.
2. Though C/NC tape is strong, it can be damaged. Handle it with reasonable care. There are specific ways to load a tape into a reader. These are shown on Custom Sheet 2 or in the operator's manual for the control. To be sure the tape is correctly loaded, look for the sprocket drive channel. It is off center on the tape. This prevents loading the tape backwards or upside down. Many laminated tape materials have a different color or finish on each side to help with tape loading.
3. Take care in handling and storing computer disks.
   A) Do not set disks near metal, magnets, or electrical devices. (This can erase or scramble the data.)
   B) Do not touch the actual disk. Dirt and scratches will damage disks.
   C) Keep the disk in a jacket in a disk file when not in use.
4. If the data is to be transferred from an external source, it is most common to prepare the receiving unit (controller) first. Then trigger the transmission from the sending unit.
5. If you must hand-enter the data, double-check each line as you enter it. You might, for example, say each line out loud as you scan the readout entered.

8-8 Loading the Raw Material into the Machine

THE PROCEDURE

The fixture or holding device is ready. The program is in the RAM. The tools are in the machine. Now the work material must be loaded into the holding fixture or setup. Here is the procedure:

1. Make sure the material is oriented correctly in the setup. The material should be right side up, with the right dimensions in the correct position. Production situations often require the grain of the material to go in a specific direction. Metal does have grain. When rolled out at the mill, the metal grains are elongated in the direction the material is stretched. This grain gives metal a strength bias similar to wood. When flexed, metal will tend to break more easily parallel to the grain than across it.
2. Make sure that the material is burred and clean. The accuracy of an otherwise good setup can be destroyed by a burr or chip under the part.
3. After loading the material, make sure that it is firmly seated in the setup. This may require a soft impactless hammer, a feeler gauge, or just a good close look. After the holding pressure is tightened, the material must be tight against the holding and locating points.
4. Make sure all clamps or chuck jaws are firm but not overtight. By overtightening, you mar the part. You also stretch the tooling, perhaps to a point where it will break under further machining load. A hint: When tightening down chucks or clamps, snug them. Then slightly release the pressure just to release any out-of-alignment condition that may have been held in the first time. Then re-tighten. You may need to do this more than once. Have your instructor double-check your first part load.

8-9 Coordinating the Tools to the Part and Setting the PRZ

At this point the program comes together with the fixturing and raw material. We need to coordinate the cutting tools to the work and setup. We then need to set the PRZ in the control. This involves moving the cutter or the spindle to the correct location in relation to PRZ and then resetting the positional registers in the control to zero.

DEPARTURE POINT

The departure point is the coordinated park position for the tool. Before we study the actual movements and procedure, we need to define the departure point for a tool. All programs assume that the tool is in some coordinated relationship to the PRZ. The place the tool starts from is called its *departure point*. The departure point can be the PRZ or some other location. However, the tool must be parked in some *known* relation to the PRZ. This concept is the same on lathes and mills. We will look at examples for both.

The PRZ Is the Departure Point

For simplicity, we will consider a program with only one tool on a milling machine. The simplest case would be where the tool starts from the PRZ. Fig. 8-19. In this example, the PRZ is 2.0″ off the surface of the part. This is shown in the setup document.

1. *Move the spindle over the corner of the part.* Using an indicator or edge finder, the operator moves the machine to the left lower corner of the part so that the spindle is right over the corner.
2. *Mount the tool, then establish the departure distance.* Next, the operator brings the tool to the part along the Z axis and just touches the surface at the correct corner. He or she then backs the tool away (Z axis) from the part exactly 2.0″.
3. *Reset the axis registers.* The operator then sets all axis registers (X,Y, and Z) to zero. This means that the tool is parked at PRZ and the axis registers in the control are set so that this position is the PRZ (X=0 Y=0 and Z=0).

A second option, rather than touching the tool to the part, would be to touch the tool to a 2.0″ gauge block that is sitting on the surface of the part.

Either method leaves the tool at the PRZ—over the corner of the part and 2.0″ off the part in the Z axis. In this example, the departure point is the PRZ. That is the way the programmer intended the sequence to work. The first tool movement in the program is from the departure point.

There are advanced methods of touching off the workpiece using electronic or laser probes.

The PRZ Is not the Departure Point

In this example, a single tool is being used on the lathe. Here the PRZ is on the centerline and right at the end of the workpiece. The departure point is not at the PRZ but is rather at the park position for the tool turret. Fig. 8-20. There are two options at this point.

Option One

The simplest option is first to touch the tool to the workpiece. Then set the axis registers to zero (X and Z). Next, back the tool away from the work to the position as described in the setup document—the departure point. Assume that this distance is 2.0″ in the X axis and 6.0″ in the Z axis. Once the tool is parked at this position, the program is ready to run. The tool is parked at the departure point, which is a known distance from PRZ. All moves in the program assume that this coordination has been made. This is the responsibility of the setup person. The axis register will read the correct position in relation to the PRZ: X=2.0 Z=6.0. This option is common on small lathes that do not use tool offsets or have an automatic park position.

The program must contain initial tool moves that are roughly correct. In this case, to get the tool to the work the programmed Z move is 6.00″ and the X move is 2.0″.

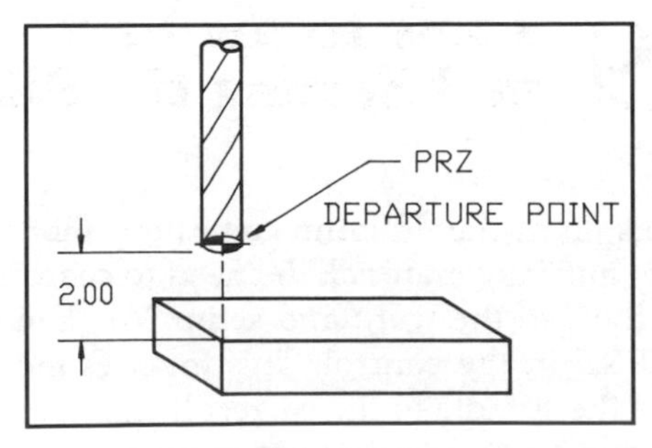

Fig. 8-19. In this setup/program, the PRZ is also the departure point.

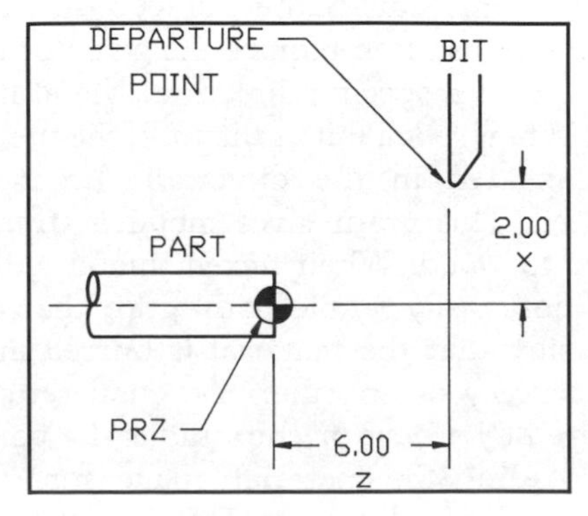

Fig. 8-20. This setup has the PRZ on the part, but the departure point is off the work at the tool park position.

Option Two

The operator touches the tool to the workpiece then zeros the axis registers. He or she then commands the tool back to the tool park position. This position could also be called machine home or tool change position. Many machines have a special command to park the tool turret at this position. The actual distance traveled from the touch-off point to the park position is on the X and Z axis registers on the control. The difference between the programmed distance and the actual distance on the registers is then hand entered into the tool axis offset registers.

This option adjusts the initial moves to include the offsets. The initial move will be roughly calculated by the programmer then fine-tuned using the axis offsets. The program distance will be shortened or lengthened in each axis according to the the amount of travel from touch to park positions.

As an example, suppose that after touching the material, zeroing the registers, and then moving to park-departure point, the registers show an actual distance traveled of X 2.35″ and Z 5.55″.

The offset for the X axis must be +.35″, added to the programmed move. The programmed move must be lengthened by .35″ in order to move to the PRZ from departure. The offset for the Y axis must be –.55″, subtracted from the initial move of 6.0″ to arrive at PRZ.

Option two is recommended. The machine will automatically compensate for the extra distance entered when the correct program callout is used for the initial movement from the departure point. This method is common. It is used extensively in multitool setups for both lathes and mills.

The initial movement callouts must include a statement that positions the tool to the gauge height and also considers the X and Z position offsets. We will look at how to do this in Chapter 21.

Multitool Setups

If the machine has more than one tool in the setup, the operator will have to repeat this procedure for each tool unless the machine is loaded with qualified tools. In that case, the first tool is the only one that needs to be tested. The relationship between tool points is known for qualified tools that are correctly mounted.

Since the programmer knows the relationship of each qualified tool point, his programmed initial moves for each tool will be adjusted to include the differences. For example if tool two is 1.0″ longer than tool one, it needs to travel only 5.0″ to touch the work. Therefore, the initial move of the tool will be 5.00″. Each succeeding tool offset will be the same as for tool one because the difference is built into the program.

Fine Tune Offsets

Fine tuning must be done for each tool to bring the part exactly into manufacturing tolerance. This is determined upon the first part run with actual measurements. Some tools, such as drills, are set up on a bench tool gauge. This gauge will allow the tool to protrude an exact amount from a bushing. When tightened in the bushing, the drill then becomes a qualified tool. A head on the bushing slips into the tool turret against a shoulder. Thus, the drill is gauged exactly. It then becomes a qualified tool in this setup.

As you can see, qualified tooling can save time. If tool two broke, it could be replaced with a qualified tool. The only difference in the size of the part would be the manufacturing tolerance in the tool itself. This could be compensated for with a slight adjustment to the offset value.

Final Adjustment

The part being machined needn't be ruined, even if the new tool takes more material. Before using the new tool, a smart operator will adjust the offset so that the tool moves away from the part more than the manufacturing tolerance. After machining a small portion with the new tool, the machine is stopped and a measurement is taken. The result is then adjusted into the offset registers so that the new tool makes the part to size.

Gauge Height

Another position often used in C/NC programs is that of the gauge height point. The *gauge point* is a minimum distance height for tool travel. A good example would be found in a drilling program. After leaving the departure point, a rapid travel move would bring the drill close to the material but not touch it. The drill would slow to the programmed feed rate and then contact the work. This safe amount held back is the gauge height point.

If you have coordinated the PRZ correctly and have the tools parked at the departure point, you need not be concerned with the gauge height. It is built into the program. It is the minimum safe distance the tool retracts to move to another location. In the previous examples, the tool wouldn't actually rapid-travel to PRZ. It would be stopped at a gauge point a short distance from the part.

The concept of *departure point* is complicated. Do you understand it?

All programs must start the tool from somewhere. This point must be coordinated between the control and the tool. The tool must be where the control thinks it is. The tool must be a known distance from the PRZ. This is the responsibility of the setup person. Without correct coordination, the part will be incorrectly machined. There will be a crash and the fixture or setup will be ruined.

It is a good idea to have someone double-check your setup and coordination. Also a cautious first part, single-step test or a dry run will reveal any errors. The next Activity will help measure your understanding.

Presetting a Program Reference Zero—G50

On lathes, there is usually a way to set the PRZ from the departure point without resetting the machine axis registers. This is done by a code or word that causes the control to reset the zero point from the present location of the tool. The PRZ is established by the G50 code followed by the distance it lies from the present tool location. This is a useful code. A setup person needs to coordinate the tool to the part but not necessarily to the PRZ. The control reads the G50 code followed by the coordinates of the new PRZ. The present location of the tool is taken as the temporary reference.

For example, a piece of raw 1.00″ diameter stock that protrudes from the collet 3″ must have the PRZ at the collet face on the centerline. The setup person does not need to move the tool to that position (which would be inside the stock). Instead, during the setup, he or she coordinates the Z axis by touching the tool to the outer face of the stock and the X axis by touching off the diameter. This positions the tool at the outer corner of the stock. The setup person then moves the tool back 3.0″ further in Z. This is the departure point, which is 6″ from PRZ at the 1.0″ diameter in the X axis.

The program then starts with this command: G50 X 1.0 Z6.0. This command informs the control that the tool is presently 1.0″ in X and 6.0″in Z from the PRZ. The control then knows that the PRZ for this operation is located 6″ toward the chuck in the Z axis and on center in the X axis. (Diameter programming is used in this example.) This is useful when non-qualified tools are being used. Each tool is touched off the stock and positioned in the same position as tool one. A G50 code is given for that particular tool.

The zero pre-set method is a useful programming technique that can also be used on mills. If a G50 technique is used, it should be mentioned in the setup document. This technique is also used on conversational machines.

MANUAL MOVEMENT OF THE MACHINE

Jog Mode

All C/NC machines have an operation mode called *jog*. In this mode, you have free movement over the axes. Each machine has specific procedures to initiate the jog mode. After we discuss some of its basic points, you will need Custom Sheet 3, Manual Movement of This Machine, or a demonstration on the machine.

To begin, let's look at the word *mode*. All controllers have many different functions. Each function in the control requires different internal programming and circuitry. The controller must be told what internal programming to bring up to accomplish what the operator wishes to do. This internal programming is called the *mode*. Find the mode select switch on your control. It probably looks like the one shown in Fig. 8-21. Some machines using interactive video may select the mode from a screen menu such as that shown in Fig. 8-22. This will be made clear to you on Custom Sheet 3.

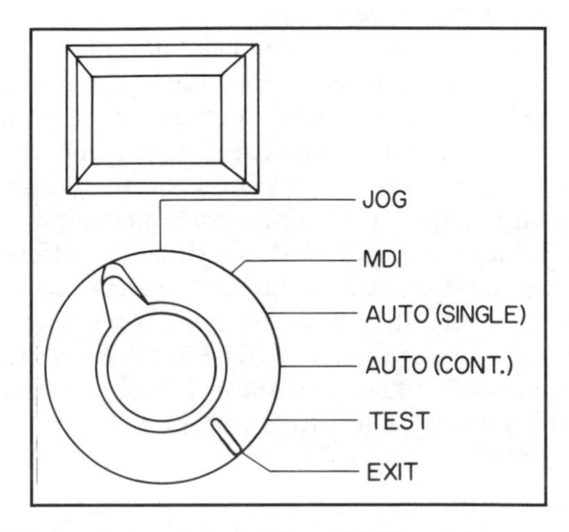

Fig. 8-21. A typical mode switch for a C/NC machine.

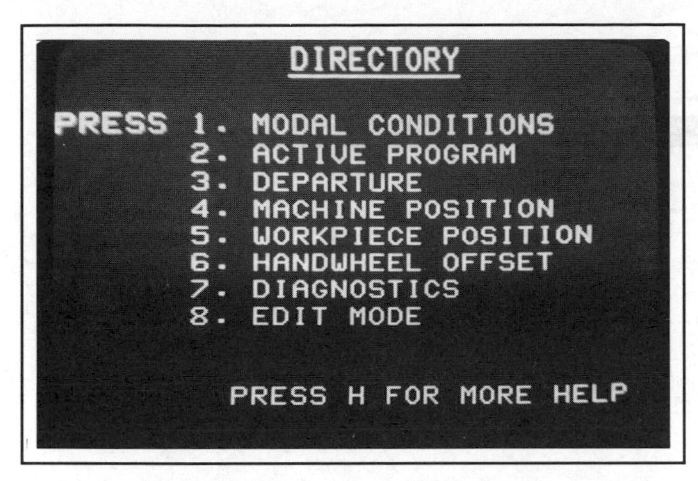

Fig. 8-22. A typical screen menu for mode selection.

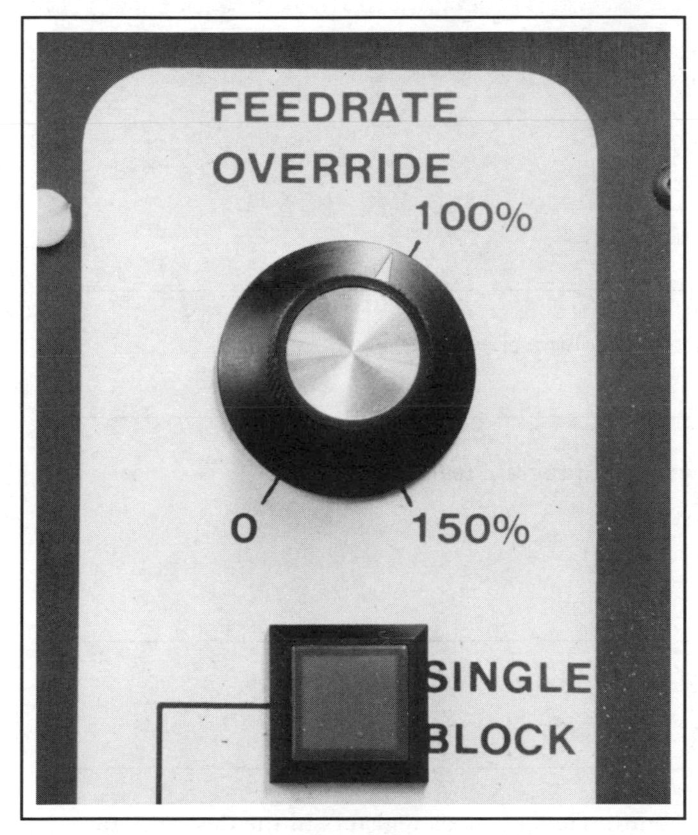

Fig. 8-23. A typical feed rate override switch.

THE PROGRAM, THE SETUP DOCUMENT

Jogging is usually done in the following three stages.

1. *Rapid continuous feed.* This selection will move the axis in the direction you have selected until you take your finger off the button, switch, or lever. Some machines allow overriding the rapid feed rate at the control with a switch that looks like the one in Fig. 8-23.
2. *Selected feed continuous.* In this stage, the machine moves at a selected rate. As long as you hold the button, the machine moves. All machines equipped with feed rate override control will allow you to adjust this speed.
3. *Delta jog.* In this stage, the machine will move a discrete amount with one push of the button. You may be able to select the amount the machine will move. The word *delta* means "small amount." On most machines the amount will be .1″ .01″, .001″ or .0001″. This amount is called the *jog resolution.*

Manual Pulse Generators or Resolvers

Some machines have a delta jog dial. Rotating this dial sends jog pulses to the machine. The farther you turn the dial, the farther the machine axis moves. Such a dial is called a manual pulse generator (MPG) or resolver.

While jogging, the positional readouts will track the movements you have made. You can reset these readouts and have some control over their usage. Also, the jog mode may or may not affect the coordination of the machine. You must know how this works on your control. This is especially important if, in a coordinated setup, you wish to move over a slight amount. All of this will be machine particular. It will be explained on Custom Sheet 3 or in the operator's manual.

MDI Movement of the Machine

Another way to move the machine involves changing modes. You will need to select the manual data insert (MDI) mode. With this, you can supply single blocks of information to the controller. If you have first told the controller a feed rate and whether absolute or incremental values are being used, you will be able to instruct the control to move the machine to an exact position.

In effect, you are providing a program one block at a time. This technique is also valuable in moving odd distances during setup. It can also be used to machine a simple part. After the spindle is turned on, the machine is commanded—through MDI mode—to move and machine a part. The part machined could be a first part and each successful move could be saved in RAM to build a program as you make a part. This requires skill. Do not try it until you gain some experience.

Ask your instructor for Custom Sheet 3, Manual Movement of the Machine. When the machine is not being used, practice first continuous jog moves. Then practice delta jog moves. Once these are mastered, try to move the machine using MDI movement. Note that on some controllers, the resolution in jog is not as fine as in MDI. If you had to coordinate to a PRZ within, say, .0003″, you might have to make the final adjustments in MDI rather than in delta jog.

Listed below are the positions used on a C/NC machine. Answer the following questions about these positions.

A) Program reference zero.
B) Local reference zero.
C) Machine home.
D) Departure point.
E) Gauge height.

1. Which position never changes?

____

2. Identify the position that is the safe minimum distance for tool travel.

____

3. The *departure point* is used as a reference point for a group of program features.

   Is this statement ☐ True or ☐ False?

   If it is false what will make it true?

____

4. Could the machine home be the departure point?

____

5. Describe the departure point. Include the terms PRZ, coordinated, and axis registers in the description.

____

____

____

6. Could the departure point and the gauge height point be the same for a vertical drill press program?

____

Mystery Part Challenge

This exercise will tell you if you understand C/NC coordination of a PRZ and manual movement of the machine. Choose either the lathe or mill exercise. Use Custom Sheet 3 as a guide.

Safety Rule. *Take your time. Think before pushing a button. Lock out, turn off, or place the spindle in neutral.*

1. Mill Exercise: Find the Size of a Mystery Block

Objective

Using an edge finder, coordinate the spindle over the left lower corner of an indicated block. Then set the tip of the edge finder 2.00" above the block. This position will be a simulated PRZ and also the departure point. *Using only the edge finder and axis register positional readouts, determine the actual size of the block in X and Y.*

Instructions

A) *Indicate the block parallel.* Set the mystery block in the machine. This can be any true rectangular block you choose. A ground parallel would be ideal. Clamp the block fingertight to the table. Fig. 8-24. Indicate the long side, parallel to the X axis of the machine. Use continuous jog and an indicator in the spindle.

B) *Clamp and check indication.* Once the block is true to the machine, *lightly* clamp it down. Check the parallelism after indication. Adjust if needed.

C) *Position over PRZ.* Remove the indicator and replace it with an edge finder. Now, one axis at a time, move over the corner of the part. You will need to understand the edge-finding technique and the position readouts of the machine.

D) *Position the Z axis at zero.* Move the tip of the edge finder up 2.00" above the surface of the work. Reset the Z register to zero. You have now completed a PRZ coordination (X0,Y0,Z0).

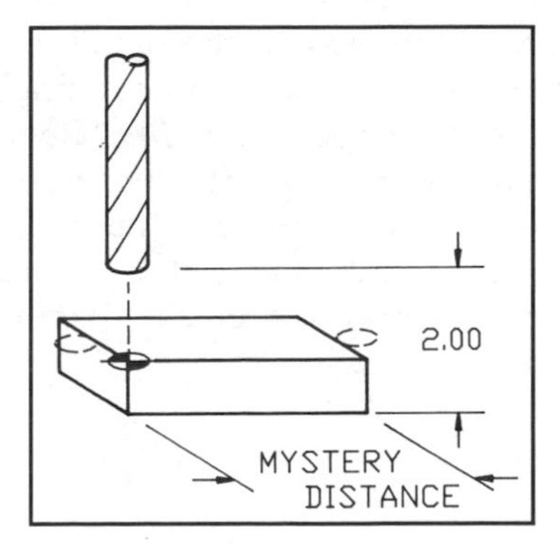

Fig. 8-24. The mystery block correctly mounted on the machine.

E) *Find the length and width of the mystery block.* Now, using only the edge finder and the position registers, determine the length and width of the block by jogging. Remember to take into account one-half the edge probe diameter in your calculations. You should be able to get the dimensions within .003". There will be some inaccuracy in using an edge finder. There also will be a small backlash factor in the machine.

X distance = Y = X = Y =

%112 %112

Your determination of the actual measured size of the mystery block:

____

2. Lathe Exercise: Determine the Mystery Shoulder Distance

Objective

In this exercise, you are to coordinate a turning tool for a PRZ at the end of a workpiece, on the centerline. Fig. 8-25. Use a turning tool to touch the outside diameter of the part. Then move toward the center, half the distance. Touch the end of the part and set the PRZ in the axis registers. The "mystery shoulder distance" will be a step turned in any round part. Any round object with a turned step will do.

Instructions

A) *True the chuck workpiece.* You can use a standard three- or four-jaw chuck, or a collet. Make sure it is running true by turning it and checking the runout with an indicator.

B) *Mount a turning tool.* The turning tool need not be a carbide tool or a sharp tool. A piece of mild steel or aluminum ground to simulate a turning tool would work just fine. For safety, you might use a rubber tool.

C) *Establish contact on the diameter of the work.* Touch the work with the tool somewhere near the end (X axis). Then move the Z axis beyond the end of the part (+Z). Now jog the tool to the centerline by moving in the X axis one-half the diameter of the piece as measured by a micrometer (–X). This is the X axis centerline. The X-axis positional registers may read out diameter figures. If they do you must command or jog the machine to center by moving the full diameter on the registers. In effect, you are moving from one diameter position to another—(zero, in this case.)

D) *Establish the PRZ.* Bring the tool toward the work (–Z) until it touches. This is the PRZ. Set the registers to zero.

E) *Find the mystery distance.* Using the tool as a probe and *carefully* jogging the tool, touch the shoulder of the part and determine the exact "mystery shoulder distance."

8-10 Entering Offset Data and Tool Geometry into the Control

Offsets need to be updated because of the following:

- Tool wear.
- Machine warmup, which causes size changes.
- Tool replacement.
- Need to change size of part. For production reasons you wish to make the part slightly bigger or smaller inside the print tolerance.
- Need to allow for cutter deflection in machining.

MANAGING OFFSETS

There may not be a need for offset data in the program you are about to run on the machine. A first part run with a single tool is easy to coordinate without offset entry. However, this is a good time to actually enter the information into the machine. Remember that some machines may not be capable of offset management. Ask your instructor if your machine uses offsets.

The Two Kinds of Offsets

The first type of offset is the adjustment to the axis moves just discussed. The second involves the tool geometry data discussed at the beginning of this chapter. These may have different procedures for entry. Make sure that you understand both entries. Many newer CNC machines have a "tool data page" on the video screen where each kind of data is easy to see and enter. Other machines may have a special memory set aside for the tool data.

Custom Sheet 4—Enter Tool Offsets

Following is the general procedure for entering tool offsets.

1. *Examine the program.* Determine that the program actually needs axis offsets. Your instructor can show you what to look for at this time.
2. *Enter the tool geometry offsets.* If you have a mill, you will also need to enter the *cutter diameter or radius.* Which does this machine accept? You must know. You must also enter the *tool length.* The tool length may be taken into account by the Z axis offset. If you have a lathe, you will need to enter the tool nose radius (TNR) and a standard tool shape from the possible set of choices and X and Z axis offsets.
3. *Determine the axis offset amount.* Find the amount needed for each axis. Do this in two stages.

   First, determine a rough amount, using the touch-off procedure just learned.

   Second, machine a part to find the fine-tune amount to be taken to bring the part into exact tolerance. Again, if you use good technique, the first part needn't be surprisingly undersize. Set the rough amount so you are sure the part will be oversize or right at the top of the tolerance. Then adjust to print size with the fine-tune amount.

8-11 Testing the Program: Running the First Part

Now that you have made the setup, you are ready to test your skill. The material is in the setup. The cutter is in the machine. The PRZ and departure points are established.

Review Program Testing in Chapter 7 to see the various testing options available. The general procedure to follow is given in Custom Sheet 5.

Here is the general procedure. A program test will probably require the "Program Execute Mode" on your machine.

1. The simplest test is a graphic test of the run on the video screen. Follow the individual machine instructions.
2. Next, perform a dry run in single step with the tool removed or backed away from the work in all axes.
3. Finally, start the machine. Set the optional stop switch to the ON position if the control and program have opt-stop or put the machine in the PROGRAM RUN—SINGLE STEP mode. Using both features would be acceptable.
4. Put on your safety glasses. Now, analyze every block of data that comes up on the "ready to execute" readout. Ask yourself, "Is this reasonable?" If it is, then press the START button. This button may be called several things. It is the signal to the control to execute the information in the readout. *Don't guess and hope. Analyze. Think first!* Remember, you are running the machine. It is not running you.
5. Take notes for the corrections you intend to make. If the program is already written, there should be few changes. However, the offset data will need to be fine tuned.

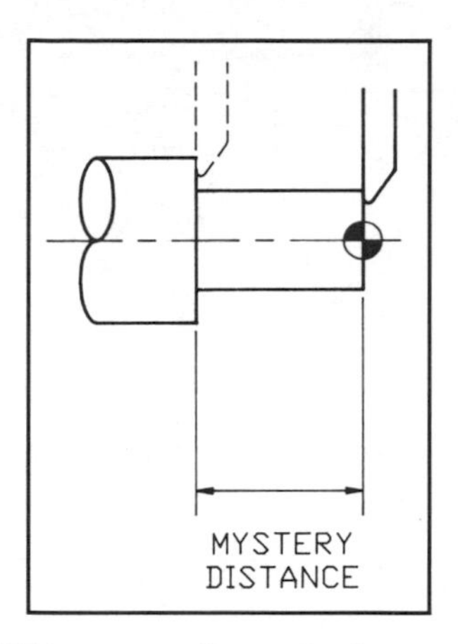

Fig. 8-25. The PRZ is on centerline at the far tip of the work.

Z distance (your determination) =

____

Z distance (actual measured distance) =

____

Note. When touching the tool to the work, it is best to use a feeler gauge. Bring the tool close to the work, but do not actually touch it. The extra distance can be compensated on the control by jogging after the touch. A piece of paper .003″ thick is always available for this purpose. The .003″ is easy to adjust in the axis register.

PRACTICE EFFICIENCY AT SETUPS

Remember that the skills discussed earlier in this chapter involved "down time," nonpart-making setup time. The skills you have just learned involved setup work that a skilled operator must be able to perform in a minimum amount of time.

An operator might also be called upon to perform the following duties:

- Organize the work area. Tools should be laid on a clean surface or shop towel—on the work bench. They should not be laid on the machine. Keep prints and documents away from the machine.
- Clean up chips that prevent production and are a safety hazard.
- If there is time, pre-burr or post-burr the parts. Measure them and stack them neatly in the pallet. This also helps with part counts for the run.
- Protect parts. Many employers have noted that new employees do not understand how to correctly handle machined parts. Here are a few guidelines.
  1. To avoid marring parts by overclamping, make sure clamps and jaws are level on the part. If necessary, place soft material such as copper or aluminum between the tooling and part.
  2. Stack completed parts in such a way that they will not scratch each other. Also, make sure that they will not fall over.
  3. Never throw or hit parts. If you need to tap parts into alignment, use a hammer softer than the work material.
  4. Although burring is a good idea, be cautious. It can be overdone. Read the print. Sometimes sharp edges must not be rounded.
  5. Take any necessary precaution to preserve a perfect finish on any part you work on.
- Monitor tolerances. As each part comes off the machine, measure it after the next part is in and running. If an adjustment is needed, put it into the control on the next part. Change tools when called for. Try to keep the machine running at all times.

Do you feel confident that you understand each task needed to set up a C/NC machine? To check your knowledge of the information in Sections 8-1 through 8-9, complete the assigned activity. Complete any "in lab" program that your instructor has assigned for this activity.

Remember that the nervousness you might feel during your first C/NC part run is common. It will go away with time. The idea is to let the nervousness disappear. You should keep, however, a healthy respect for the machinery.

A major goal of most C/NC machinists is to write their own programs. This unit is designed to teach you the basic part of programming in both EIA coded language and conversational input. We will prepare and write a program for a lathe part or a mill part.

Programming is more than just entering a series of commands to make a part. Several steps must first be taken to insure a safe and efficient program. Most student programmers try to short-cut these planning steps and move right to the actual program. This can result in an unsafe program.

This unit will trace two jobs from initial planning to the production program. Study the development of each plan closely. You will be called upon to complete certain parts of the plan.

CONTENTS

CHAPTER 9

PREPARING TO WRITE A PROGRAM

Many new programmers fail to plan the whole sequence of events. Chapter 9 will show you how to successfully plan a safe, efficient, and accurate program.

OBJECTIVES

After studying this chapter, you will be able to:

- Plan the whole setup.
- Demonstrate how the part will be held and machined.
- Choose successful reference points.
- Write an operator setup document.
- Select the best cutting tools for the job.
- Select correct speeds and feeds for the cutting tools and material.
- Prepare a drawing of significant points for the program.
- Plan an efficient and safe program with good sequencing.

9-1 Planning the Whole Setup

THE BASIC QUESTIONS

In operating a standard machine, have you ever machined yourself into a corner? Has the sequence of operations made it difficult or impossible to complete a part. Poor planning will make it difficult to complete a part. The same problem arises in C/NC. To avoid this problem, never begin to write a program without planning. You must first think through all aspects of the program and find the answers to these questions.

1. How is the part to be held?
2. What will be the basic steps in the overall process?
3. Where will the PRZ be located?
4. What tools and cutting methods will be used?
5. Can the part be measured?
6. How will this program affect future programs or operations?

Related Facts Call for Careful Planning

Notice that some of the decisions are related. For example, the choice of the PRZ may affect the fixturing or chucking you will do to the part. Because some answers affect others, the best procedure is to review the list after making the initial decisions. See if any changes will make the program more clear. Check the safety and efficiency of the program and setup.

Double-check your plan by asking these questions:

- Can this setup be made quickly and accurately?
- Is this a safe, solid setup and program?
- Is the program as efficient as it can be?
- Can parts be loaded accurately and quickly?

In the following examples, notice that even for simple parts all the decisions affect the success of the job. You must think the job through completely before you try to write a program.

PREPARING TO WRITE A PROGRAM: LATHE PROBLEM

Refer to Appendix A Drawing 4. The overall planning for this lathe problem requires careful study of each of the following five planning points.

Planning Point 1: Holding the Part

Is the part to be held in a chuck or a collet? This is an important decision but on this part, it is not very difficult. Since the center portion is 1.00″ in diameter, a piece of 1.00″ diameter cold-rolled material could be held in a collet.

Planning Point 2: The Overall Process

How is the part to be completed? Examine the drawing closely. The steps in the test pin get smaller at both ends. That creates a production problem. If you try to make the part in one single program, it will be weak on the end closest to the chuck. It is possible to make the part by holding the excess stock on the .5625″ diameter end and then parting it off. This, however, would require a tailstock operation, and not all C/NC lathes have tailstocks.

Two separate setups might be the answer. Although this is only one answer, we will choose this one. We will need to turn the part around one time. In a production situation, that might not be a good decision because of the extra part handling time required. For this exercise, we will start with 1.00″ diameter material cut to the 6.00″ length. Bear in mind that there are no definite answers to these questions. This answer will work for the purpose of this demonstration. Fig. 9-1.

For the first operation, we will turn the right end of the part first (.5625″, .625″, .6875″, and .75″ diameters). Then, for the second operation, we will turn it around and machine the left end. For the first operation, the part must protrude from the spindle 2.5″, plus a small amount (say, .25″) to avoid hitting the collet. The total, then would be 2.75″. In a production situation, this would be done using a stock stop from the tool turret so that each part protrudes the same amount.

After roughing and finishing the four right-end steps, we will remove the part and turn it around for a second setup. For the second operation, we can hold the part on the .750″ diameter and index the 1.00″ diameter shoulder against the collet. This will be a reliable surface from which to start the second operation. Think as though you were doing more than one part. Putting the parts against this shoulder would be a quick, solid, foolproof way of indexing them.

Planning Point 3: Locating the PRZ

The centerline of the part is almost always the X axis PRZ for lathe work. Locating the Z axis is not such an easy decision. Notice the symbol -B-. This means reference datum B on the part. *Datums* are surfaces or other features from which reference is taken in dimensioning and tolerancing modern blueprints. Datum A is the centerline. If the print uses baseline dimensioning, the baseline will be the same as a datum. It is always good practice to reference the PRZ to the intersection of datums.

Implied Datums

Not all prints use datum reference, but the dimensions on all part prints must be referenced to something. Measurement and relationships between features of a part mean nothing without a reference. If a part is 1″ thick, that dimension is referenced to some other feature. If a hole is located from an edge of a part, that edge is the reference. Every dimension or feature control on the print must refer to some point. This is called an *implied datum*. Even though the drawing does not show formal datums, they are still there. Without formal datums, look for common features of the part that are used as a reference for other measurements and tolerances. Then use these to locate the PRZ.

First PRZ and Departure Point

For the first operation, we will choose datum -B- as the PRZ, but the departure point will be the tip of the part plus 1.00″ more. Fig. 9-2. By choosing the PRZ at datum -B-, the mathematics of programming will be simple. We will use absolute values from the print. By looking at the drawing, you can see that the PRZ is inside the collet. This is not wrong as long as you, the programmer, remember that an axis callout of absolute Z 0.0 would cause a crash.

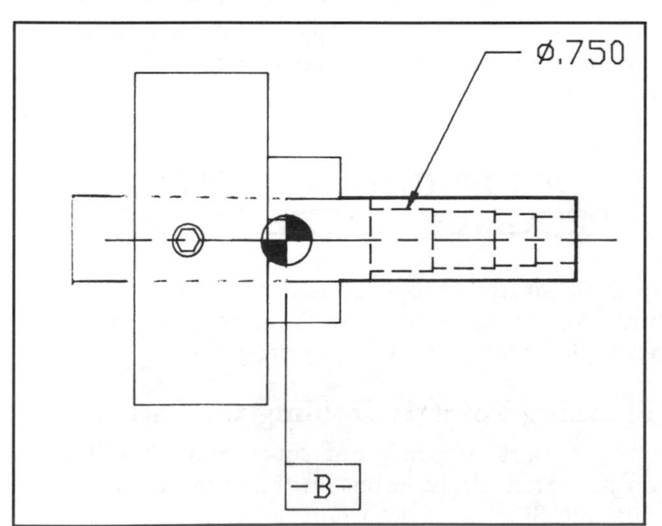

Fig. 9-1. First setup — turn right end of part. Part is protruding 2.75″.

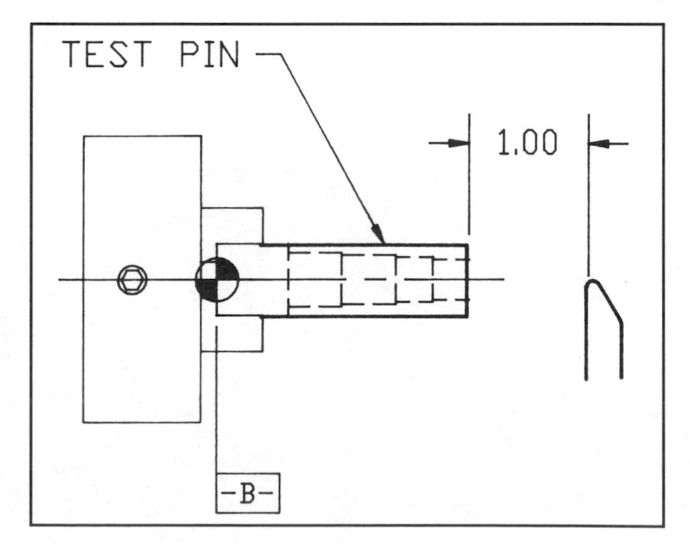

Fig. 9-2. The PRZ is Datum -B- inside the chuck. The departure point is the tip of the part plus 1.00″.

Set Limits

There are soft limits and hard limits.

Soft limits — programmable work envelope. Some CNC machines have programmable software limits. This means that you can write, at the start of the program, a set of travel limits. By defining this software work envelope, you can prevent the tool from traveling in the Z direction any further than Absolute Z + 1.500″. The tool could not reach the PRZ because it would hit a programmed limit. You will learn how to program soft limits in Chapter 19.

Hard Limits. Some machines have adjustable limit slides. These can be positioned to trip a limit switch when the tool is commanded beyond a safe distance. The operator sets these at initial coordination.

Second Operation PRZ and Departure

For the second operation, the PRZ will be datum -B- again. This time, however, the departure point will be the other tip of the part plus a 1.00″ safety margin. Since the PRZ is the shoulder against the spindle nose, there is less danger of hitting the machine. Still, it is a good practice to use limits if you have them. The part will protrude 2.00″ plus 1.50″, or *3.50″* from the collet. There may be some doubt about chatter and strength, but you are holding on a .75″ diameter, and turning three small diameters. If the machine is so equipped, use the tailstock.

Planning Point 4: Cutting Tools and Methods

By looking at Appendix A Drawing 4, you can see that there are no special angles, radii, or features that would limit the shape of the tool. The tool we use must be able to get into a square corner. It must have a nose radius no larger than 1/32″. A right- or left-hand turning tool with an 80° cutting angle and a 1/32″ nose radius is the choice. The choice of right-hand or left-hand will depend on whether the lathe has the tool in front or back. Because the part is already cut to length, there is no need for a parting or facing operation.

Planning Point 5: Measuring the Part

Micrometers are easily used to measure turned surfaces. A depth gauge or depth micrometer could be used to verify the shoulder lengths if the tailstock has been pulled back. In measuring the part, it is essential not to disturb it in its setup. This is important.

Planning Point 6: The Program's Effect on Future Operations

This is often a tricky decision that is linked directly to Planning Point 2. Will what we have just planned make it easy or difficult to perform future operations? In this case the answer is no. This part is complete with the second operation.

RECHECK PLANNING

The plan is ready to go. Let's recheck the points outlined earlier.

Is the setup quick and accurate?—Yes.

Is the setup safe and solid?—There is some doubt about the overhang of the part. A tailstock is recommended.

Is this the most efficient plan?—Yes.

Is part loading efficient?—Yes.

This may not be the best plan for all situations. It does, though, demonstrate how a plan is constructed. It also shows that a C/NC plan is seldom simple. We will follow this plan with caution.

PREPARING TO WRITE A PROGRAM: MILL PROBLEM

Refer to Appendix A Drawing 3. You will need to consider the following planning points. Planning points 1 and 2 will be considered together.

Planning Point 1: Holding the Part, and Planning Point 2: The Overall Process

Because the material is thin, use of a vise is out of the question in machining the outside perimeter. For this exercise we will not drill the holes. In this setup, we will only profile-mill the part. Can we do it in one operation?

A Fixture

The use of a simple fixture might be a good idea. Since datums -A- and -B- are square to each other, they could be premachined. We could square up these two sides of the rough material on a manual milling machine. If the fixture and clamps were spaced correctly, the .25″ corner radius at the lower left could be put in and the whole part would be complete. For this part, assume that we have drilled an undersize hole at the .625″ diameter hole location.

A capscrew could be used to hold the part to the fixture through this hole. One other clamp would hold the part. The fixture would need two locating pins on one square side, and one locating pin on the second square side. Fig. 9-3.

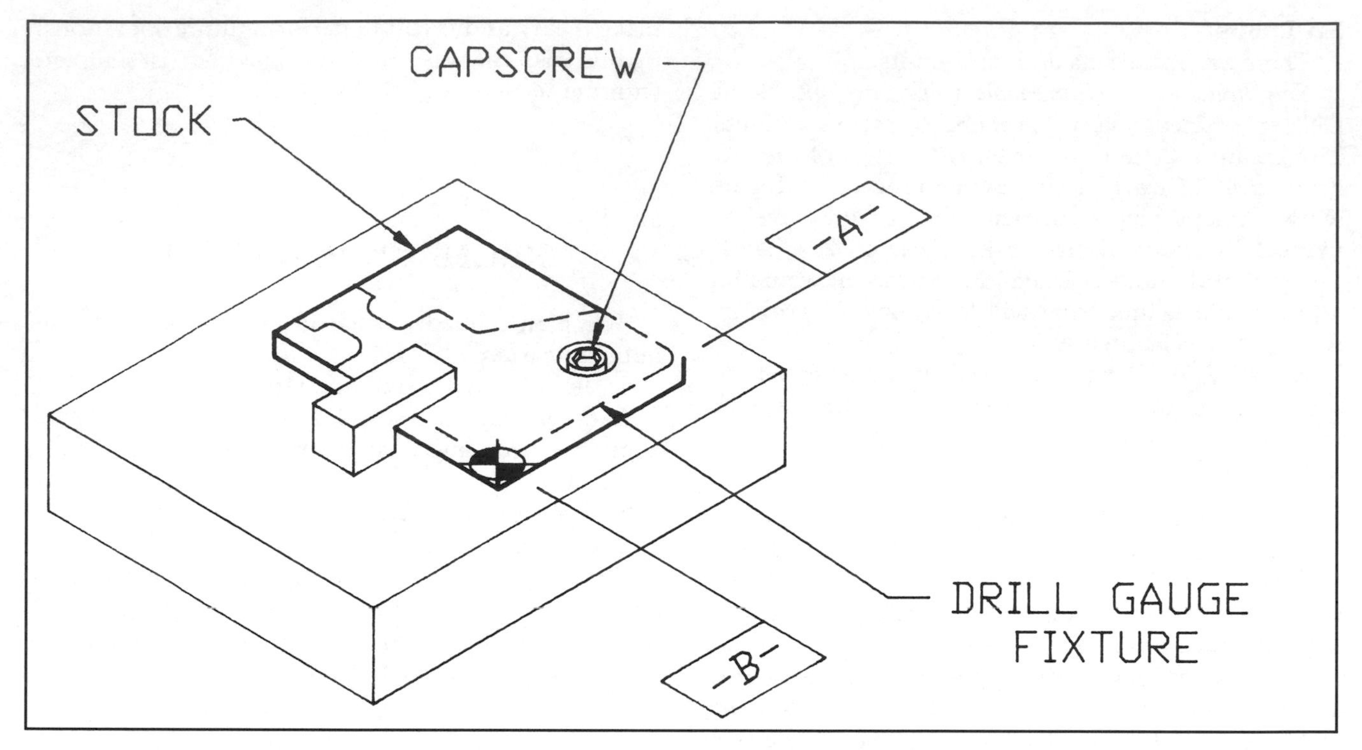

Fig. 9-3. A simple shop-made mill fixture to hold the drill gauge.

Note that Planning Points 1 and 2 had to be considered simultaneously. They were so interrelated that they could not be separated. At this point, you have designed the fixture too. This would be a good time to get it into fabrication. Then it will be ready when the job is ready to go.

Planning Point 3: The Location of the PRZ

A 1/4″ tooling hole bored in the fixture right at the intersection of datums -A- and -B- would be an excellent way to set up the X-Y PRZ. The Z axis could be 2.00″ off the material surface. The departure point will be the PRZ.

Planning Point 4: The Tools and Cutting Methods

We will profile-mill the part using climb milling. As we discussed earlier, the internal corner radii are the limits on maximum cutter size. A .375″ cutter would be used. We will not generate the corner radii. This does cause a secondary operation. The radius in the corner of the 59° angle must be machined away in a second operation. With some reservations, a single pass around would probably be acceptable, especially if the workpiece was aluminum and the feed rate was slowed to help eliminate cutter deflection.

Planning Point 5: Measuring the Part

Calipers and micrometers can measure the straight cuts. The angle will need to be checked with a protractor. This will not be easy in the setup, but it is possible.

Planning Point 6: The Program's Effect on Future Operations

The program will affect future operations. Again the cutter radius is left at the 59° corner. This is a minor problem. The fixture used to hold the part for milling could be used for C/NC drilling. The bolt would be replaced by a second clamp.

Note. An improved plan might be to halt in mid-program and change clamps and tools and then drill the part in a single setup. If the machine had an automatic tool changer, this would be easy. We will not plan this at this time. Another plan sequence might be to drill the holes first while the part is a squared-up rectangle in a drill fixture. The finished holes could then be used for clamping screws.

Recheck Planning

Recheck the plan by asking the following questions.
Can the setup be made quickly and easily? Yes.
Is the setup safe and solid? Yes.
Is the setup the most efficient? No (see above note).
Can the parts be loaded quickly and accurately? Yes.

Again, this is not a bad plan, with some reservations. We will use it.

These two exercises demonstrate the need for careful preplanning before writing a C/NC program. Planning the entire job saves many lost hours of down time due to errors. Careful planning also makes the job safer.

9-2 Planning the Tool Path

EFFICIENCY AND SAFETY

The path along which you plan to drive the tool is important. Properly planned, the machining operation will:

- Take the minimum amount of time.
- Require the least number of tool changes.
- Be the safest.
- Produce a quality part.
- Produce the part with maximum efficiency.

It is obvious that you couldn't do this task until the overall planning is complete. You need to know the departure point, PRZ, tooling obstacles (such as clamps), part and/or cutting tool changes, and the overall sequencing of events. Without this information, planning a tool path would be premature and dangerous. Once the whole setup and overall process is planned, you are ready to plan the tool path and choose the significant points on the part. Then you can write the program.

FOLLOW YOUR OVERALL PLAN

Look at the print and draw a rough sketch of the setup. This sketch will also be in your setup document. Use it as a guide for the tool path plan. There were two setups in our lathe plan, so two sketches would be needed. We will examine the first setup. Fig. 9-4 (Setup Document) shown on page 132.

THREE KINDS OF TOOL PATH RECORDS

There are three ways to record a tool path. It is acceptable to draw the tool path on your sketch if the path is not too complicated. If the path is complicated, you may describe it in words. Refer back to Fig. 7-1. In this setup document, the programmer has described the overall path in words in the sequencing comments. If the control has the capability, the tool path might be drawn on the video screen of the controller. Even so, simple sketches should still be in the document.

The tool path for the lathe job can be straightforward, a rough cut, and a finish cut. You will need to think out the tool path for the mill job in Activity 9-1.

TOOL PATH CHECKLIST

There are many decisions to make in writing an efficient and safe tool path. Follow these guidelines.

1. Are there clamps, fixture components, or chucking tools that could be hit by the cutter?
2. Are the moves restricted to the least possible distance. (See the paragraph following this list.)
3. Is climb milling used whenever possible?
4. Do I make turning passes toward the chuck when possible? (On the lathe straight cuts should be directed at the chuck.)
5. Are the machining forces directed at solid holding components, such as the solid jaw of a vise?
6. Does the plan require the material to be roughed and then finished?
7. Do all moves that can be done safely occur in rapid travel?
8. Do the rapid moves stop short of the material and then feed into the workpiece at the minimum safe gauge height?
9. When repositioning the cutting tool, does the tool clear by a safe minimum distance? (Don't back away 20″ when 2″ would be acceptable.)
10. If more than one tool is used, does the tool have to be changed more than once? (Perhaps this could be avoided.)
11. When repositioning the tool, are diagonal moves used whenever possible? Take shortcuts in rapid if such shortcuts are safe.
12. Are there times when no material is being cut that could be eliminated from the program?
13. Is there an operation that is causing a burr or other mismatch that will require secondary finish work? If so, a small change in the program can save hours of handwork.

Large Part Runs

Careful planning of the tool path is important when large numbers of parts are to be run. A time saving of 1 minute per part on a batch of 2000 parts can save a great deal of money. If there are only a few parts, don't spend too much time removing every unnecessary move from the plan. Also, remember that safety applies for both large and small batches.

C/NC PLANNING AND SETUP DOCUMENT

F.O. #4 - TEST PIN	FITZPATRICK
PART NAME OR NUMBER	PROGRAMMER/PLANNER NAME
12" CNC LATHE	CNC - SIMPLIFIED
MACHINE TO BE USED	CLASS DATE

****

SEQUENCE OF EVENTS

	EVENT TOOL #	LOCATION	OFFSET #	COMMENTS
1.	Ø1 ROUGH	Ø1	Ø1	CARB 1/32" INSERT RAD #TANR-164B ALL RT DIAMETERS LEAVE .03 EXCESS
2.	Ø1 FINISH	Ø1	Ø2	SAME TOOL

SKETCHES AND COMMENTS

1" Ø COLD ROLL

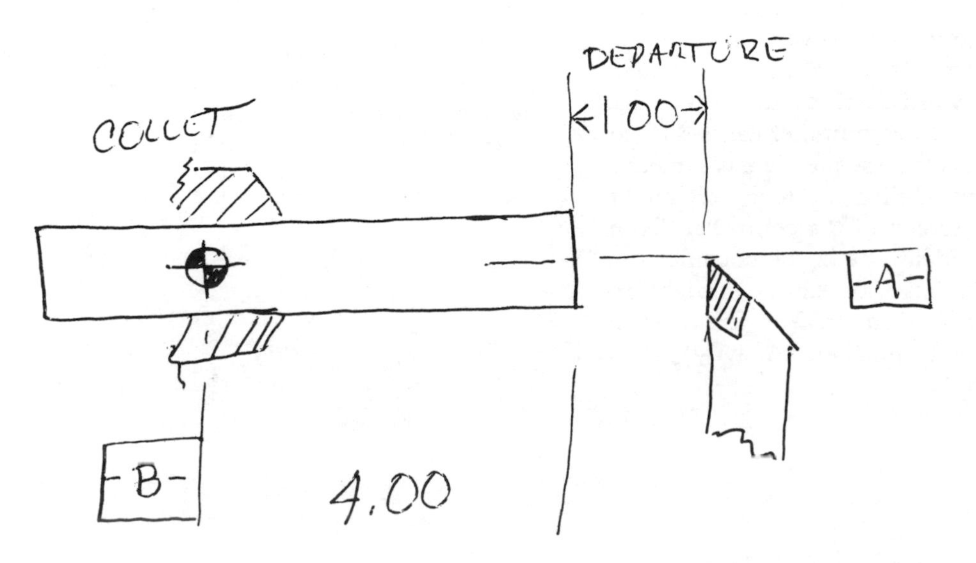

Fig. 9-4. An example setup document for the lathe job.

9-3 Selecting Efficient Speeds and Feeds

PREPLANNED SPEEDS AND FEEDS

Your study of basic machining should have included information on speeds and feeds. We will not discuss the selection of correct RPM and feed rates here. This section will emphasize the necessity of preplanning correct speeds and feeds from the start of a program.

If you need to review RPM calculation and inches per minute or inches per revolution, ask your instructor for supplementary study material. Many C/NC controls allow speed and feed override by the operator. Do not depend on this, however. Write correct programs from the start. In general, you must consider the following:

- The type of cutter.
- The size of the rotating cutter mill or, in a lathe, the size of the rotating part.
- The rigidity of the setup.
- The kind of machining operation.
- The availability of coolant.

Correct cutting speeds and feed rates are essential for long tool life, good machine finish, maximum program efficiency, and safety. *Calculate the speeds and feeds as you plan your program.*

9-4 Writing a Setup Document

A setup document helps everyone involved to do a good job. You will need a setup document even if you are the only person involved in the process. The setup, program, and part produced will be better if you plan and record your plan. Fig. 9-4.

The following page shows a blank setup document. Using this blank document and the guides in Unit 4, Chapter 9, develop a setup document for the overall plan for the mill job we have already developed. Appendix A Drawing 3.

1. Sketch the setup and show:
   A) The location of the PRZ and departure point.
   B) The location of the clamps.
2. Develop a tool path and show it on your sketch. Use solid lines for cuts and phantom lines for rapid travel moves.
3. Be sure to fill in all information, such as information about tool size, material to be cut, part name, and machine name.

C/NC PLANNING AND SETUP DOCUMENT

PART NAME OR NUMBER	PROGRAMMER/PLANNER NAME	
MACHINE TO BE USED	CLASS	DATE

SEQUENCE OF EVENTS

EVENT TOOL # LOCATION—OFFSET #	COMMENTS

SKETCHES AND COMMENTS

CHAPTER 10

WRITING A CODED LANGUAGE PROGRAM

The logic patterns, organization, and axis moves used in machine tool programming are similar, no matter what language is used to enter the program into the control memory. However, programming in the standard coded language is common in industry today. Invented many years ago, this system standardized the machine commands used in numerical control machining. This chapter will show you how to program using codes.

After a brief introduction, we will look at a simple cutter centerline program used to define a shape on the mill. Examples of the program will be given for Bridgeport and Fanuc controls. You will then go on to write a program for a lathe or mill as assigned by your instructor. This exercise will limit you to using only simple commands.

OBJECTIVES

After studying this chapter, you will be able to:

- Combine letters and numbers to form alphanumeric code words.
- Use the basic coded words in a program.
- List the fundamental codes used on machines in your shop.
- Describe the difference between G and M code words.

10-1 Coded Words

ALPHANUMERIC CODING

Coding that combines a letter prefix and numbers is called *alphanumeric*. It is easy to adapt to coded input. You simply need a language list from which to work. The best way to learn codes is to write a program. As you need a machine function, look it up in Appendix B.

There are many different kinds of words. The category into which they are placed is determined by the letter prefix. The most common words are G, X-Y-Z, M, F, S, and N.

G Code Words—Preparatory Commands

A common word is the G code word. G words always define a mode or operation. They define the way in which the next action is to be interpreted. Several G words cancel certain actions. According to EIA standards, there are 100 G words for milling and 100 G words for turning. Many of these words are the same in both turning and milling. G words are used in four ways:

1. The G command may be an actual movement such as:

   G00—rapid travel movement
   G01—linear interpolation (move in a straight line at feed rate)
   G08—accelerate the tool in a smooth transition

2. The G command may define the movement or mode:

   G82—drill cycle with a dwell at the bottom
   G90—go to absolute value mode
   G91—go to incremental value mode
   G71—go to metric value mode
   G70—go to inch value mode
   G17—working in the X-Y plane
   G99—feeds are in inches per/revolution (turning only)

3. The G command may cancel some modes:

   G40—cancel tool radius compensation
   G80—cancel last cycle mode

4. Unassigned EIA G words may be used as custom words by the manufacturer:

   G22—define safety zone limits (Fanuc control command)
   G98—Cycle return to PRZ (Allan Bradley Bandit control)

X-Y-Z Code Words

X-Y-Z words are common in programs. They define the axis to be used in the G command. For example, a block might read: N005 G91 G00 X3.75

This indicates that at sequence number 5, using incremental values (G91) and at rapid travel (G00), the X axis should move 3.75″.

M Code Words

The next most common words found in programs are the miscellaneous (M) words. There are 100 EIA M words. M commands are used for all machine functions other than axis movement. They are also used for certain other program operations.

M00—program stop
M09—coolant off
M19—stop spindle at an oriented position
M06—tool change
M01—planned optional stop

Other words are used in basic programming. These include:

F—define feed rate.
S—spindle speed command.
T—tool number word.
I,J,K—define center of an arc (a form of local zero).
I,K—also used as threading parameters.
A,B,C—secondary rotary axes.

Special Custom Words

Special words are designed by the individual machine manufacturer. These words will be found in the programming manual of the machine.

10-2 Coded Language in Industry

There are advantages and disadvantages to programming in coded language.

ADVANTAGES

1. Coded language is standard throughout the United States and parallel with the International Standards Organization (ISO) language used worldwide.
2. Coded language is common in United States machine shops, especially those using NC.
3. Coded language can be quickly used by those familiar with this input system.

DISADVANTAGES

1. Coded language is slightly longer to learn because of the need to memorize a library of codes.
2. Coded language is not typical of user-friendly computer systems. No other application of computers today is coded.
3. Of the 100 preparatory (G) code words assigned for milling and turning, about half are unassigned by the EIA. This allows custom commands to be set up for individual machines. The disadvantage is that many codes are *not* standard from one machine manufacturer to another.

It is difficult to memorize the codes for milling and turning and variations from machine to machine. Memorize the basic commands. Know what your C/NC machine is capable of doing. Then find the code that causes the action. Each machine control comes with a language library.

Ask your instructor for Custom Sheet 6, The Language Library. This will cover the machine you are going to program.

Tape or staple Custom Sheet 6 inside the back cover of this book. It will be your language library for the next exercise.

Review Custom Sheet 6 for the more common G and M words. Be familiar with what the machine can do. For now, memorize the abilities of the machine, not the entire word list. Test yourself by answering the following questions.

1. What code turns on the coolant on your machine?

   ____

2. How is the feed rate programmed?

   ____

3. Write a coded command that calls for a feed rate of 20.5 IPM.

   ____

4. How many assigned G words are on the list?

   ____

5. What is the command to cause a rapid travel movement of –3.74″ in the X axis?

6. What is the code for Spindle On?

7. Write a command that will cause a linear interpolation of X 1.5 and Z–2.0.

8. What command signals Program End on your machine?

9. Does your machine have an optional stop?

   What is that code?

10. What is the program stop command used for a tool or clamp change?

10-3 Writing a Coded Program

There are three sections to a program: Initial Commands, Program Body, and Program End.

INITIAL COMMANDS

All programs require initial commands to get going. These commands fall into four categories: remarks, safety cancel lines, mode and condition commands, and special custom functions.

Remarks

Remarks define what is going to happen. Not all machines can have remarks in the program but if they do they are ignored by the machine and used only for your reference. Example: **REM (PROGRAM TO MAKE 65B-32405 HINGE)**

Safety Cancel Lines

If the last program to be run used certain modes, they can be retained in the control. But they should be cancelled to avoid surprise. Examples: G40—Cancels Tool compensation, G80—Cancels Any Automatic Cycle Used Last.

Mode and Miscellaneous Commands

These commands include any modal command such as feed rate, rapid travel, or circular interpolation. Miscellaneous commands include Spindle On, and any program condition mode you want to use. Examples include *G70—inch programming, G91—incremental value mode,* and *G01—linear interpolation.*

Special Custom Functions

A custom function on a particular machine can be actuated by one of the EIA unassigned M words. Examples include *M10—Close Automatic Clamps,* and *M08—Turn On Secondary Coolant (Mist).*

PROGRAM BODY COMMANDS

The words that cause the actual machining. Axis moves, tool changes, rapid positioning commands, linear and circular interpolation are only a few examples of the program body.

END COMMANDS

All programs must have certain command words to properly end the program. The program and command may have one or all of the following three functions.

1. It stops all machining and turns off functions such as spindle and coolant. This command is usually M2 or MØ2.
2. It moves the tool to the departure point or some other position to change parts and prepare for the next part cycle. This command is usually MØ6 or a custom word.
3. It signals the control to stop reading data and rewind the tape or set the sequence number to the first block of the program. This command is usually M2 or MØ2.

After reading the following coded programs, you will write a cutter centerline program for the machine assigned by your instructor. (You will need to read Chapter 12, Manual Data Entry of Programs before entering and testing your program.)

These programs assume the availability of a CNC machine. We will concentrate on CNC from this point forward in this book.

CNC made special formats obsolete. Remember, we are using a form of interchangeable format here. The same program may appear slightly different on your control. We will use variable length blocks. Commands may be arranged for convenience of reading. Also, the decimal point will be shown. There will be no zero management. Zero management means that the number 1.87 will look just like that. It will not be written as 018700, as it would have been in NC formats. Computers have eliminated the need for special entry formats. Unless your C/NC machine requires special zero management, enter numbers as you see them. Adding is acceptable if the decimal point is in the right place. I prefer some zeros to keep the decimal point in a column. The entry of numbers in modern CNC is algebraic, meaning that you enter what you see. This is similar to making common calculator entries.

Program 1

The program shown in Fig. 10-1 will produce the drill gauge shown in Appendix A Drawing 3. However, there will be no curves, only straight lines. It will simply trace the perimeter with a pen in the spindle, without allowances for tooling. *This is an exercise only.* Follow the program along the blueprint. Fig. 10-2.

The PRZ is the intersection of datums -A- and -B- and 2.00″ off the work surface. The tool is a plotter pen. We will trace the perimeter, going clockwise from the PRZ. Place your pencil on the drawing and follow the program.

Sequence Number	Command	Comments
REM: PERIMETER OF DRILL GAUGE		
N001	G90 F20.0 G70	absolute mode, feed rate 20 IPM, inches
N002	G00 Z–1.95	rapid travel, drop pen near paper
N003	G01 Z–2.00	linear feed mode, feed to paper
N004	Y3.150	feed to top left of groove
N005	X0.687	feed into groove
N006	Y3.650	feed to top of groove
N007	X0.393	out of groove
N008	Y4.00	
N009	X1.393	
N010	Y3.65	
N011	X1.099	
N012	Y3.150	
N013	X1.786	
N013	Y1.7803	a trig calculation for top of angle cut
N014	X2.875 Y1.126	linear interpolation down the angle
N015	Y0.0	where does this command place the tool?
N016	X0.0	where does this command place the tool?
N017	G00 Z0.0	
N018	M02	what does this command do?

Fig. 10-1. Drill gauge program.

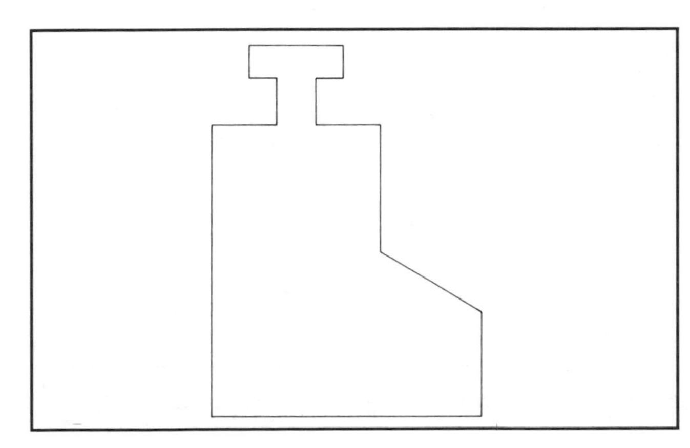

Fig. 10-2. The simplified drill gauge shape produced by example programs.

Program 2. Bridgeport Control. Drill Gauge.
Refer to Fig. 10-3.

Program 3. Fanuc Control Lathe. Producing a Test Pin Foldout.
This program produces the test pin from left to right (turned around), chucking beyond the big diameter end on excess material and using an end support from the tail stock. Fig. 10-4.

Now it's your turn. Your instructor will assign a shape to program into the control. If you do not have a coded control, move directly to Chapter 11. If you have a mill, trace the outline with a pen. If your machine is a lathe, you can devise a plotter or you can turn a piece of soft material or test material. Note. Before you write your lathe program, find out whether the lathe is programmed in radius or diameter figures for the X axis. Programming is common using diameter input.

Ask for Custom Sheet 7—Practice Shape 1.

```
                          REM:  FIGURE 10-3 DRILL GAGE
                          REM:  PRZ IS LOWER LEFT CORNER 2.00" OFF PART
                          REM:  PROGRAM IS CUTTER CENTERLINE

N001 G75 G90              REM:ABSOLUTE VALUES
N002 G00 X0.Y0. T01M6
N003 S2000 M03
N004 Z.1                  REM:RAPID TO WORK
N005 G01 Z-.25            REM:GO TO LINEAR FEED RATE DROP INTO MATERIAL
N006 X2.875
N007 Y1.126
N008 X1.7855Y1.7806 REM:ANGULAR LINEAR INTERPOLATION
N009 Y3.15
N010 X1.099
N011 Y3.65
N012 X1.393
N013 Y4.
N014 X.393
N015 Y3.68
N016 X.687
N017 Y3.15
N018 X0.0
N019 Y0.0
N020 G00M25               REM:RAPID TRAVEL TO HOME POSITION
N021 M02
```

Fig. 10-3. Coded program to machine the drill gauge appendix drawing. Program postprocessed from screen image by MasterCam Software.

```
(REM  Figure 10-4)
(REM  To machine the test pin appendix part)
(REM  Machined reverse of the drawing)
(REM  Single pass - no roughing)
(REM  PRZ is C/L at Datum B)
(REM  For Fanuc lathe control)

     G90 G40 G71     (REM safety and cancel lines, inch mode)
N001 G50 X0.0 Z0.0
N002 G50 X    Z      (REM to be established at setup)
N003 G00 M25
N004 M05
N005 G00 X.250 Z2.0  (REM note diameter callouts for X)
N006 G01 Z1.5
N007 X0.375
N008 Z1.00
N009 X0.4375
N010 Z0.500
N011 X0.500
N012 Z0.000
N013 X1.000
N014 Z-1.500          (REM note Z axis beyond PRZ toward chuck)
N015 X0.7500
N016 Z-2.250
N017 X0.6875
N018 Z-3.000
N019 X0.6250
N020 Z-3.000
N021 X0.5625
N022 Z-4.000
N023 G00 X1.0 Z3.0     (REM rapid to clear part)
N025 M03
```

Fig. 10-4. Coded program to machine the test pin appendix drawing. Program postprocessed by MasterCam Software.

CHAPTER 11

PROGRAMMING IN CONVERSATIONAL LANGUAGE

Conversational input differs from coded input in several ways. First, the standard upon which it is based is not set forth by a regulating group such as the EIA. Instead it is based upon the English language. Secondly, using conversational input and output allows the computer to combine symbols (words). This means that the microprocessor in the controller has a larger "vocabulary." By being conversational, the controller can communicate more freely with the user. Modern conversational software is "user friendly." It works with the user to avoid mistakes and speed up the learning process.

This chapter will explain and give examples of conversational programs. If you have a conversational control in your shop, you will be given a program to write, enter in the memory, and run.

Programming organization, logic, and actions do not change. Only the input language used for storing the machine moves in memory are different.

OBJECTIVES

After studying this chapter, you will be able to:

- List the various ways in which controls prompt.
- Respond to prompts correctly to build a program.
- Prepare a program in conversational language.
- Be able to compare conversational input to coded input.
- Enter a program in a conversational control (if available).

11-1 Conversational Language in Industry

C/NC machines perform similar activities no matter what input unit is used. It is simply a matter of adapting a concept to the correct input unit on a specific machine. This is analogous to microcomputers. Most microcomputers can be programmed in BASIC language, yet there are differences in the word symbols used from computer to computer. The logic and syntax of BASIC are the same on all machines, but the actual words may differ.

Conversational, user-friendly controllers work with the user in two basic ways. The programming of a controller is interactive or prompted.

INTERACTIVE PROGRAMMING

The controller analyzes MDI input data and tries to eliminate errors. This means that the controller is active, not passive. A passive controller simply accepts everything you send it. An active controller analyzes the data. For example, a controller may not accept an axis move that is too big. It will read back "TOO BIG." A controller may not accept an entry that is not in the correct form.

There are some errors that a controller cannot analyze on input. An example would be a logic move where the programmer steps out of the main program to call on a sub program. The programmer plans to return to the main program, but forgets to do so. To find this type of mistake, the controller must analyze the entire program after completion. The controller cannot see the length of a tool or whether you have coordinated it correctly. Crashes can still happen, even with programmed soft limits.

PROMPTED

Conversational controls ask what you want to do next. They then ask questions or present spaces to be filled based on what actions you wish. The questions they ask are called *prompts*. Your responses are stored and become the program. The data stored may be translated into several forms, but English is the input language used.

Your responses to prompts fall into five categories.

- A word typed in from the keypad.
- A number typed in from the keypad.
- A letter or symbol typed in from the keypad.
- A simple answer—yes or no.
- A menu response—a key that represents a choice. This is similar to general menu-driven computer software.

Some of the words used will be abbreviated. In the prompts and the correct responses, we find differences between conversational controllers. The symbols can be very different, although they initiate exactly the same action at the tool.

The term *machinist shop language* is sometimes applied to conversational input because words such as HALT are substituted for the coded word M00, FAST for G00, and SPINDLE ON for M04.

11-2 Conversational Comparison

There are advantages and disadvantages to conversational programming.

ADVANTAGES

- Relates to words we already know; input words are learned more quickly.
- "User friendly," allows computer to communicate with user.
- Interactive, can help to eliminate errors in the program.
- Programs are easier to read back.
- Complicated logic programming can be easier in conversational programming because the power of the computer in the control is opened up to work with the user.
- Expands the vocabulary of the controller.

DISADVANTAGES

- The English language is the only standard for words or symbols.
- Because it is new, there are fewer conversational units in industry.
- Some conversational units can be slower to program once the mastery level has been reached. Answering all the prompts can slow down a skilled programmer. Some controls allow the programmer to tab through the questions or skip them altogether once the programmer no longer needs prompts.
- Conversational language is leading edge technology. As such, it is changing in industry. It is difficult to stay up with the changes and to learn very many languages at the same time.

As a new student of CNC, you should learn about both coded and conversational systems. You should then become proficient on whatever unit you are assigned in industry. The transfer from one programming system to the other is easy if you understand the logic and organization of programming, the fundamental concepts that are universal for all CNC.

EXAMPLES OF CONVERSATIONAL PROGRAM

The following program will produce the drill gauge shown in Appendix A Drawing 3. This program is from a conversational control. It will plot the outline of the part using the cutter centerline as the part outline. The part will have no corner radii. Follow this program as it traces the shape of the drill gauge.

Note that sequence numbers 17 through 22 were using polar coordinates to move down to the top of the angle. Sequence numbers 17, 18, and 19 set up a local reference point at the bottom of the angle. Sequence numbers 20 and 21 moved the tool to the top of the angle from the top of the chip groove. Sequence number 22 moved the tool to the center of the new reference point down the angle. This was linear interpolation using a polar destination.

If you have a conversational control, answer the following questions about your control. Then write a program of your own. Your instructor will assign the shape to be programmed from Custom Sheet 7 (practice shape).

There are three possible input formats for a conversational control. Your control will have one or more methods of inputting the program data. Suppose you select a drilling routine that includes pecks of the drill in and out. You wish to drill 2″ deep with 6 pecks at 1250 RPM.

1. Does your control prompt (ask questions) that are answered from a specialized keypad? Your responses are entered by depressing individual buttons. Prompt example: What is the total depth? Your response would be 2.00″. Example: Do you wish pecks? Your response would be yes.
2. Does your control allow direct keypad entry? Example: You wish to enter a spindle speed of 1250 RPM. Your entry would be made on a keypad and it would be the letters and numbers "SPEED 1250."
3. Does your control feature screen menu responses? For example, you may select the auto cycle page. A screen appears that asks, "Which Cycle?" Your response will be one of several cycles available. You will select by touching a number or letter or by moving a cursor to the drill cycle. A new screen page appears with the drill cycle selections for depth, speed, and pecks.

If you have a conversational control, write a program for a shape your instructor assigns. To enter the program, you will study Chapter 12 next. Then you will enter and test your program. To do this you will need Custom Sheet 8 (the language library for your machine) and Custom Sheet 7 (the shape to be programmed).

DYNA Mechtronics—2400 Vertical Milling Machine

SEQUENCE	DATE	COMMENT
001	START INS 75	start in inches, file number 75
002	TD = 0.0	tool diameter
003	FR = XYZ 10.0	feed rate for X,Y, and Z = 10 IPM
004	setup ->zcxyu	setup block for PRZ during prog run
005	GO f Z–0.95	GoAbsolute fast (0.05″ gauge height)
006	GO Z–1.00	GoAbsolute feed to work
007	GO Y 3.15	
008	GO X 0.687	
009	GO Y 3.650	
010	GO X 0.393	
011	GO Y 4.00	
012	GO X 1.393	
013	GO Y 3.65	
014	GO X 1.099	
015	GO Y 3.15	
016	GO X 1.786	
017	ZERO AT	change reference point to:
018	X 2.875	right edge
019	Y 1.126	end of angle (new local zero)
020	GO a 149	GoAbsolute angle 149 (90 + 59)
021	r 1.271	radius 1.271
022	GO r 0.0	GoAbs to center r = 0.0
023	GR Y–1.126	GoRelative Y–1.126 to base
024	GR X–2.875	completes shape
025	End Newpart	reset program, up spindle, move to PRZ

MANUAL DATA PROGRAM ENTRY

All CNC units have some form of programming at the control. This is known as manual data insert (MDI) programming. MDI programming consists of making single line data entries at a controller. Although not strictly MDI programming, if your class has an offline computer as a programming station, this is the time to learn to use it. This chapter is designed to teach you the skill of data entry from a keypad. It does not matter whether the control is conversational or coded.

There are two stations in CNC where a program may be entered into a storage device. These stations are a control keypad and an offline computer. The primary objective of this chapter is to give practice on data entry from a CNC controller keypad. A secondary exercise is that of entering the program at the keypad of a microcomputer and then downloading it to the controller. If necessary, review the procedure for initial system turn-on by referring to Custom Sheet 2.

OBJECTIVES

After studying this chapter, you will be able to:

- Select the correct mode for MDI data entry.
- Enter data from C/NC control keypad.
- Enter program data into offline computer (if available).
- Store program on data storage device, either a control or a computer.

12-1 Manual Data Insert at a CNC Control

At this point, the control is "up and ready." Your program has been written and you are ready to place it in the controller memory. At the end of this process, you would store the program in a permanent storage device, if available. You would then test-run the program as discussed in Chapter 8.

Here is the general procedure for MDI programming at the controller:

1. Switch on the control. Refer to Custom Sheet 2.
2. Make sure that the data in the control is not important. If it is, save it to a data storage medium. This is important. Do not erase a program without knowing what you are erasing. You may also enter the program elsewhere in the RAM. Most controllers have enough storage lines in RAM to allow storage of several programs at one time. You may need to switch up to an empty section of the RAM before storing your program.
3. Erase or clear the blocks that you will need for your program. Many controllers have specific codes or commands that allow erasing of large sections of the memory. If yours does, refer to the programming manual. Again, use this function cautiously. Do not erase information you might need.
4. Select the MDI mode on the control.
5. Select the correct line of the RAM for the first line of your program.
6. Begin the entry one line at a time. On some controls, you will need to link entire commands together. For example, in a program that requires a linear interpolation, the entire command would be G01 X1.5 Y0.75. This may be entered on one line on some controls. On other controls, it must be entered separately, then linked together as it is stored in RAM.
7. Store the program on a permanent-storage device, such as computer disk, magnetic tape, or punched tape.
8. Verify and store the program. At the end of this process, return the RAM to the first line of your program. Then step through the program one line at a time and verify each entry. Edit any incorrect entries. The programming manual will provide information on this.

PROGRESSIVE STORING

Most MDI entry programming is raw data, which is data being entered for the first time. The program does not exist on any storage device. It is good practice always to backup your work by making a permanent copy of the program. If possible, always store your work on a permanent-storage device.

Such storage of materials is called *progressive storage*. In the case of a power failure, the stored program would be safe. Only the work being done in the active RAM. Many CNC controls have a battery backup on the active RAM. Thus, in a power failure, the program is not lost. Most microcomputers lack this feature. When writing programs on a microcomputer, it is good practice to store them every five minutes.

This activity covers MDI program entry. If MDI program entry is familiar, you may be able to skip this activity.

1. Ask your instructor for Custom Sheet 9, MDI Entry of Data.
2. Follow the above general procedure. Enter your program into the controller RAM.
3. Have someone double-check the program entry.
4. Test-run the program. If you have a mill, use a pen in the spindle. If you are using a lathe, cautiously test the program in the single-block execute mode. Then run the program on test material.

12-2 Offline Computer Program Entry

OFFLINE COMPUTERS ARE EFFICIENT

The use of a microcomputer console for data entry is common in industry. The computer may be removed from the workplace. This frees the CNC control to run parts. Programs may be developed at the same time as parts are being run. This increases up time.

DATA MANAGEMENT NETWORK

If your lab uses offline computer programming for C/NC, this is the time to learn to use it. The process is usually simple. Software loaded into the computer prepares it to create a C/NC program. This software will be specific for the computer you are using. It may have the ability to translate a program for one control into a program for another control. This ability, called *postprocessing*, will not be used at this time unless required by the software. Some large offline programming computers accept programs in only one form. They then postprocess the data into specific programs for individual controllers.

Our objective is to enter the program you have developed and then download it to the CNC controller. As you write the program lines, remember that spacing and decimal points are important. Here is the general procedure.

1. Make the computer system operational.
2. Load the specific software for C/NC programming from hard or floppy disk.
3. Enter the program commands, one line at a time. There may be a difference in the way large lines of data are arranged on a computer as compared to the controller. Usually, microcomputers are able to accept more data per line than a C/NC control. You will need to find this out from Custom Sheet 10, Offline Computer Data Entry.
4. Save the program to disk or to tape. Save as you go.
5. Edit the program for errors.
6. Save the edited program to disk.
7. Download the program to the controller.

Activities

This activity covers offline program entry.

1. Ask your instructor for Custom Sheet 10, Offline Computer Data Entry.
2. Write your program, edit it, and save it on disk.
3. Download the program to the CNC controller. (You do not need to test run this program.)

If the equipment is available, this would be a good time to practice uploading and downloading programs. If your shop has offline data management, you should be familiar with the basics of data management. You will have one or more of the following capabilities in your shop network. Check those that you do have. Note which controls in your lab have which capabilities.

1. Control MDI programming.
2. Offline programming.
3. Downloading programs from computer to control.
4. Uploading programs from control to computer.
5. Saving programs to permanent control storage.
6. Saving programs to computer disk.
7. Saving programs on punched tape.
8. Saving programs on cassette tape.
9. Printing out a hard copy of the program.

CHAPTER 13

EDITING A PROGRAM

You may need to change a program after it has been written to RAM. For example, the program may have errors or it may need to be updated. Improvements to the efficiency or safety of the program can also prompt editing. Another reason would be to improve the quality of the part produced. A basic programming concept is to improve and update a program once it is written.

Once a program has been sent to the control and tested, the control finds errors that may not be detected during program entry. These will appear only after the PRZ is established and the program is run. Such errors include hitting a limit switch or stating a command that is not possible mathematically. The control will inform you of such errors through words or symbols called *error messages*. You need to be able to interpret error messages to edit your program.

OBJECTIVES

After studying this chapter, you will be able to:

- Determine errors in programs.
- Determine improvements in the process of the program.
- Insert blocks into programs.
- Delete blocks from programs.
- Edit existing blocks in programs.
- Make permanent changes to programs in RAM and on permanent storage media.
- Interpret error messages from the control.

13-1 Finding Errors in Programs

THE FIVE STAGES OF ERROR DETECTION

Program errors are detected in a five-stage process. In this chapter, we assume that you wrote the program and are testing it, too. You are responsible for the correctness of the program. The five stages of error detection are:

1. After initial writing, review the program.
2. After initial entry, review your entries in the control.
3. Test the program. (This includes video testing,too.)
4. Measure the part produced.
5. Observe the program during a part run.

We will look at each stage.

Stage 1—Reviewing the Program

At this stage, you have not entered the data. It is still on your notepaper. Some errors will be difficult to find. Inaccurate entries in the program may seem correct every time you review them. One way to test for errors is to use a pencil test. Here, you use a pencil just like a cutter. Read a program line. Then move the pencil exactly as the program commands. Do not think in terms of what you intended the program to do. Instead, move the pencil exactly as the program line commands. This technique is especially useful for short programs. For long programs, this stage might be omitted. Graphics, plotting, and testing can then be used to find errors.

Stage 2—Reviewing the Entries

This stage usually finds only a single type of error, such as a miskeyed number, letter, or symbol. After entering the program into the computer or controller, return to the beginning and scan each line. Compare the entries to the written version, making sure they match. Look for decimal, axis, and positive/negative sign errors. Program errors that you missed in stage 1 will probably not be detected here. You might form the habit of saying the line as you enter it.

Stage 3—Test the Program

Here, you will find major errors. At this stage, graphic representations of the program are useful. With or without graphics, the program needs to be tested or plotted on a plotter. An image is the result. Errors detected at this stage include major math problems, incorrect sign values on axis entries, and incorrect axis callouts. These errors will produce a part of the wrong shape. These errors are easily detected on graphics or the initial test.

Stage 4—Measure the Part

Small math errors are eliminated at this point. Perhaps the error is not in the program. For example, it might be an offset or setup coordination adjustment. Such fine tuning will produce a quality part.

Stage 5—Observe the Program

After the program has been run successfully, watch for machine moves that can make the program more efficient. You may see shortcuts that could be written differently. A comparison of the number of parts to be run and the amount of wasted time will determine if you will need to make such corrections. For a short part run, making the correction might take more time than making the part run.

In the following program, there are several errors that should be edited. Read the program and detect the errors. This program uses standard coded language and defines the right end of Appendix A Drawing 4, Test Pin.

Planning Parameters

A) One roughing pass leaving .030″ excess, then a finish cut using diametrical dimensions for the X axis.
B) PRZ is on centerline and at datum -B-.
C) The part is end supported.
D) The departure point is 6.00″ in Z and 4.00″ in X from the PRZ.
E) The part is chucked on excess material to the left of datum -B-. This program should machine only the right end of the part up to the 1″ diameter.
F) The part is mild steel with a carbide turning tool. The recommended surface speed is 300 FPM.
G) The part has been faced and center drilled. If you see a code you do not know, look it up in Appendix B at the back of this book.

```
N001  G90   G71 G99
N002  F.004 S1500. (speed and feed are acceptable)
N003  G00   Z4.1 X5925
N004        Z3.53
N005        X.655
N006        X3.53
N007        X.7175
N008        Z2.28
N009        X1.1 (end of roughing pass) faces
            1" shoulder
N010  G00   Z4.1
N011  G01   X.5625
N012        Z3.5
N013        X.625
N014        Z3.0
N015        X.6875
N016        X2.25
N017        X.75
N018        Z1.50
N019        X1.1
N020  G00   X0.0 Y0.0
N021  M2
```

1. How many errors did you find in this program?

___

2. How many incorrect G codes did you find?

___

3. Although the spindle speed was defined, was the spindle turned on?

___

4. Were there any missing G codes?

___

5. Was the part completely machined? Were there any missing steps in either the rough or finish cuts?

6. Why was there a crash in sequence number 20?

7. Why was the spindle not turned off at the end of the program?

8. How long will the corrected program be? (You will need to know how many blocks need be inserted.)

9. Were there any decimal point errors?

10. What was the problem between blocks 3 and 10?

11. Is this a safe program? Explain.

12. How many blocks were deleted?

Here is the corrected program. The corrected entries are italicized. Does this program match your corrected program?

N001	G90	*G70* G99	inch programming, not metric
N002	F.004	S1500. *M04*	"spindle on" code added
N003	G00	Z4.1 X.5925	decimal point missing
N004	*G01*	Z3.53	left out linear feed rate code
N005		X.655	
N006		*Z3.03*	axis and number were wrong (2 errors)
N007		X.7175	
N008		Z2.28	
N009		*X.780*	insert axis move
N010		*Z1.53*	insert axis move
N011		X1.1	
N012	G00	Z4.1	
N013	G01	X.5625	
N014		Z3.5	
N015		X.625	
N016		Z3.0	
N017		X.6875	
N018		*Z2.25*	wrong axis letter
N019		X.75	
N020		Z1.50	
N021		X1.1	
N022	G00	*X4.0* *Z6.0*	not X0.0 Y0.0 (3 errors)
N023	M2	turns off all functions	

[12 errors]

13-2 Maximizing a Program

There are three reasons to improve a program that is making acceptable parts. A program may be improved to:

- Make the program more time efficient.
- Improve part quality.
- Improve the safety.

MAKING A PROGRAM MORE TIME-EFFICIENT

Once the program is running, you can always see shortcuts you would like to incorporate. Watch for the following problems. Causes and solutions are also given.

1. Axis moves that waste time.
   A) Indirect path from one point to the next. This usually results from not using diagonal moves.
   B) Too much travel. The gauge height is too great or an axis move could be shortened for part load or tool change.
   C) Failure to insert rapid travel on an axis move.
   D) Tool path rearrangement could reduce passes and pecks.
2. Tool change problems.
   A) Selecting the wrong tool at the wrong time can increase tool change time.
   B) Selecting the wrong tool can cause unnecessary axis moves.
   C) The process plan requires that the same tool be used more than one time.
3. Improve the machining process.
   A) Too many roughing passes. The part run can be accomplished by taking more material in the roughing phases.
   B) Forming radii takes less time than generating and might work.
   C) Take a small deburr pass. This will not save time now, but it will prevent extra secondary work.
   D) A new tool would work better, though it might require new offsets.
   E) Optimize programmed speeds and feeds.
   F) If burrs or other rough edges can be corrected during machining, hand work will be reduced.

IMPROVING PART QUALITY

Most programs seldom make the best part possible on the first run.

1. Tool action problems.
   A) Tool path is not cutting correctly. It is not climbing, or it is cutting toward the chuck. Chatter destroys finish. Reduction in programmed RPM.
   B) Burrs are not acceptable. Their elimination requires a different cutting action.
   C) Coolant is unable to reach the cut, such as in deep drilling. This may require more pecks to get chips out and coolant in.
   D) Chatter requires different tools, speeds, and feeds.
2. General process.
   A) Speeds and feeds do not produce optimum finish.
   B) Not enough or too much material is left for the finish cut.
   C) The tooling required to hold for machining is marring the part. Lighter holding is required.
   D) Use other tools. For example, replace a drill cut with a center drill. Drill, then ream.

IMPROVING PROGRAM SAFETY

1. Change cutting action.
   A) Increase gauge height.
   B) Reduce speeds or feeds. Chuck speeds are dangerous in turning operations. Some machines have programmable upper limits for chuck RPM.
   C) Tool change position is too close to work.
   D) Part change position is too close to tools.
2. Change general process.
   A) Reduce amount of material taken per pass.
   B) Lathe tools index at the wrong time or place. This is not possible on many lathes because they must be parked to index the turret.
   C) The tool path comes too close to the tooling or the part.
3. Tooling is inadequate.
   A) Part needs more support. It slips or vibrates.
   B) Clamping or chucking is in the wrong place.
   C) Cutting tool selection is too weak. Breakage is imminent.

13-3 Insert and Delete Data in Controls and Computers

Once improvements or errors have been identified, the program must be changed. Programs are changed in three ways. Information is added and or inserted, deleted, or changed. You will need to know how to perform each type of change. Ask for Custom Sheet 11. This information can also be found in the operator's manual. The operations of adding, deleting, and changing CNC program data are similar on CNC controllers and microcomputer keypads. Here are some general procedural hints:

- Before making axis move changes, make sure you know what effect the change will have on further moves. This is more important when using incremental values.
- Be aware of how the control deals with line number management. When we add or subtract lines, does it affect other line numbers in the program? Editing can sometimes introduce new problems by changing the total lines of information in the program. We will look at this later, during our study of logic.
- Store the edited version of the program.
- Be conservative in maximizing. Too much feed or speed can be dangerous. It can also reduce tool life and lower part quality.
- In industry, *never* edit a program without permission from your supervisor.

ADDING AND INSERTING PROGRAM LINES

This editing capability allows the programmer to add lines to the program or to insert information at a specific point. The procedure consists of the following:

1. Select the correct mode or software for editing.
2. Select the line on which you wish the new information to reside.
3. Write the new line of program commands.
4. Insert the information.

DELETING PROGRAM LINES

This editing capability allows you to erase a particular part of the program. Use caution in erasing data.

1. Select the correct mode.
2. Select the line number you wish to erase.
3. Delete the line.

Most controls will then reorder the remaining lines and close the gap created by the deletion. You must understand how your control deals with line number management. Refer to Custom Sheet 11.

CHANGING PROGRAM LINES

This editing capability allows you to rewrite a single line in the program. Probably the most common of the program editing procedures, changing lines is simple.

1. Select the correct mode.
2. Select the line you wish to change.
3. Rewrite the line.
4. Save the changed line over the top of the old line.

While this procedure is the most common, some controls might require you to clear out the old information first.

There will be no activities for editing. You are encouraged to read the programming manual and your Custom Sheet. Then you should practice on a program that is not needed. Make sure that you develop skills in all three editing procedures.

13-4 Error Messages from a Control

THE TWO TYPES OF ERRORS

Frustrating, confusing and seldom welcome, error messages appear during program testing. Sometimes they appear during a test of an already proven program. What do they mean? The type and number of errors detected by a particular control depends on the intelligence of its software. As a beginning programmer, you will need to be able to interpret error messages from controls. Not all controls will detect all the error types listed below. Controls find errors at two different times.

Immediate errors are detected during program entry. Sequential errors are detected during the test run of the program. The control detects a different type of error at each stage. Each control will have a set of error codes. If the control is conversational, it may simply state the problem. Some controls use both words and codes for error messages. You should already have an error message list on the language card given you. Here are the error detection stages:

Stage 1. The immediate errors detected during program entry include:

A) Entry form errors: (decimal point problems, wrong letter, incomplete command, wrong combinations of data).
B) Missing parameters: (incomplete cycle information, center of arc not identified, unspecified feed rate, speed range, or RPM).
C) Beyond capacity of equipment: entries are too big, too small, too fast, or impossible to carry out.

The detection of capacity errors at Stage 1 depends on the intelligence of the control.

Stage 2. The sequential errors detected during the program run or test run include:

A) Incomplete program (failure to include program end, tool offset data, or program commands needed for callout).
B) Limit switch errors that will not be detected until the PRZ is set. These errors can occur in an already proven program due to the fixture placement or logic errors. (We will study these in Chapter 19.) These errors include (missed return from subroutines; attempt to move through soft limits without authorization, loops that do not lead back to parent program, subroutines that have no end, and overnesting of logic moves).
C) Math errors (miscalculation of center points for arcs). Note that parts of the wrong shape or size cannot be detected by the control.

COMPLETE AXIS DISABLE DRY RUN

Many controllers will perform a whole program test run without actual machine movements. This is called a complete axis disable dry run. This is requested through the keypad by the operator. Testing the program this way will detect logic errors and math errors. The control may or may not detect hard limit switch violations.

LEARNING ERROR MESSAGES

When an error is detected by the control, you must interpret the message and make the correction. Consider the following guidelines. They offer hints as to what is wrong, even without the error message.

- *Errors will occur at a specific sequence number line.* Look at the general program. Check especially those parts of the program where there are many calculations or logic moves.
- *Plots show errors.* Where errors occur, a plot of the part, usually shows funny shapes. By analyzing these, it is possible to detect the errors. For example, a program going from left to right may produce a strange little jump as the tool passes over a quadrant line. This jump was caused by an incorrect arc center placement for the first arc, not the second. This is a math error. The line in which the error occurred would indicate the problem.
- *Errors fall into categories.* You may tend to make certain types of errors. Look for a certain error pattern in your programs.

Activities

The following questions are based upon the errors in the program in the Activities following 13-1. The line numbers refer to the corrected version of the program.

1. Would the error at line 003 have been detected at data entry or during program testing?

2. The error at 004 would have been detected at:
   A) Entry of data.
   B) Test of program without axis movement.
   C) Not detectable until real tryout.

3. Refer to number 022. Assume a hard limit had been set up to prevent hitting the chuck and assume that the Y error was exchanged for a Z. When would the crash have been detected in the original program?
   A) Entry of data.
   B) Test of program.
   C) Not detectable until real tryout.

4. Refer to number 022. Assume that a soft limit had been set up in the program and assume that the Y problem was corrected. When would the rapid into the chuck have been detected?
   A) Entry of data.
   B) Test of program.
   C) Not detectable until real tryout.

5. How many errors in this program would have been detectable during data entry. (To answer this, assume that the error in line 001 was found. If left in metric mode, because of decimal placement, all entries would be wrong. Correcting that error would make all further entries in the correct form for inch programming.)

From this brief activity, you can see that many errors will be undetected until the dry run or the visual evaluation of the shape produced. Not many of these errors were visible to the control. Even the disastrous crash at the end was invisible to the control without soft or hard limits. Soft limits would have shown the crash during dry run, axis disabled, testing. Hard limits would have shown the crash during the actual test.

In Your Lab

Becoming proficient at error detection is a matter of practice. Scanning your language custom sheet for the error messages possible is helpful. However, the best experience will come from writing and improving programs at the machine. Follow these guidelines.

1. Scan the language card for the machine you are learning to operate. Refer to Custom Sheet 6.
2. Try to memorize only the major error messages. Do not attempt to memorize all of them.
3. Do not seek help immediately. Ponder the error. Finding errors yourself is good self-instruction.

Even the very best programmers make errors. Be cautious, use good planning, recheck your work, and cautiously test new programs.

CHAPTER 14

USING POLAR COORDINATES IN PROGRAMMING

The use of polar coordinates is a programming tool that is extremely helpful in C/NC work. Parts that are dimensioned using angles and radii are always easier to program using polar coordinates. Over half of the trigonometry calculations programmers make are due to the need to convert basic polar dimensions into Cartesian coordinates.

Many new controls have polar capability. As computer integrated manufacturing becomes more common, polar coordinates become more important. Polar coordinates simplify much of the complex geometry and trigonometry needed to define high tech parts. CAD units use polar coordinates in their part generation software.

Polar CNC controllers require more internal programming. Due to development costs, manufacturers have been slow to add the full range of capabilities.

As a student of CNC, you must be aware of polar programming. Persons trained in the use of polar coordinates will find increasing job opportunities.

OBJECTIVES

After studying this chapter, you will be able to:

- Identify significant points using polar coordinates.
- Select correct coordinates based upon drawing.
- Be able to identify relative and absolute polar values.
- Know when to use and when not to use polar coordinates.

14-1 Polar Grid

A POLAR GRID IS CIRCULAR

Polar coordinates do the same task and follow many of the same principles as Cartesian coordinates. They identify a point in a grid. The difference lies in the appearance of the reference grid. A polar grid is a series of circles concentric to a reference point. Fig. 14-1.

POLAR COORDINATES

Polar coordinates are identified by reference to radius and angle. The mathematical symbol for radius is R. The symbol for angle is Ø (theta). A point is identified by the radius distance from the center of the grid and the angle from a base line extending from center. The base line extends from the center parallel to a primary axis, usually the *plus X axis*. This line is known as the *polar reference line (PRL)*. In Fig. 14-2, point A has a location of radius = 1.25″ and angle = 52°. Point B is R = 2.0; Ø = 105.

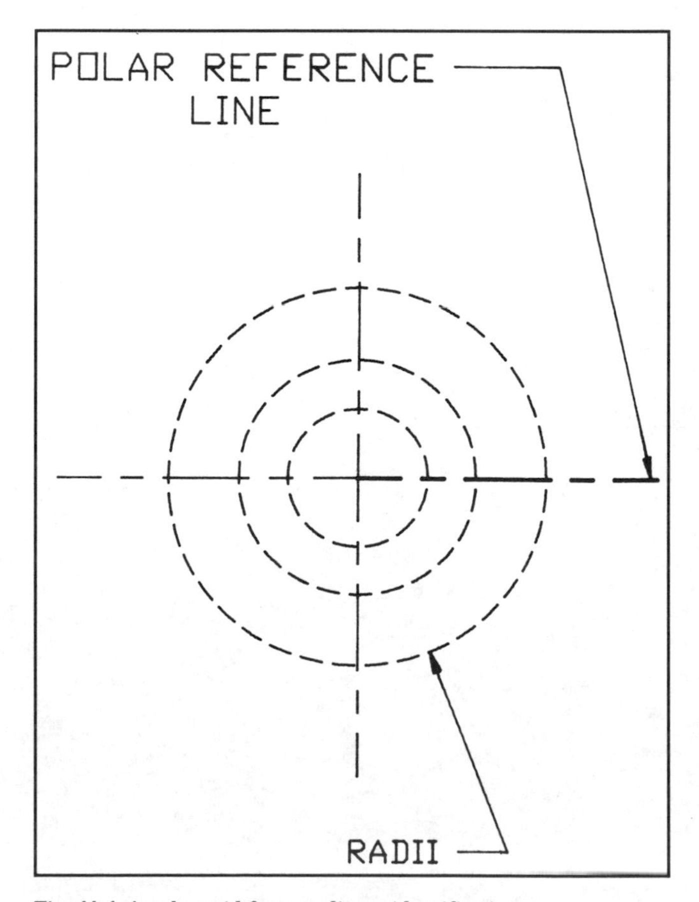

Fig. 14-1. A polar grid for coordinate identification.

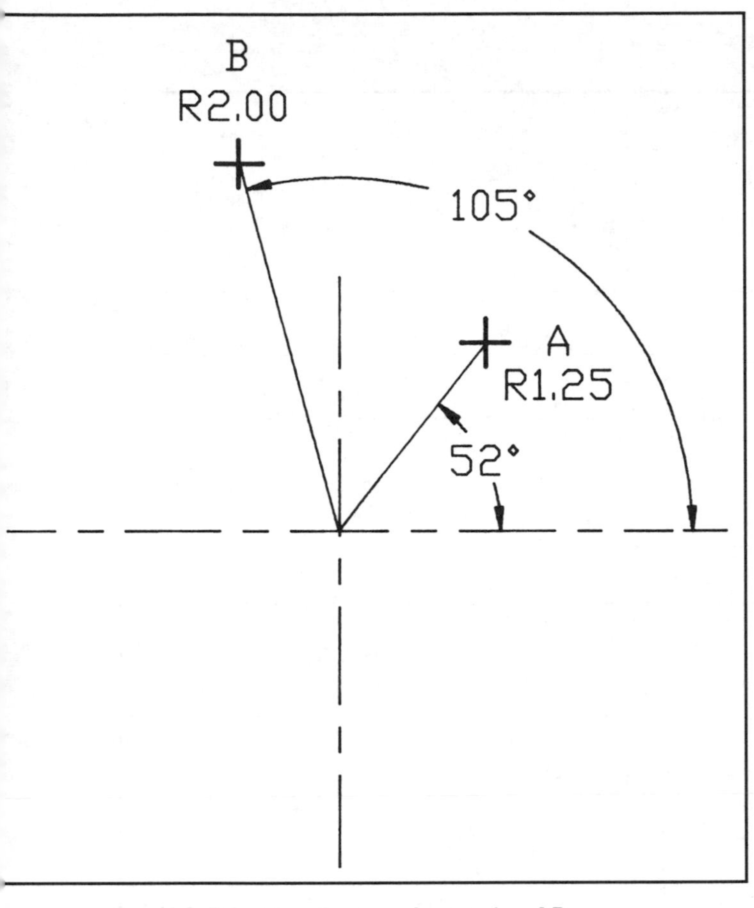

Fig. 14-2. Polar identification of points A and B.

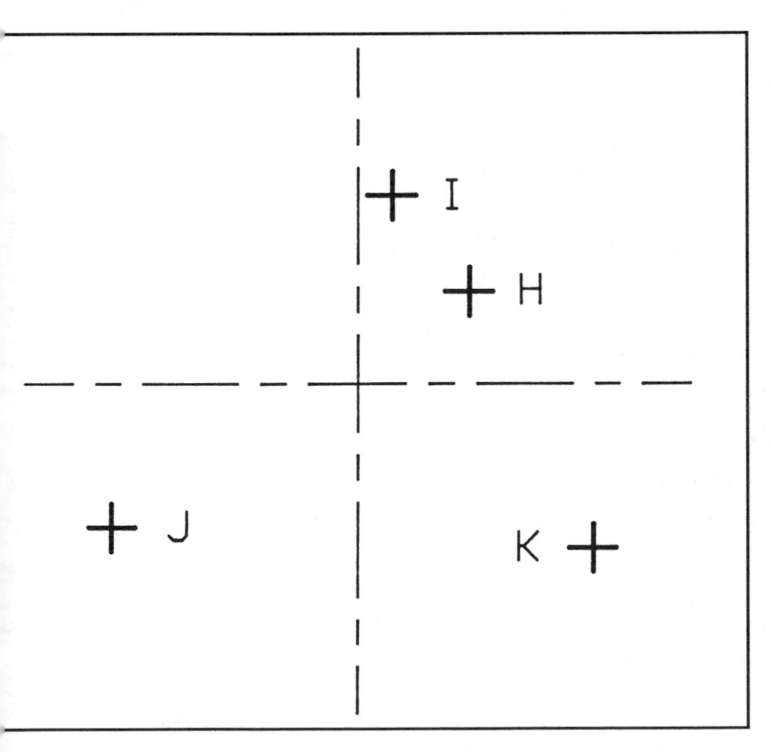

Fig. 14-3. These points can be identified using polar coordinates.

POLAR CONVENTIONS 1 AND 2

Here are two conventional guides for polar. We will learn more later.

1. *Grid not shown.* This is similar to working with Cartesian coordinates. The grid is normally not shown when discussing points.
2. *Order of identification.* A second convention is to identify the radius first, then the angle. Some controls will accept either R, Ø or Ø, R. This is similar to Cartesian entries in which coordinates should be entered as X,Y,Z but would be accepted as Z,Y,X.

The polar coordinates for each point shown in Fig. 14-3 are

(H) Radius	.75, Angle	40°
(I)	1.0	80°
(J)	1.5	210°
(K)	1.5	325°

14-2 Absolute Values in Polar Coordinates

As in Cartesian coordinates, there are both absolute and incremental values in a polar coordinate system. We will look at each in turn.

The concept of absolute value is similar to Cartesian coordinates in that the coordinate values always refer to the center of the grid and the polar reference line. Absolute value polar coordinates identify a point from the center of the grid in terms of radius. The angular coordinate is referenced from the polar reference line (PRL).

POLAR CONVENTIONS 3 AND 4

3. *Any point in a plane.* Any point in a plane may be identified by the absolute radius and angular distance.
4. *Absolute positive and negative angles.* Both positive and negative angles are used for absolute value polar coordinates.

The positive circle begins at the PRL and goes counterclockwise, while the negative circle goes clockwise from the PRL. Fig. 14-4.

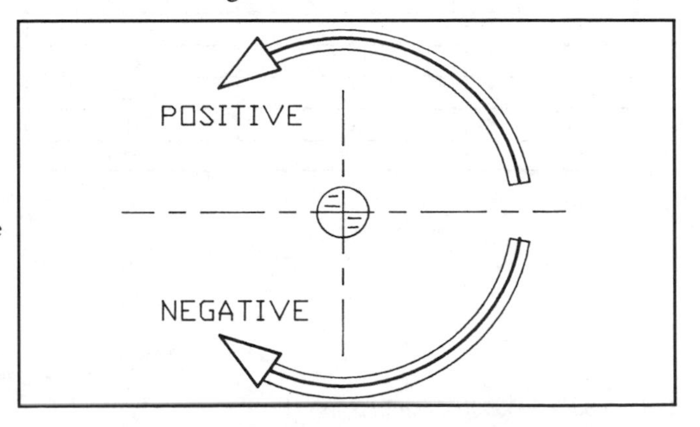

Fig. 14-4. Two polar circles. Positive goes in a counterclockwise direction, negetive goes clockwise.

Activities

Absolute Value Polar

1. Refer to Fig. 14-5. Identify the polar coordinates for points J through N.

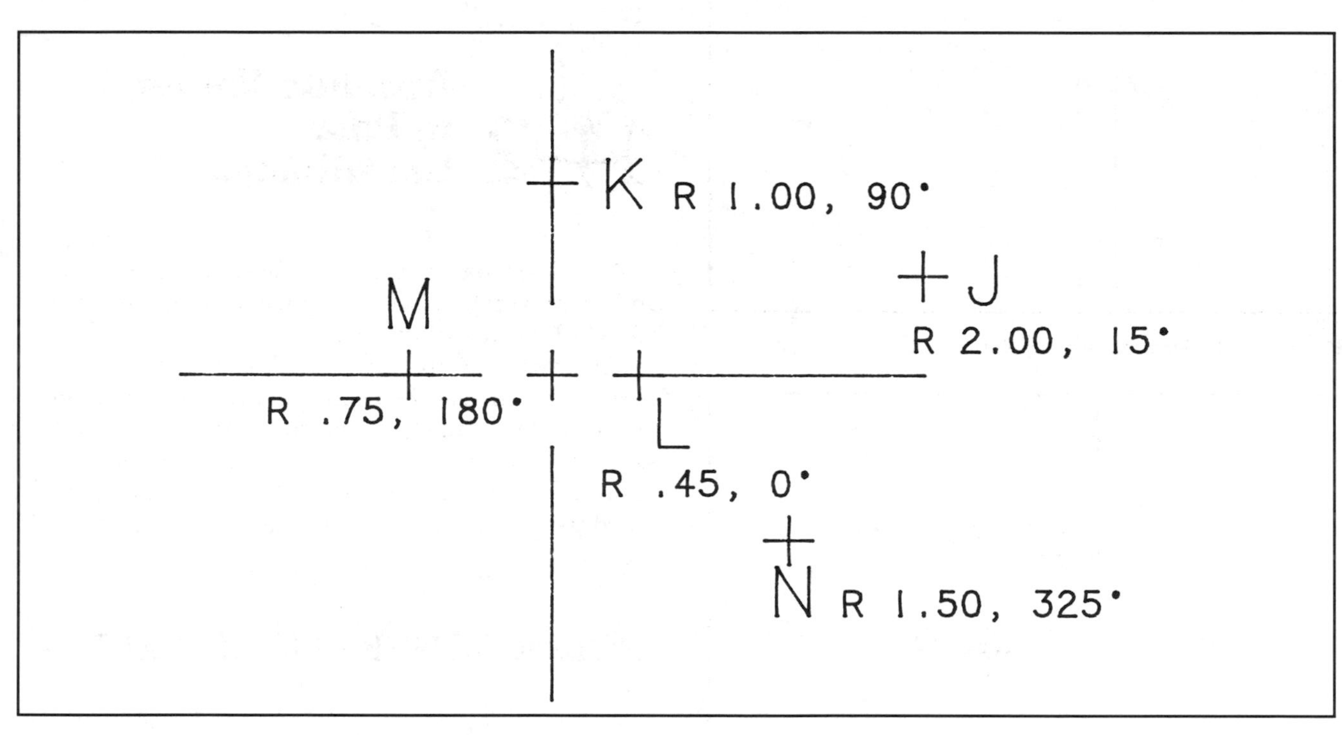

Fig. 14-5. Identify points J through N using polar coordinates.

2. Locate the following polar point locations on the grid below shown in Fig. 14-6. Note the use of positive and negative angles. Sketch accurately. Use a protractor and ruler.

Point	Radius	Angle
A	2.00	–270
B	2.00	360
C	.50	45
D	.50	–135
E	2.00	270
F	.50	–225
G	2.00	180
H	.50	315

Fig. 14-6.

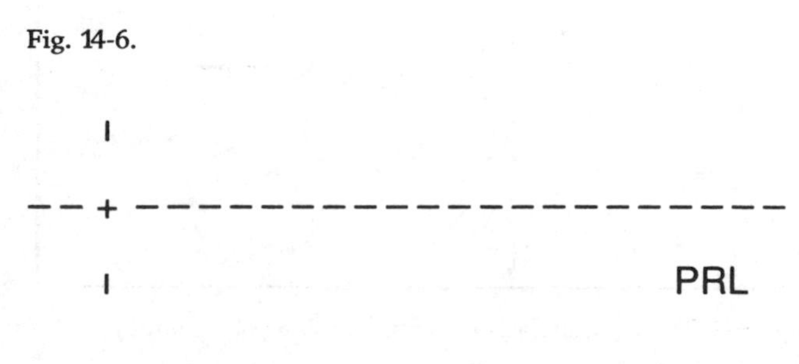

3. Connect the points located in Problem 2. Correctly located, they form a four-pointed star.

14-3 Incremental Values in Polar Coordinates

Using polar coordinates, a point may be identified using another point as a reference. This is relative identification, also called incremental identification. The relative *difference* in the radius and angle from one point to the next will determine a position. Incremental point values depend on the distance, in terms of radius and angle, from one point to the next. The previous point is taken as a temporary reference. This concept also applies in incremental Cartesian values.

INCREMENTAL EXAMPLE

In Fig. 14-7, point B is identified by the relative difference from point A. Using point A as the reference, the incremental coordinates of point B are radius = .50 and angle = 73°.

Point B has .50″ more radius and 73° more angle than point A. This is the relative difference. Point A was used as the basis for the new identification. Point B also has an absolute value: R = 1.5 and A = 108.

POLAR CONVENTIONS 5 & 6

5. *Direction is determined by the "rule of thumb."* The direction of the new incremental polar coordinate is indicated by a sign value (+ or –). Counterclockwise angles are positive and clockwise angles are negative. (To review the "rule of thumb," refer to Section 3-2, Fig. 3-11.)
6. *Radius out is positive.* A move outward from center on the radius produces a positive incremental value for the radius coordinate. A move toward the center (smaller radius) produces a negative value coordinate.

As an example, refer to Fig. 14-7. If we were to reverse the identification for the previous example (identify A from B), the relative coordinates would be radius = – .5″ angle =– 73°.

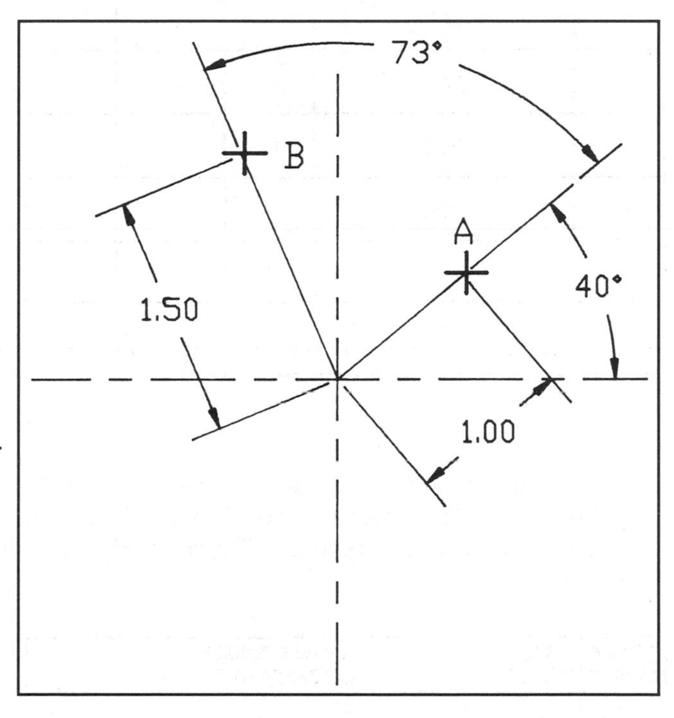

Fig. 14-7. Point B is identified incrementally from Point A.

Activities

Incremental Values

1. Give the incremental radius difference and angular values from one point to the next for the following points shown in Fig. 14-8. Note that you will need positive and negative angular coordinates and that you may need to add or subtract angles and radii. Use the rule of thumb.

FROM POINT	CLOCKWISE OR COUNTERCLOCKWISE DIRECTION	TO POINT	RADIUS	ANGLE
A	CCW	B		
A	CCW	C		
D	CW	A		
E	CW	C		
F	CCW	C		
C	CCW	F		
F	CW	B		

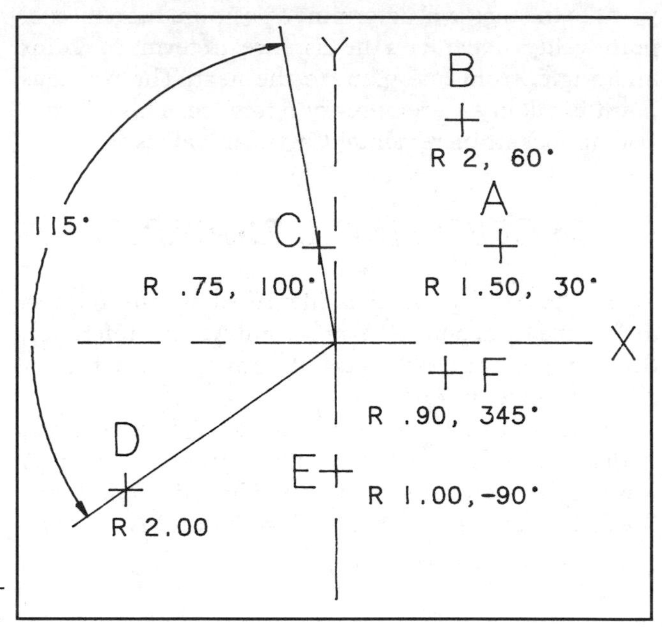

Fig. 14-8. Give the incremental (relative) polar value of each point relative to the given point. The direction will determine which arc to use.

2. On the grid shown in Fig. 14-9, locate the following incremental polar coordinate values. You must have a polar point from which to start. The absolute value of this point (A) is radius = .50 angle = 45°. The plus signs are shown as a reminder. Normally they would not be shown, for the absence of a sign value indicates positive.

FROM POINT	TO POINT	RADIUS ANGLE COORDINATES	
A	B	+1.5	+45
A	C	+1.5	–45
C	H	–1.5	–225
C	D	–1.5	–45
H	G	+1.5	+45
H	F	0.0	+90 (no change in radius)
C	E	0.0	–90

Fig. 14-9.

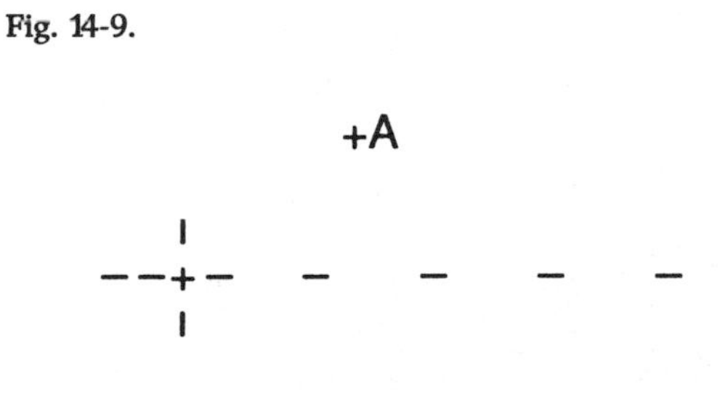

3. Connect the points around the outside of the figure. Do not connect them in letter order. They should form the same star pattern formed in the last problem.

14-4 Using Polar Coordinates to Define Geometry in Programs

PART PROGRAMMING EXAMPLES

This section discusses some of the ingenious ways polar coordinates can save math and time in programming. In each example, notice that the part geometry to be produced was defined and dimensioned using mostly radii and angles. This indicates that polar coordinates would be the best point values to use. Controllers that fully use polar coordinates can move freely between Cartesian and polar values.

The following example demonstrates a programming technique called a *smart loop*, which is also called a *Do Over LOOP*, or *DO loop*. A DO loop repeats a series of commands a given number of times. It then skips the loop and moves to the next commands. The program examples may seem unclear now. After studying program flow and DO loops in Chapter 19, you might review these programs. They might then be clearer to you.

Example 1. A Bolt Circle Drill Pattern

This useful application of polar coordinates is common on many CNC controllers. A bolt circle is a repeating pattern of holes or other features that are concentric to a center reference. Note that Fig. 14-10 is dimensioned using polar parameters, not Cartesian coordinates. To program each location using Cartesian coordinates, one would have to solve a trigonometry problem for the X and Y coordinate value for each hole. Most C/NC drilling and milling machines have a "canned cycle" that will handle this programming problem.

Bolt circle parameters: 10 holes, equally spaced 18° to first hole from the X axis.

Radius of B/C = 1.5″

For these examples, I will use the programming language of the DYNA control, which is conversational. For clarity, start-up blocks have been skipped.

Program parameters: PRZ is on center of part
Departure is PRZ = 1.00″ above part
Gauge height is .10″ above
Hole depth is .25″ deep

	Block	Comment
	004 SPINDLE ON	
	005 GO r 1.5	Position at first hole
	006 a 18.0°	*Absolute Polar Cords.*
	007 GO f Z–.9	Drop spindle to gauge height (f = Fast).
	008 REPEAT 10	Define loop start. Repeat loop 10 times.
L	009 GO Z–1.25	Drill to depth.
O	010 GOf Z 1.0	Clear to gauge height.
O	011 GR a 36	Move tool 36° *relative* angle.
P	012 REPEAT END	Loop end. Return to beginning of loop.
	013 END	

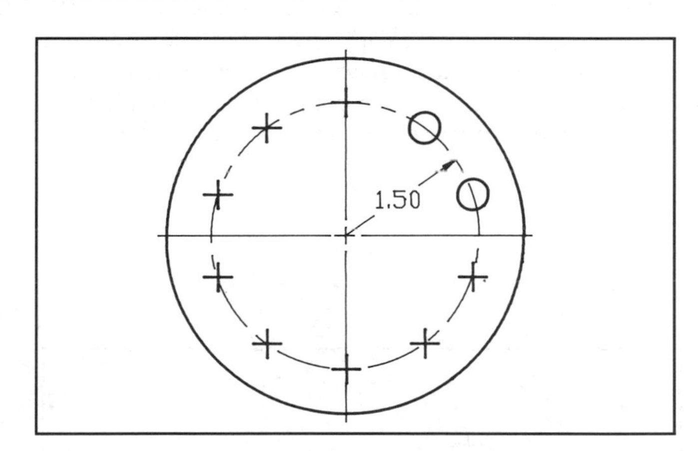

Fig. 14-10. A bolt circle pattern is easy to program using polar coordinates.

Can you see how this program might easily be reworked to drill 180 holes? Where would you make changes? How many changes?

Answer: There would be only two changes:
008 = REPEAT 180 Causes 180 cycles
011 = GR a 2.0 Each cycle is 2°

Example 2. Part Geometry Problem

Here, we will look at two approaches to the same program. We will define a significant point (point C) on the part, using Cartesian and then polar values. Refer to Fig. 14-11 on page 162. You will see that because the part has polar dimensions, the point becomes significantly easier to program using polar coordinates.

This drawing shows a typical industrial part. The part is solidly defined. The height is 2.0". The width (3.0655") is fixed by the intercept of the 40° line with the base.

To program this part using Cartesian coordinates, one needs to do some trigonometry to find the X and Y values of point C. Fig. 14-12.

Cartesian Solution

Going either way around the part from the PRZ, using Cartesian coordinates, eventually you need to arrive at point C. To find the Cartesian X and Y value of point C, we make the following calculations based upon the triangle in Fig. 14-12.

X' = Sine 40 × .5"
.64278 × .5 = .3214"

Y' = CoSine 40 × .5"
.7660 × .5 = .3830"

Point C Absolute value;
X total = .5 + .3214" = .8214"
Y = 1.5 + .383 = 1.883"

Using these calculations, you could then write the total program. You would need, however, some information on how to write partial arc callouts in Cartesian values. This information is given in Chapter 15. We will not write the actual program because there are still some complications in writing the arc callout using Cartesian coordinates. This program is relatively complicated due to the trigonometry and arc callouts.

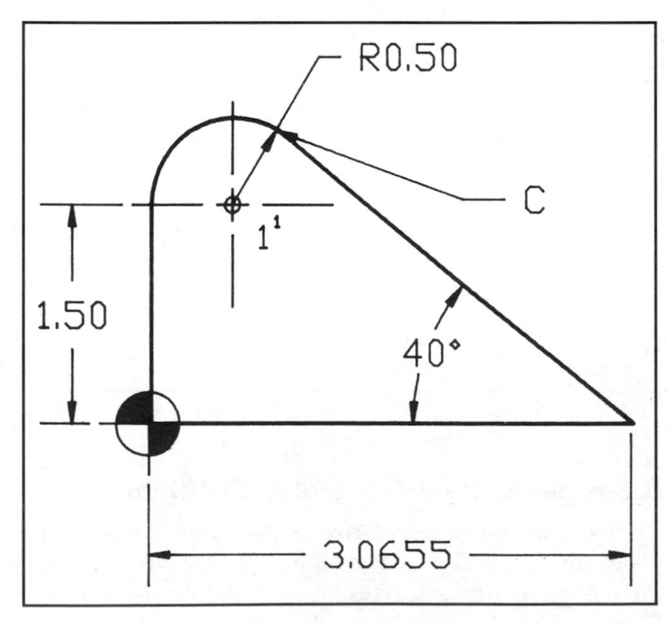

Fig. 14-11. A typical part geometry that is dimensioned using polar information.

Polar Solution

The polar solution to this problem will identify point C using polar coordinates without any trigonometry. Here is the solution program. It is a plot of the shape.

Polar Solution Program Refer to Fig. 14-12.

005	GO Z– 1.0	Bring pen to paper on PRZ.
006	GO X–3.0655	GO Absolute to point B.
007	ZERO AT	Blocks 7,8 and 9 fix a local zero
008	X .5	(from PRZ) at center of arc.
009	Y1.5	
010	GO r.5	Blocks 10 and 11 move pen from
011	a 50	point B to point C. Absolute 50° from PRL and .5" from center.
012	GR a 130	Plot ARC from point C to point A.
013	GR Y–1.5	Complete shape.
014	END	

We could have chosen the center of the arc as the PRZ and thus saved blocks 7,8 and 9. However, this was not done so that we could demonstrate the defining of a local zero point in polar coordinates. You will come to appreciate the programming of the arc, too. Arcs are a great deal simpler in polar.

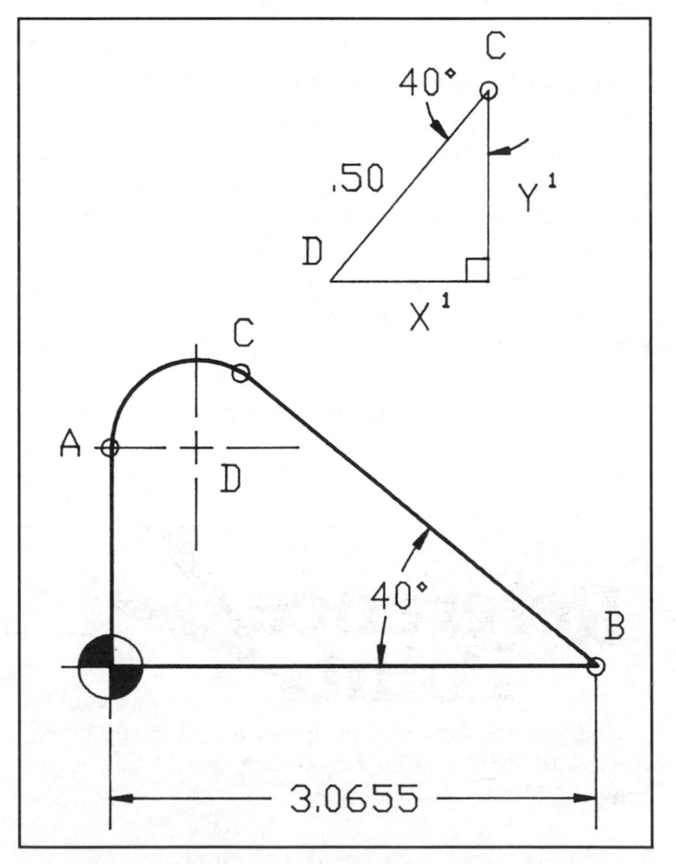

Fig. 14-12. This triangle must be solved for X' and Y' to identify point C. This is then added to the drawing values program. This shape using Cartesian coordinates.

14-5 When and When not to Use Polar Coordinates

Polar coordinates are not right for every point in a program. In using Cartesian coordinates, if you wish to identify a point, use the coordinate value that is known on the drawing. If you find yourself doing a lot of arithmetic, you are probably not using the right point values. This is also true in using polar coordinates.

LET THE DRAWING DICTATE

If you know the Cartesian value of a point, use it. If you know the polar value of a point, use that. (This assumes that you are using a Cartesian/polar control.)

Refer to the example program in Figs. 14-11 and 12. The first move to the right was given in Cartesian values (X 3.0655) because it was on the drawing. Polar would have been wrong. So too for the final move (Y–1.5) there was no need for polar values.

Custom Sheet 12 Polar Coordinates

Custom Sheet 12 explains the use of polar coordinates on the control you will be using in your lab. Staple this custom sheet in the back of this book as a guide. The more polar-capable a machine becomes, the more conventions you must learn. Learning full polar ability requires more study than the basics we have discussed. There are many differences between controls that use polar coordinates.

If your control uses polar coordinates, you will receive Custom Sheet 12 for a polar programming problem.

There is much to learn about machining straight and curved lines using polar coordinates. You will need to study some of this information on your own, using the programming manual of the machine in your lab.

UNIT 5

Intermediate Programming Concepts

The following five chapters discuss the programming techniques that will allow you to manipulate the program like a professional.

CONTENTS

Chapter 15. Programming Arcs—This chapter discusses the three ways to program arcs. You will learn to calculate the I,K or I,J values in programming arcs. Also, you will study radius and polar arc methods.

Chapter 16. Program Flow—Branching Logic—Programs do not have to start at the beginning and move sequentially directly to the end. You can manipulate the sequencing of events inside the program. This can save programming work and improve part quality. This skill is called *branching logic*. Millions of combinations of manipulations are possible. Once understood, the art of arranging the logic in the program flow is challenging and enjoyable.

Chapter 17. Programming with Canned Cycles—Fixed cycles, more commonly called canned cycles, are the labor-savers on most controls. Canned cycles use parameters to completely program long operations. Once cycle programming is mastered, entire programs can be built entirely from canned cycles. You will also learn about building your own canned cycles.

Chapter 18. Threading on a C/NC Lathe—This chapter discusses the main steps in threading on the lathe. A C/NC lathe is able to thread at nearly the same speed as turning.

Chapter 19. Advanced Programming Concepts—We will learn a couple of tricks here. This chapter discusses techniques such as a mirror image part from the same program. It also discusses shrinking and enlarging a part through programming. Safety crash zone programming and constant velocity machining are also discussed.

CHAPTER 15

PROGRAMMING ARCS

You may have noticed that the programs discussed thus far had no generated curves. Programming arcs in C/NC requires further training, especially when the points in the curve must be defined in Cartesian coordinates. This chapter will teach you to generate circular cutter paths on both Cartesian and polar controls.

OBJECTIVES

After studying this chapter, you will be able to:

- Given the parameters of a curve, write a curve program in coded language and write a curve program in conversational language.
- Identify the three significant points in a curve.
- List the four parameters of a curve.
- Calculate X,Y, or Z and I, J, or K points using the Pythagorean theorem and trigonometry.

15-1 Curve Parameters

THE NATURE OF CURVES

All curves discussed in this section will be circles or parts of circles. All controllers need four parameters to program a curve. These parameters are:

1. Center point identification.
2. Start point identification.
3. End point identification.
   (Identified with Cartesian coordinates or an angular distance.)
4. Radius length.

Refer to Chapter 4, Fig. 4-24, which illustrates the need for the four parameters.

THREE METHODS OF PROGRAMMING CURVES

Three methods are used to program curves. The methods differ in the format and the combination of parameters entered. The different methods are known by the entry parameters you must enter into the program.

As you will see, all four parameters are not always entered in the program. Given certain information about a curve, the controller automatically knows other facts. These secondary facts do not need to be entered.

The three methods are the IJK method, the radius method, and the polar method. We will look at all three methods in turn. But first let's look at the start point, which is the first parameter that is always automatically identified in all three methods. The start point is the present tool location. When a curve begins, it starts at the location of the tool. *When a curve command is read, the control takes the present location as the start point.* This is true for all three methods of curves and controllers. This eliminates one of the parameter entries. The start must be identified, but you do not need to enter it. In the program, it is known automatically.

IJK Method

The required entry parameters of the IJK method are the end point and the center point.

In the IJK method, if the controller knows the *start point* (automatically known), *center point*, and *end point*, the radius is also automatically known. When told to generate a curve, the control knows that the center is the radius distance away from both the start point and the end point. Your entries must yield equal radii between center-end and center-start. If they do not, you will be given an error code.

Coded controls require a G02 code for clockwise circular interpolation and a G03 code for counterclockwise interpolation. The I,J, and K are used as an alternate set of X,Y, and Z coordinates that identify the center of the arc. This method is commonly called an *IJK curve* or a *partial arc format* for Cartesian controllers. On coded, Cartesian controls, the G02 or G03 code must also be present in the entry. This initiates circular interpolation. We will write actual circular programs in the next section.

Radius Method

The required entry parameters of the radius method are the end point and the radius distance.

Since the control already knows the start point and is now given the end point and radius distance, it can reason that the center point is in one of two locations and that there are four possible curves. Can you see these in Fig. 15-1?

To signify the correct curve, this combination requires that the curvature either clockwise or counterclockwise be specified. This is indicated by a G02 or G03 entry for coded controls or by a direction sign value (+,–) on polar controls. This entry tells which curvature is needed between the start and end point and initiates circular interpolation on coded controls.

Polar Method

The required entry parameters of the polar method are arc distance and center coordinates.

In the polar method, the radius is automatically identified. The control knows that the arc refers to the center reference point. Knowing the start point and the arc distance in degrees, or knowing a coordinate for the end point and working from an assigned center reference point, the control can calculate the radius. Note that the only information you must enter in the curve command is the arc distance in decimal degrees and the center point.

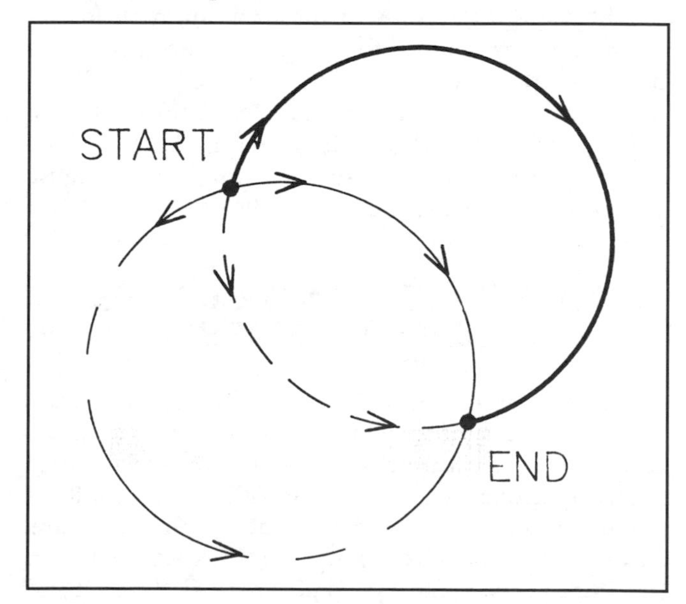

Fig. 15-1. Four possible curves from the given information.

Some controls require that the center point be preidentified as a local reference zero. Then the actual program arc command is simply the arc distance and a sign value for clockwise or counter-clockwise movement. This is useful for curves sharing the same center point, such as spacing around a bolt circle. Some controllers use a version of this method in their special cycles for curves and angular spacing.

Here, you are asked to make three sketches. These will help you understand the three methods. In each sketch, use a pencil to mark each parameter point that must be entered. Sketch each combination to help you see why the remaining parameters are automatically known. Then with the given parameters only, draw the curve as indicated, using a compass.

In every combination, you will automatically know the start point. Use graph paper and make any mark for the start point.

Sketch 1 IJK Method
Given: 1. Start point (automatically known).
2. End point is X+1.5, Y+.927 (from start point).
3. Center point is I+1.5, J–.75 (from start point).

Sketch 2 Radius Method
Given: 1. Start point (automatically known).
2. End point is X+1.0, Y+.382 (from start point).
3. Radius is 1.50. With this information you can draw several curves. See if you can sketch four possibilities.
4. Now the curvature is counterclockwise (G03) and less than 180°. This method is used for curves less than 180° to eliminate incorrect curves.

Sketch 3 Polar Method
Given: 1. Start point (automatically known).
2. Center point X0.0, Y–1.5 (identified as center of polar grid).
3. Radius (automatically known). The distance from the start point to the center point is the radius distance. Therefore, it is known by the controller.
4. The arc is –45° (rule of thumb). This is the only entry required for this type of arc. The direction is determined by the rule of thumb. A minus value arc would be clockwise.

What is the radius of the above curve?

____

15-2 Programming Curves—Formats

There are differences in the programming format for curves from control to control. However, each uses one or more of the three methods just outlined. If you understand these, you can transfer from control to control with no difficulty. In general, Cartesian controls require more math to program curves because curves are inherently polar in nature.

CURVATURE AND G02-G03 CODES REQUIRED

Some symbol is usually needed by the control to indicate circular interpolation and the type of curvature. Is the curvature clockwise or counterclockwise? Coded controls use the G02 or G03 code to initiate circular interpolation and to establish the curvature. Polar controls require a plus or minus sign value on the angle callout. The sign is plus for CCW curves and minus for CW curves according to the rule of thumb.

Plane of Curve

Each significant point must be identified by a pair of coordinates. These coordinates will depend on the plane in which the curve is to be programmed. The plane will be X-Y for vertical mills and X-Z for lathes. If the machine is capable of circular interpolation in more than one plane, it may require a program entry to signify the plane in which the circle is to be produced. For uniformity, all examples will be in the X-Y plane. However, Z could be interchanged for Y with no change in the concept.

PROGRAMMING COMBINATION 1: IJK CURVES OR PARTIAL ARC CURVES

This type of program curve is commonly called an IJK curve because of a need to separate coordinates within the command block. The letters I, J, and K are incremental values used for this purpose. This type of curve is also called a *partial arc curve*. On Cartesian controls, this is the way to program arcs of less than 90° and arcs not divisible by 90°.

IJK—Alternate Coordinates

In this method, we first identify the end point, then the center point. Each point requires a pair of coordinates. The control needs a way of separating the coordinates for the end point from the center point. This is done by assigning an alternate form of X, Y, or Z. The alternate of X is I. The alternate of Y is J. The alternate of Z is K.

X, Y or X, Z are used for end point coordinates and I, J or I, K are used for center point identification. The combination used depends upon the plane in which the circle lies.

Since the radius and start points are automatic, Combination 1 requires the identification of two significant points, the *end point* and the *center point.*

The IJK programming format is:

1. Circular interpolation code (G02 or G03).
2. Identify end point.
3. Identify center point.

Example 1. Refer to Fig. 15-2. The program command would be G03 X–3.0 Y3.0 I–3.0 J0.0.

Explanations

G03	Counterclockwise interpolation.
X and Y	Identify the end point from the start point.
I and J	Identify the center point from the start point. Note that the J coordinate was zero. There was no J distance. This could be eliminated as a null entry, but it is best left in for clarity. Notice the sign values on the X and I coordinates, they indicate the direction from the start point to the end and center. X-Y identify the end point; I-J identify the center point.

Example 2. A program command for Fig. 15-3 would be G02 X1.4202 Y–1.25 I–.5 J–2.0.

Explanations

G02	Clockwise interpolation.
X and Y	Identify the end point from the start point.
I and J	Identify the center point from the start point.

Incremental and Absolute Modes

There is a slight difference in programming IJK curves in absolute value mode. The examples we have discussed have been incremental.

Incremental Mode in Curves

In the above examples, the X, Y, I, and J coordinates were given in incremental values from the start point. This is easiest to deal with. The dimensions deal with the curve only. You do not need to add or subtract any distance to relate the end point to the PRZ.

The X-Y or X-Z coordinates identify the end point in reference to the start point. I-J or I-K coordinates identify the end point from the start point.

Absolute Mode

If you are in the absolute mode, then the X and Y or X and Z coordinates would identify the end point with *absolute values* from PRZ. Even so, I,J, and K still are incremental for curves. In either absolute or incremental modes, *I, J, and K coordinates are always incremental.* For a new programmer, it is easier to program IJK curves in incremental mode. It is not difficult to use absolute mode. For now, however, we will consider all examples in incremental. This is a recommended practice until you feel secure in programming IJK curves.

In coded language, enter G91 in the program to switch to incremental mode if at all possible. This way your entry will not be a mixture of values. The math will relate only to the general area of the problem, not back to a reference zero point.

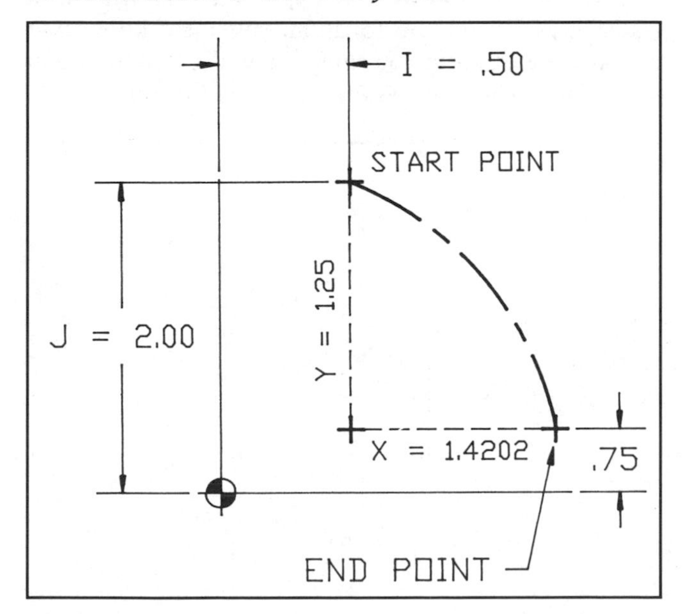

Fig. 15-2. A 90°, 3″ radius, counterclockwise curve. The I, J, K method was used.

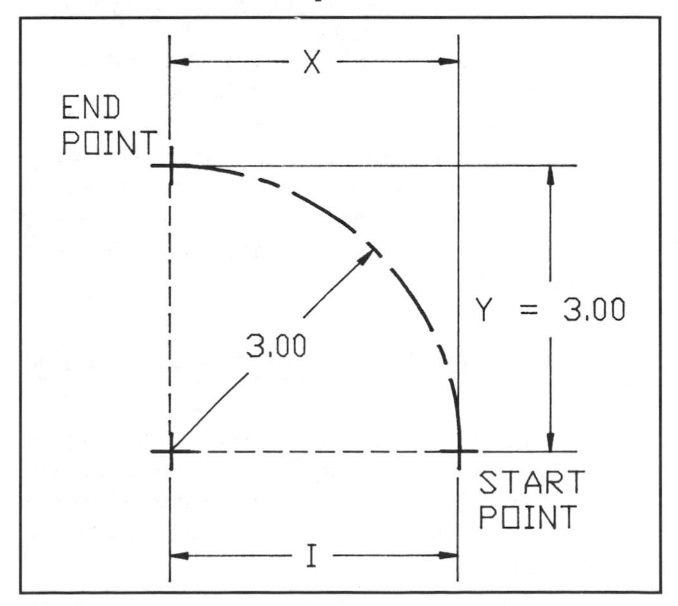

Fig. 15-3. An arc of less than 90°. Note the sign values on the program entries.

Custom Sheet 13

Before you read further, find out which methods will be used to program curves in your lab. If your control does not use the IJK curve method, your instructor might ask you to skip the rest of this development. He or she may tell you to move to Programming Combinations 2 and/or 3.

IJK = 1 Radius method = 2 Polar = 3

You will need to establish these facts:

1. For circular interpolation does your control use Combination 1, 2, or 3? (IJK, radius or polar)

   ____

   How many combinations does it use?

   ____

2. Will your control do circular interpolation across a quadrant line?

   ____

3. What is the default feed rate for circular interpolation?

   ____

4. Will your control do circular interpolation in more than one plane?

   ____

5. For most controls that are capable of circular interpolation in more than one plane, need to know in which plane the circle lies: G17 for X-Y, G18 for X-Z and G19 for Y-Z. Probably, your control will not need these to do simple circular interpolation. For vertical mills, the base plane X-Y is usually assumed unless otherwise specified. This is called the *default parameter.* This parameter is the one the control assumes unless told otherwise.

15-3 The IJK Method of Calculating Partial Arc Curves

The IJK method should be practiced as a common arc program command. The objective is always to find the X,Y or X,Z coordinates for the end point and the I,J or I,K coordinates for the center point. Each problem will be slightly different. You will always need to identify the end point and the center point. The two math tools needed to accomplish this are the Pythagorean theorem and trigonometry.

THE PYTHAGOREAN THEOREM

The Pythagorean theorem is used for finding the third side of a right triangle, given two sides. The theorem states that the sum of the squares of the two lesser sides is equal to the square of the hypotenuse. The formula is $A^2 + B^2 = C^2$.

In this formula, A and B are lesser sides adjacent to the right angle; C is the hypotenuse. Fig. 15-4. The hypotenuse of a triangle is always the longest side. Therefore, A and B are lesser sides.

You will be finding the hypotenuse or finding a lesser side. To use the formula, you must know that one angle is 90° and you must know the length of two sides. There are two variations of the formula.

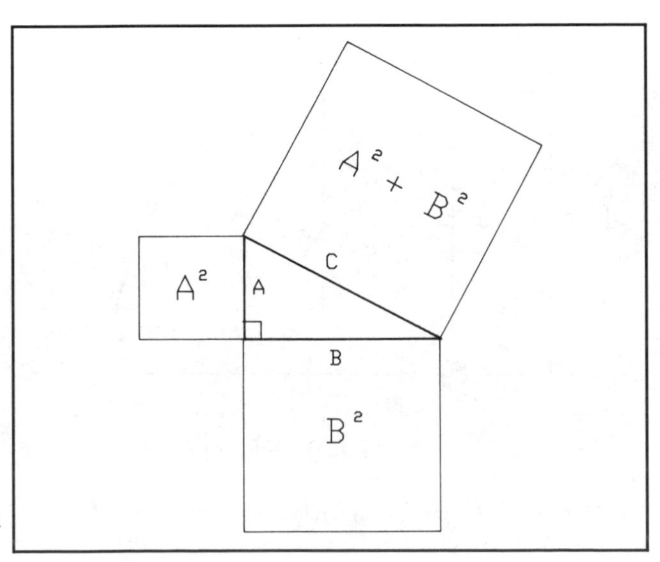

Fig. 15-4. A right triangle can be solved using the Pythagorean theorem.

The following formula is used for finding the length of a lesser side given the hypotenuse and one lesser side. A and B are interchangeable. The lesser value inside the square root sign is the known lesser side.

$$A = \sqrt{C^2 - B^2}$$

The following formula is used for finding the hypotenuse when you know the two lesser sides.

$$C = \sqrt{A^2 + B^2}$$

For example, you may need to find the hypotenuse (C), when you know the two lesser sides (A and B). Refer to Fig. 15-5.

The formula is: $C = \sqrt{A^2 + B^2}$

This problem is easily solved, using a standard scientific calculator.

Calculator Entry	Display
3	3
X^2	9
+	9
4	4
X^2	16
=	25
$\sqrt{X}$	5

The hypotenuse is equal to 5.

In another problem, you may need to find the length of a lesser side (A or B), when you know the length of the hypotenuse and one side. Refer to Fig. 15-6.

The formula is: $B = \sqrt{C^2 - A^2}$

Calculator Entry	Display
.35	.35
X^2	.1225
–	.1225
.25	.25
X^2	.0625
=	.06
$\sqrt{X}$	.244949 = .2449

Side B is equal to .2449.

Rounding Off

Notice that we rounded off the number to the fourth place to the right of the decimal. That is standard policy in programming math. Because controls are commonly set to a resolution of .0001", our math has to be that accurate also. This can make a difference. Never round off the number until the final answer.

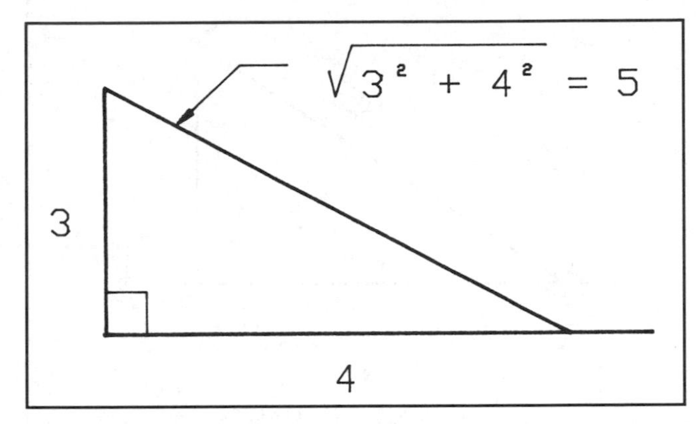

Fig. 15-5. Solving a right triangle for the hypotenuse using the Pythagorean theorem.

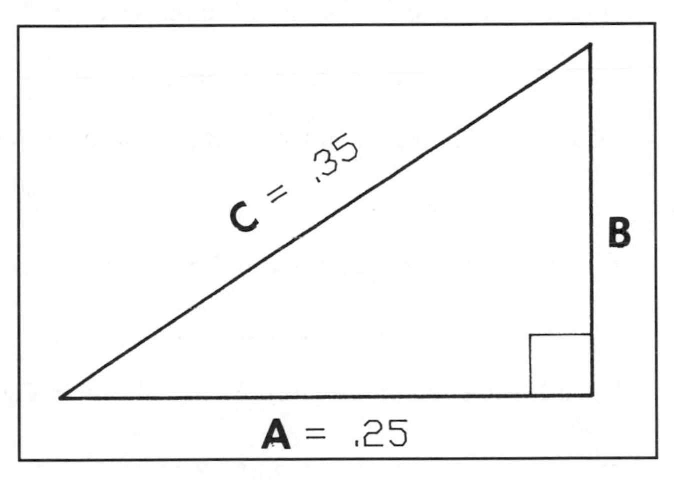

Fig. 15-6. Solving a right triangle for a lesser side.

The Pythagorean Theorem

1. Given two sides of the following right triangles, find the third side.

Triangle	(Hypotenuse) Side C	Side A	Side B
1.	.75	.25	
2.		1.065	.78
3.		.025	.090
4.	6.03		.50

2. Refer to Fig. 15-7. Find the missing dimension.

Radius = 3.5. J = 1.0. I = ?.

____

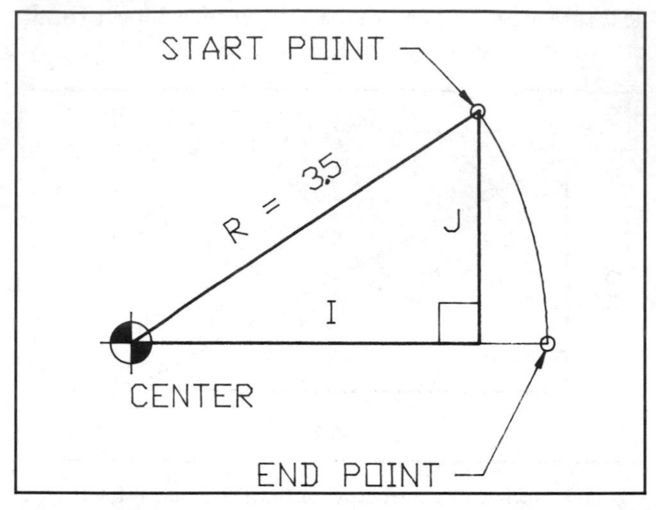

Fig. 15-7. Solve this arc problem and triangle for the missing I and X dimensions.

3. If you were programming problem 2 above, would "I" be positive or negative? Remember, both XYZ and IJK coordinates are incremented in this chapter.

____

4. What is the X dimension for the shape shown in Fig. 15-7?

____

Is this a positive or negative coordinate value?

____

5. What is the Y dimension for the shape shown in Fig. 15-7?

____

Is this a positive or negative coordinate value?

____

6. Refer to Fig. 15-8 on page 171. Find the missing dimension (J).

____

Is J positive or negative?

____

7. Refer to Fig. 15-8. What are the X and Y values?

X = ______________ Y = ______________

Is X positive or negative? Is Y positive or negative?

The answers to the above questions are given in the following Reference Point.

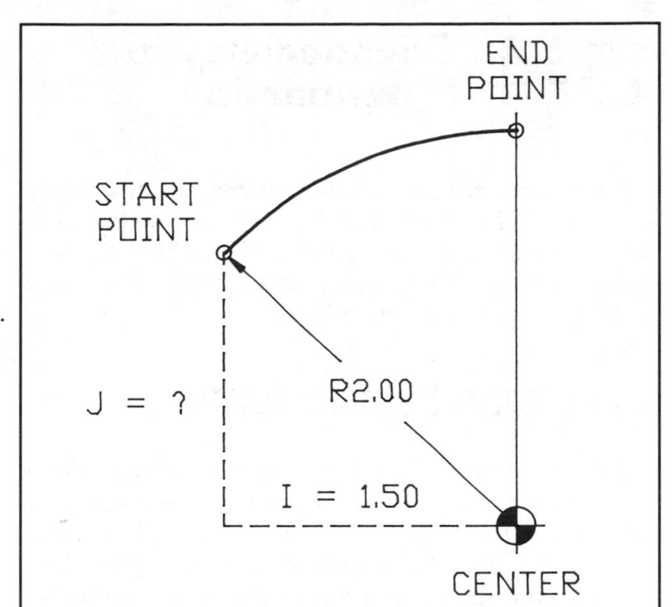

Fig. 15-8. Solve this arc problem for the missing parameters.

Do you understand the Pythagorean theorem? Here are the answers to the preceding activities.

1.

	C	A	B
1	.75	.25	*.7071*
2	*1.3201*	1.065	.78
3	*.0934*	.025	.090
4	6.03	*6.0092*	.50

2. I = 3.3541.
3. Minus (–).
4. X = .1459 (3.5 – "I").
   Plus (+).
5. Y = 1.0 (the same as J).
   Negative (–).
6. J = 1.3229.
   Negative (–).
7. X = 1.5 Y = .6771 (2.0 – J).
   Positive.
   Positive.

The Pythagorean theorem is basic to programming math. If you had a great deal of difficulty with these activities, ask your instructor for some reinforcing material from a math book.

15-4 Trigonometry in Programming

A complete review of trigonometry is not possible in this book. This section will briefly review the basic concepts and will present some formulas. If you do not understand them, ask your instructor for reinforcing material from a shop math book.

EXPRESSING RATIOS

Plane trigonometry is based on ratios. Ratios are a comparison of one number to another. Ratios may be expressed as fractions, such as $2/3$ or $1.5/5.7$. If the ratio is divided as the fraction indicates, the ratio may then also be expressed as a decimal version. Examples include:

$$2/3 = .666667$$
$$1.5/5.7 = .2631579$$

In each example above, the decimal version is compared to the number 1. In other words:

$$2/3 = .666667/1$$

The number 2 compared to 3 is the same as .6666667 compared to 1, and:

$$1.5/5.7 = .2631579/1$$

The number 1.5 compared to 5.7 is equal to .2631579 compared to 1.

When using decimal ratio expressions, the 1 is not placed in the statement, but you should understand that it is implied. Therefore, $2/3$ is correctly expressed as .66667 and $1.5/5.7$ as .2631579.

FACTS ABOUT TRIANGLES

The following facts about triangles are essential to trigonometry.

1. Given any triangle, the larger the angle, the larger the opposite side it subtends will be. Fig. 15-9.

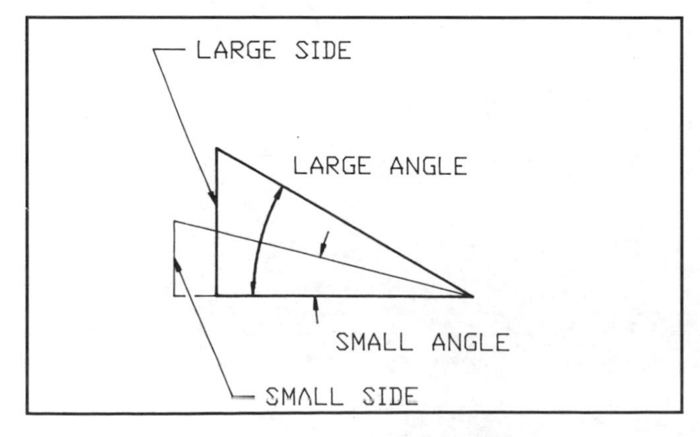

Fig. 15-9. The larger angle subtends the larger side.

2. Given any two right triangles of exactly the same shape but of a different size, the ratio comparison of corresponding sides will be the same.

   $A/B = a/b$ *and* $A/C = a/c$ *and* $B/b = C/c$
   $A/a = B/b$ *and* $A/a = C/c$ *and* $B/b = C/c$

3. Based upon one of the nonright angles, there are three names for the sides in a right triangle—hypotenuse, opposite, and adjacent. Fig. 15-11. The *hypotenuse* is always the longest side. The *opposite side* is opposite the angle being considered. The *adjacent side* is next to the angle being considered. (The adjacent side is never the hypotenuse.) Again, these names are based upon an individual angle in the triangle. Either you know the angle or you need to know it.

4. There are six possible combinations of ratios between the sides of a triangle:

Name	**Comparison**
Sine	= Opp/Hyp
Cosine	= Adj/Hyp
Tangent	= Opp/Adj
Cotangent	= Adj/Opp
Secant	= Hyp/Adj
Cosecant	= Hyp/Opp

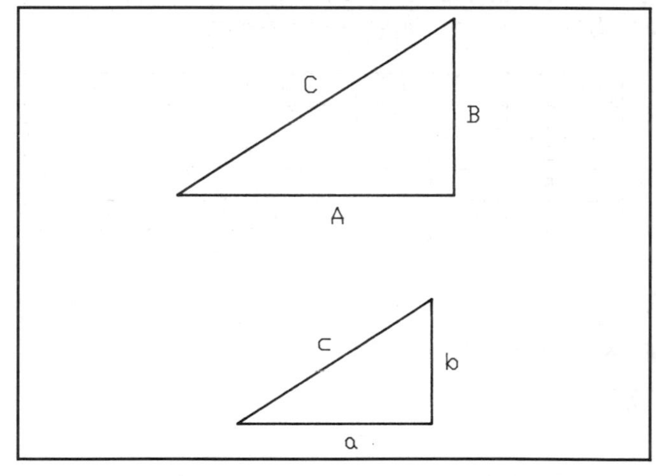

Fig. 15-10. In right triangles of the same shape but different sizes, corresponding ratios of corresponding sides are equal.

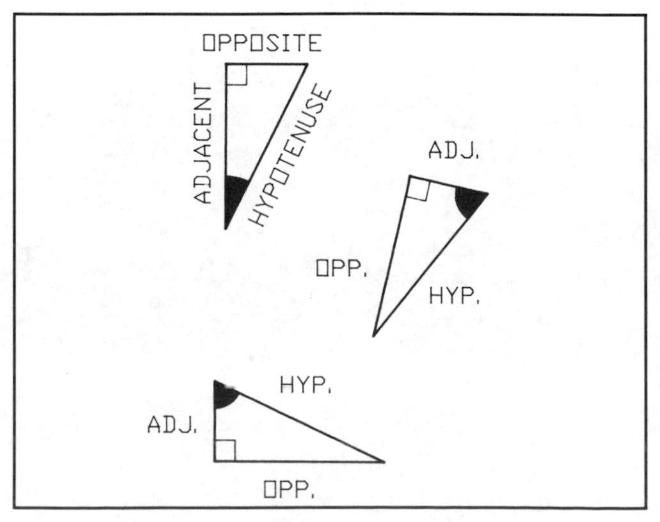

Fig. 15-11. The names of the sides in a right triangle. The sides named opposite and adjacent depend upon the given angle.

TRIG RATIOS

A trigonometry table contains decimal ratios that compare one side to another side of a right triangle based upon a particular given angle. To verify this example, sketch a right triangle with a hypotenuse of 3.00″ and a given angle of 30°. The sine ratio is equal to .50000. Look this fact up in a trig table or calculator. Sine 30 = .50000.

The sine ratio is a comparison of the opposite side to the hypotenuse. The opposite side compares to the hypotenuse as .5000 would compare to 1.000. We see that the opposite side is one-half the hypotenuse. In our triangle the hypotenuse is bigger than 1.00″. It is 3.00″. To find the opposite side, then, we multiply the hypotenuse by the ratio: 3.00 × .5000 = 1.5000. The solution for the opposite side is 1.5000″.

In this 30° triangle the sine ratio is .5000. The sine ratio would be different if the angle were different. Each possible angle has a sine ratio. The same is also true for the remaining five trigonometry ratios. We use these ratios to find sides and angles in triangles.

To use trig ratios, we need some given information.

1. We need to know that there is a 90° angle. This is a right triangle.
2. We need to know either the length of the two sides + 90° angle or one angle and one side + 90° angle.

These triangles are said to be either side side angle (SSA) or angle side angle (ASA).

Finding an Unknown Side

To find an unknown side, use the ASA given information. Multiply the indicated ratio for the given angle by the given side.

Finding the Indicated Ratio

Place the name of the side you wish to find over the name of the side you do know.

$$\frac{\text{unknown side}}{\text{known side}}$$

This will tell you the name of the ratio you need corresponding to the known angle. You must look up this ratio in a trig table or use a calculator. The ratio will correspond to a known angle. This ratio will tell you how the unknown side compares with the known side. It will be the decimal you receive from the calculator as compared to 1 but the 1 will not be stated. This would be the exact size of the unknown side if the known side was equal to 1, but it is usually not. So we multiply the ratio by the known side to find what the unknown side should be.

Refer to Fig. 15-12. In the triangle shown, we need to find the opposite side. We know ASA: a 90° angle, a 1.5″ side, and a 30° angle.

Procedure:

1. Name the sides of the triangle.
2. Select the indicated ratio (unknown side/known side).
3. Look up the ratio corresponding to the known angle.
4. Multiply the ratio by the known side.

The ratio we need is OPP/ADJ (the *TAN ratio*). Multiply the indicated ratio by the given side.

TAN 30 × 1.5

Calculator Entries	Display	
30	30	Must be decimal angles
TAN	.5773503	Tan ratio of 30°
×	.577350	Multiply
1.5	1.5	Known side
=	.8660	Calculated opposite side

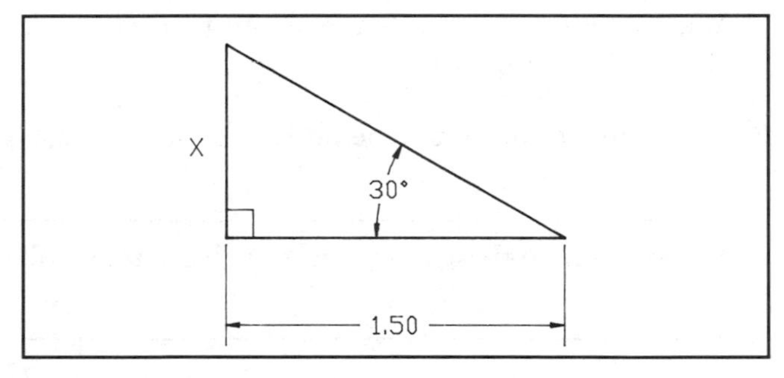

Fig. 15-12. To find the opposite side in this 30° triangle we use the tangent ratio.

ASA Practice

Procedure Review. To find an unknown side when you know another side and an angle in a right triangle, *multiply the indicated ratio by the known side.* The indicated ratio corresponds to the known angle. Thus, Ratio × Known side = Unknown side.

To find the correct indicated ratio place the *name* of the *unknown side* over the name of the *known side.*

$$\frac{\text{Unknown Side}}{\text{Known Side}}$$

Then follow this general procedure:

1. Name the sides of the triangle.
2. Select the indicated ratio (unknown side/known side).
3. Using a trig table, the ratio corresponding to the known angle.
4. Multiply the ratio by the known side.
5. Round your answers to five decimal places.

For use in a calculator, all angles must be decimal. You may have to convert if the drawing is dimensioned using degrees, minutes, and seconds for angles.

1. Solve for Y. Refer to Fig. 15-13. (Hint: $40° 30' = 40° + \frac{30°}{60} = 40.5°$.)

2. Solve for X. Refer to Fig. 15-14.

3. Solve for the Y coordinate in Fig. 15-15.

4. If we were entering this Y dimension in a curve program, would we enter it as positive or negative?

5. Solve for the X coordinate in Fig. 15-15 on page 175. (Hint: X will be the difference in the horizontal side of the triangle and the 4.0 radius.)

   You could also use the Pythagorean theorem on this problem because you have already solved for a second side of the triangle.

6. Would the X coordinate be positive or negative?

The answers to these questions are given in the following Reference Point.

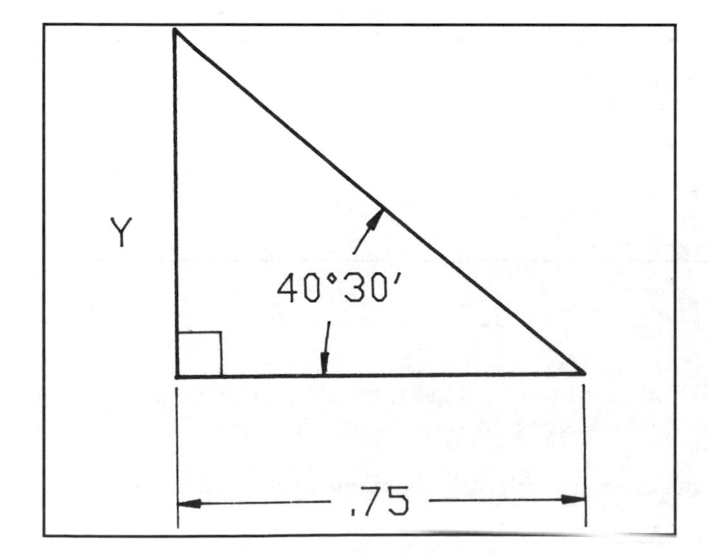

Fig. 15-13. Solve this triangle for the Y side.

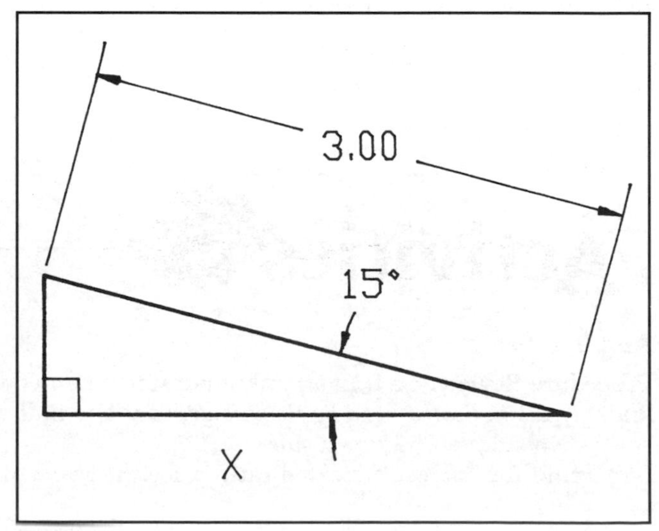

Fig. 15-14. Solve this triangle for the X dimension.

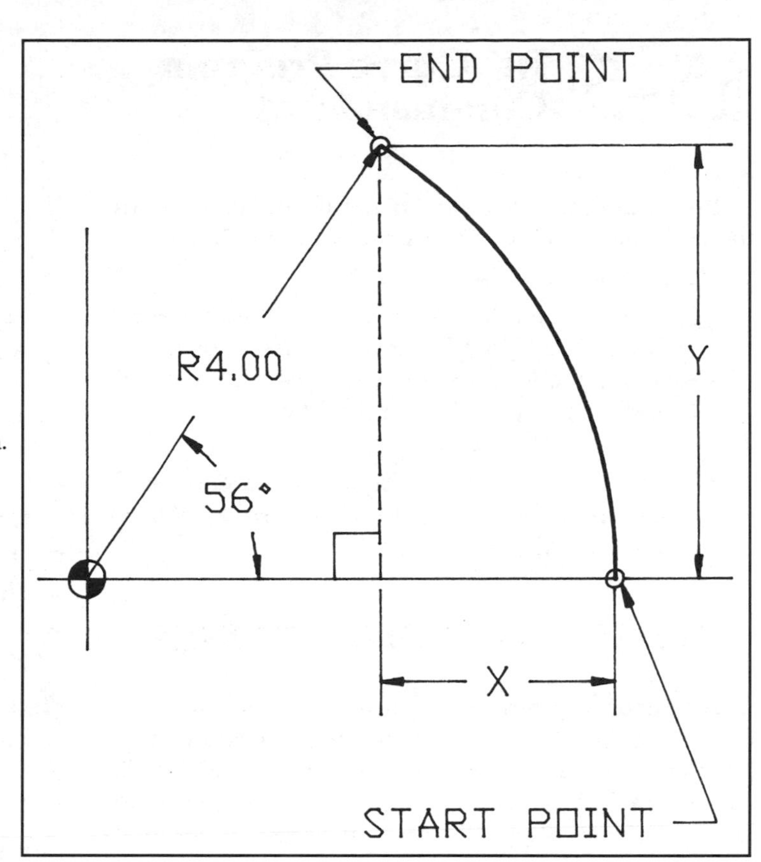

Fig. 15-15. Solve for the Y coordinate in this arc problem.

A sound knowledge of trigonometry is a prerequisite to studying programming. You should already know trigonometry. This section was only a brief review of trigonometry's important concepts. Here are the answers to the problems in the previous activities. Check your understanding.

Answer	*Explanation*
1. Side Y = .6406.	(Tan 40.5 × .75) .8540807 × .75 = .6406
2. Side X = 2.8978.	(CoSine 15 × 3.0) .9659258 × 3.0 = 2.8978
3. Y Coordinate = 3.3162.	(Sine 56 × 4.0) .8290376 × 4.0 = 3.3162
4. Positive.	
5. X Coordinate = 1.7632.	(CoSine 56 × 4.0) .5591929 × 4.0 = 2.2367716 = X′ 4.00 − 2.2367716 = 1.7632

First find X′ in the triangle, then subtract from 4.0 radius.

6. Negative.

If you had difficulty with these problems, ask for some reinforcement material from a shop math book. We will not cover solutions of ASA-type trig problems at this time because these problems rarely occur in IJK curves. ASA solutions will be reviewed when we study tool path compensations.

15-5 IJK Curve Program Commands

Program entries are similar for all IJK curves. Both the end point and center point need to be defined.

1. Define the end point.
   - Incremental mode. Define the end point in terms of X and Y or X and Z from the start point.
   - Absolute mode. Define the end point in terms of X and Y or X and Z from the PRZ.
2. Define the center point.
   - Define the center point in terms of I and J or I and K from the start point. (These are always incremental values.)

CROSSING QUADRANT LINES

No quadrant lines were crossed in any of the previous examples. Some machines will program arcs through a quadrant, and some will not. If yours will, you will be told so on Custom Sheet 13—Programming Curves in Your Shop. If your control uses more than one arc method, you will also receive Custom Sheet 13-A and Custom Sheet 13-B.

If your control will handle curves across a quadrant line, the program entries follow the two-step process outlined above. If not, you will program one block to the quadrant line and a second block to complete the arc.

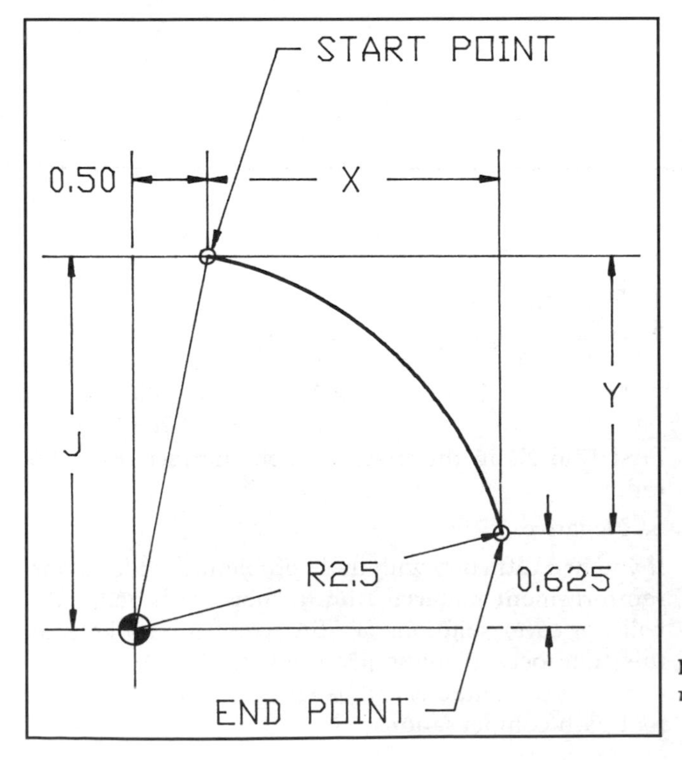

GENERAL APPROACH

We will continue to use incremental values for all coordinates. Every part print that requires a partial arc curve will be solved differently, but the same math skills are used each time. After setting up a sketch, find any coordinate: X,Y,I or J, whatever shows up first. Then find the remaining coordinates. Remember to:

1. Place the sign on negative direction coordinates.
2. Add the G02 or G03 for circular interpolation.

EXAMPLE

Refer to Fig. 15-16.

Find I

The first coordinate visible is I = *–.50*. (Because the center is left of the starting point, this coordinate is a negative value.)

Find J

Calculate the J coordinate using the I. J = *–2.4495*. (Use the Pythagorean theorem; remember the minus sign.)

Calculate the X coordinate

First find the horizontal distance in the .625 triangle:

X′ = 2.4206. Then subtract the I distance: 2.4206 –.5000.

X = *1.9206.*

Calculate the Y coordinate

Subtract .625 from J: 2.4495 – .625.

Y = *–1.8245.*

Note the minus direction.

The program command block would then be:

G02 X1.9206 Y–1.8245 I–.50 J2.4495

ORDER OF CALCULATIONS

The coordinates are not always calculated in the order used in the above example. But they will always be found using the Pythagorean theorem and trigonometry. Always start with the most obvious coordinate. Then build on that to solve for the remaining three. Once all four entries are known, write the program command block.

Fig. 15-16. Solve for the coordinates in any order. However, you need X, Y, I, and K to make a complete program entry.

Arc Programs Using the IJK Method

In the following problems, write a complete program command to define a partial arc using the correct G code and the X,Y and I,J coordinates. A very good suggestion is to draw an oversize sketch to some exact scale. After making the calculations, double-check by measuring the sketch. This is what programmers do on the print or on a sketch.

1. Refer to Fig. 15-17.
   Radius = .75
   X = .60

2. Refer to Fig. 15-18.
   Radius = 1.5
   Arc = +40°
   G03 curvature

3. Refer to Fig. 15-19.
   Radius = 2.5
   Y′ = 2.0

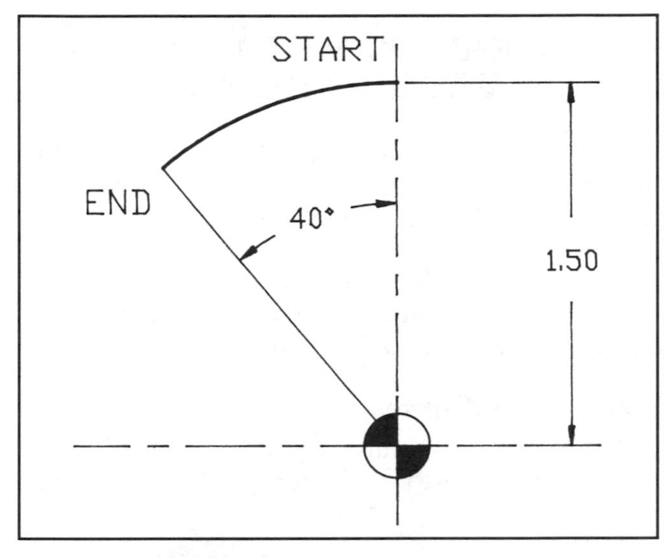

Fig. 15-18. Write a complete I, J, K program command for this arc.

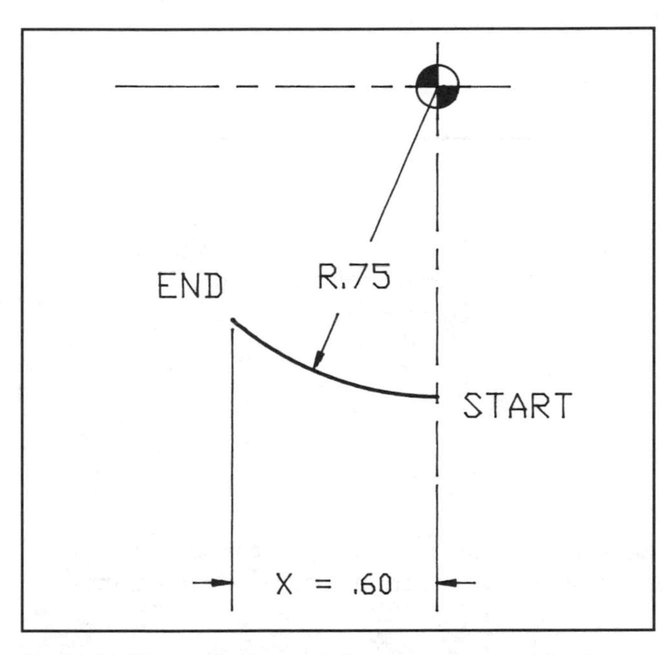

Fig. 15-17. Write a complete coded program command for this arc using the I, J, K method.

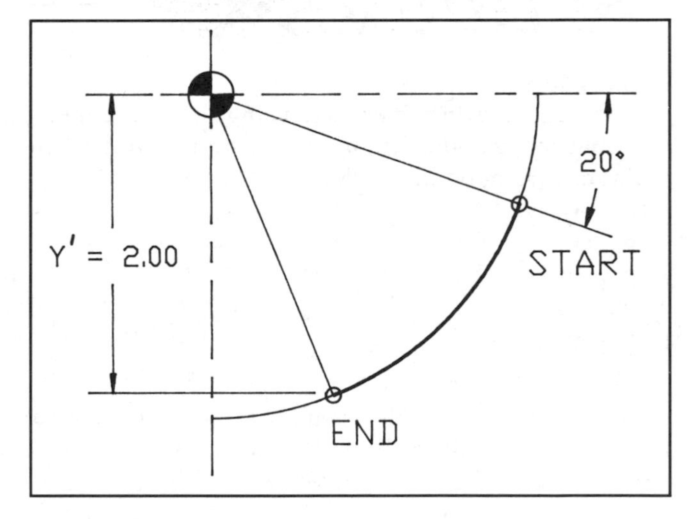

Fig. 15-19. Write a program command for this arc.

Here are the program commands for the arcs required.

1. G02 X–.600 Y.30 I0.0 J0.75
2. G03 X–.9642 Y–.3509 I 0.0 J–1.5
3. G02 X–.8492 Y–1.1449 I–2.3492 J.8551

15-6 Programming Arcs Using the Radius Method

This is method two of programming arcs. In this method, you are given the end point, start point, and curvature code.

RADIUS METHOD PROGRAMMING

This curve format is often called a *radius method* because the radius length is supplied as one of the entry parameters. If the start point is automatically known and the end point is given, the control can calculate the center point. The curvature entry eliminates the problem of determining the center point and the direction to proceed around the curve.

Fixed Cycle Curves

A version of the radius method is also called a *fixed cycle method* because it allows for specific arc programming with little or no math. Arcs such as 90°, 180°, and 360° are simplified with this method.

In a coded program format, instead of the I,J or I,K, the center coordinates are replaced with the radius entry. The letter used is **R.** The R word signifies the radius of the curve.

Example 1. Refer to Fig. 15-2.

We return to Fig. 15-2 to compare this second method.

G03 X–3.0 Y3.0 R3.0

Explanation

G03	Counter-clockwise curvature—circular interpolation.
X-Y	Identify the end point relative to start point.
R	Identify the radius.

Example 2. Refer to Fig. 15-16.

Compare the IJK curve program developed for Fig. 15-16 with the radius method entry below.

G02 X1.9206 Y–1.8245 R2.5

Explanation

G02	Clockwise interpolation, curvature.
X-Y	Identify end point.
R	Identify radius.

Radius Method

If the curve radius is known, using the radius method will save some calculations. In the following activity, note that the actual calculations are similar to those needed for IJK curves. Write radius method C/NC program command blocks for:

1. Fig. 15-17
   Radius = .75
2. Fig. 15-18
   Radius = 1.5
3. Fig. 15-19
   Radius 2.5

Note that many of the calculations were the same and that you need the same skills. For curves that start and end on a quadrant line, this method is very easy. Using the radius method from quadrant to quadrant is sometimes called a "cycle curve format."

4. Fig. 15-20
   Radius = 3.7
   G03 Curvature
5. Fig. 15-20
   Radius = 3.7
   G02 Curvature
6. A) How many degrees would the following command sweep? G02 X–1.5 Y–1.5 R1.5

   ____

   B) How many degrees will this command block sweep? G03 X–1.5 Y–1.5 R1.5

   ____

7. How many degrees will this command sweep? G03 X–1.2375 Y–.5125 R1.75

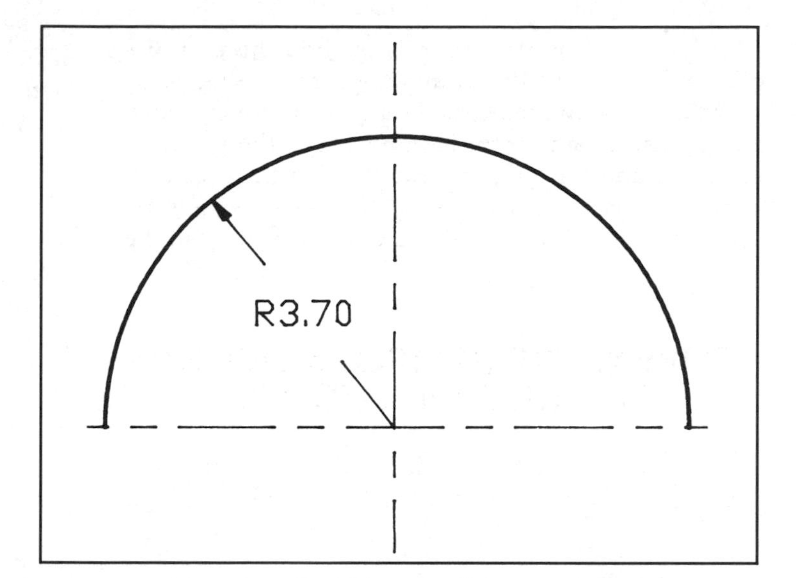

Fig. 15-20. Write a radius method program command for this arc.

Here are the correct program entry commands for a coded machine:

1. Fig. 15-17. G02 X–.6 Y.3 R.75
2. Fig. 15-18. G03 X–.9642 Y–.3509 R1.5
3. Fig. 15-19. G02 X–.8492 Y–1.1449 R2.5
4. Fig. 15-20. G03 X–7.4 Y 0.0 R3.7 (Y 0.0 is a null—left in)
5. Fig. 15-20. G02 X+7.4 Y 0.0 R3.7
   Notice that the only difference was that the curvature was opposite that caused by the G02. If you saved your calculations for the previous activity, you could simply fill in the programs above. Always save notes and calculations that support programs.
6. A = 90° B = 270°
7. 45°. To solve this you would have drawn a triangle with a hypotenuse of 1.75″. The opposite side would be 1.2375″. Using the sine ratio (1.2375/1.75), you would find that the angle would be 45°.

   A second way to solve this problem is to take the Y factor from the vertical radius. Notice that it, too, is equal to 1.2375″.

The opposite side equals the adjacent side. This can be true only for a 45° triangle. See the following sketch.

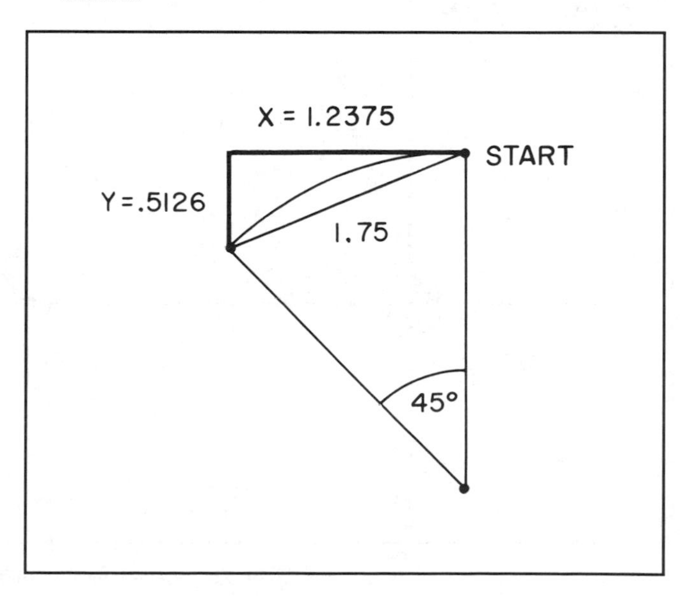

15-7 Polar Curve Programming

Polar curve programming is method three of programming curves. Programming curves using polar coordinates is simpler. Usually, a curve is dimensioned on the print, using information such as the radius of the curve and the length of the arc. The polar control is able to generate a curved path knowing the start point, center point, and the number of degrees in the arc.

CURVATURE SYMBOL CHOSEN BY "RULE OF THUMB"

The symbol used to indicate clockwise or counterclockwise curvature, whether it is a code or a sign value, is chosen according to the "rule of thumb." The G02 or G03 code or a plus + or minus – symbol indicates the curvature.

At this time, there is a great deal of divergence in the form of a polar arc entry from one control to the next. Here is an example of arc callouts using the polar method. If your controller uses the polar arc method, you will find further examples on Custom Sheet 13.

POLAR ARC EXAMPLE

Refer to Fig. 15-21. This program will plot an outline for the shape shown in Fig 15-21. The PRZ is lower left. Departure is PRZ and 1.0″ above the paper. Refer to Fig. 15-22 on page 181.

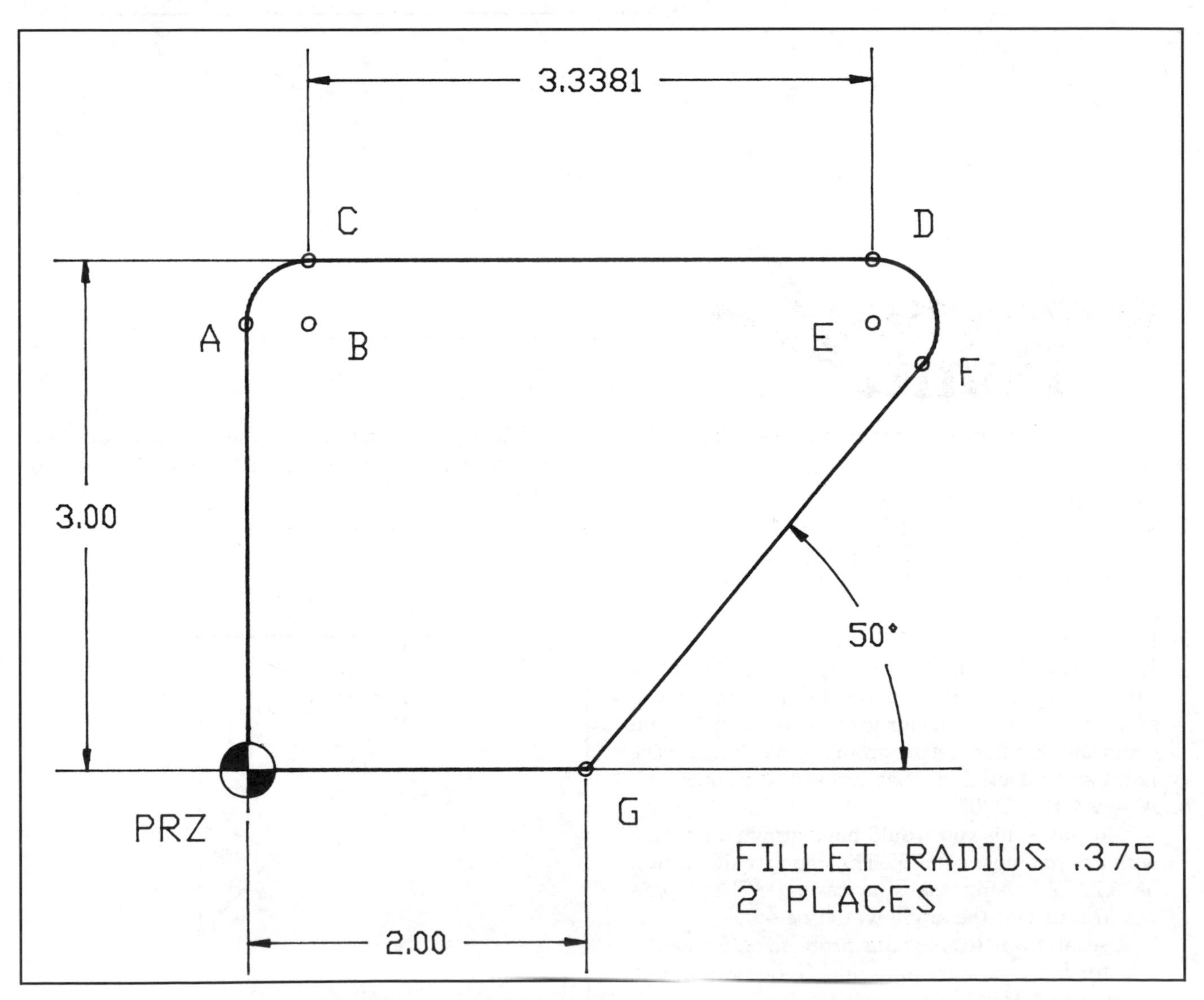

Fig. 15-21. This shape is easily programmed using a polar method arc.

Dyna 2400 Control

SEQ	ENTRY		EXPLANATION
000	REM: PLOT OF 15-20		
001	START INS 85		START IN INCHES—FILED NO 85
002	TD = 0.0		TOOL DIAMETER = 0.0
003	FR XY = 25.		FEED RATE X and Y AXIS = 25
004	FR Z = 2.0		
005	GOfZ –.95		GO ABSOLUTE TO GAUGE HEIGHT—RAPID
006	GO Z –1.0		TOUCH
007	GO Y 2.625		TO POINT A
008	ZERO AT	I	CENTER IDENTIFICATION
009	X .375	I	IDENTIFY NEW LZ AT PT B
010	Y 2.625	I	SHIFT REF FROM PRZ TO LZ
011	GR a –90		RELATIVE ANG—90 DEG (CCW), PT C POLAR CURVE COMMAND
012	ZERO AT		
013	X 3.3381		SHIFT LZ TO POINT E
014	GR X 3.3381		RELATIVE MOVE TO POINT D
015	GR a –130		RELATIVE ARC TO POINT F (90 + 40 = 130) POLAR CURVE COMMAND
016	REF COORDS		CANCEL LZ—REFER TO PRZ
017	GO X 2.0	I	
018	X 0.0	I	LINEAR INTERPOLATE TO POINT G
019	GO X 0.0		
020	GO Z 0.0		
021	END		END PROGRAM

Fig. 15-22. Here is a program to define Fig. 15-21 using the polar arc method.

Write a Program Using Curves

1. Using one of the methods outlined on Custom Sheet 13, which deals with the control in your shop, write a program for either Appendix A Drawing 1 or Appendix A Drawing 3.

   Appendix A Drawing 1, The Ball Peen Hammer Head

   A) Write program including spindle and feed commands.

      Program profile from 1.00″ diameter head.

      Chuck on excess to left of head.

      Center support is not required.

      Do not part off in program.

   B) Plot program on graphics or plotter.

   C) Test program on test or real material.

   Appendix A Drawing 3, The Drill Gauge

   A) Write a plot program for the perimeter only.

   B) Optionally, include a plot of the holes.

   C) Plot program on graphics, or on a plotter (if available).

   D) Plot program on machine.

2. Write a plot program for the shape shown in Fig. 15-21. If you have a Dyna or Heidenhain control, write the program progressing in the opposite direction of the example.

A SUMMARY OF ARC PROGRAMMING METHODS

The table below compares the three arc methods. The information you supply in the command blocks is shown first. Program commands that are automatically understood from the given commands are also shown. Notice that all four parameters are known to the control in each method.

Arc Programming Methods

	IJK Method	Radius Method	Polar Method
Program Commands	End Point Center Point	End Point Arc Radius	Center Point Arc Degrees
Automatic Commands	Start Point Arc Radius	Start Point Center Point	Start Point Arc Radius

Each method requires a curvature indicator to start circular interpolation and tell whether the curve is clockwise or counterclockwise. This will be either G02 or G03 or a plus or minus sign.

Notice that in all three methods the start point was automatically known. It is always the present location of the tool.

CHAPTER 16

PROGRAM FLOW—BRANCHING LOGIC

Program flow deals with the arrangement of the program—the logic path that the control follows as it progresses through the program. Programs need not go directly from the beginning command directly to the end. This is called *linear logic.* Within a CNC program, you may use branching logic and logic statements that shorten and simplify the program.

Using logic is sometimes the only way a geometry can be generated. Once you understand the possibilities, there are literally hundreds of different ways to manipulate a program to create the same geometry on the machine. This creative process of arranging logic statements is called *syntax.* The more you know about programming logic, the more combinations of syntax you will see.

We will learn that there are different kinds of program flow statements—*loops* and *subroutines.* We will learn that both loops and subroutines can be contained within other logic statements—called *nesting logic.* We will see how a main program can have sub-programs nested within it, and these subprograms could in turn, have still others and so on.

Finally, we will investigate a programming labor saver called *fixed cycles,* or *canned cycles.* This cycle is built into the control to simplify complicated groups of program commands. Fixed cycles in themselves are not part of the syntax, but they can be incorporated into loops and subroutines.

OBJECTIVES

After studying this chapter, you will be able to:

- Write a program containing subroutines.
- Write fixed cycles in a program.
- Use smart and dumb loops in a program.
- Draw a logic diagram of program syntax.
- Understand computer ram data management.

16-1 Subroutine Logic

SUBROUTINES ARE "MINI-PROGRAMS"

The easiest logic statement to understand is the *subroutine,* a mini-program sometimes called a subprogram. Subroutines are called up within the body of the main program, but are actually a separate group of commands that are usually outside the main program.

Subroutine Example

Refer to Fig. 16-1 on page 184. One example of the application of a subroutine would be drilling twenty-four deep holes in the part shown. We know that drills must be withdrawn from the hole to clear chips. These in-out steps are called *drill pecks.* To save time, the drill need not peck until it has drilled a certain depth. This custom routine of up-and-down action would require several lines. You would not want to rewrite it in the program twenty four times.

The group of commands repeated at each hole are written once and called a subroutine. The custom peck-drill subroutine is called out by the main program each time it is needed. After locating the tool each time, the drilling subroutine is called up by the main program.

Keep this example in mind, we will use it again to illustrate further concepts in logic and syntax.

CNC Logic Diagrams

When we study program flow, we study the path that the program takes. The logic statements are arranged to cause the flow to branch to the subroutine then back to the main program. Because this can be complicated to envision, we will use a line drawing to represent program syntax.

Refer to Fig. 16-2 on page 184. A logic diagram is used by computer programmers to make sure that their syntax makes sense. A logic diagram uses lines to follow the program flow and word statements to denote the actions taken at a particular branch of the logic. Figure 16-2 shows two holes being drilled using the drill subroutine. The twenty four hole example would be repeated twenty two more times in exactly the same form.

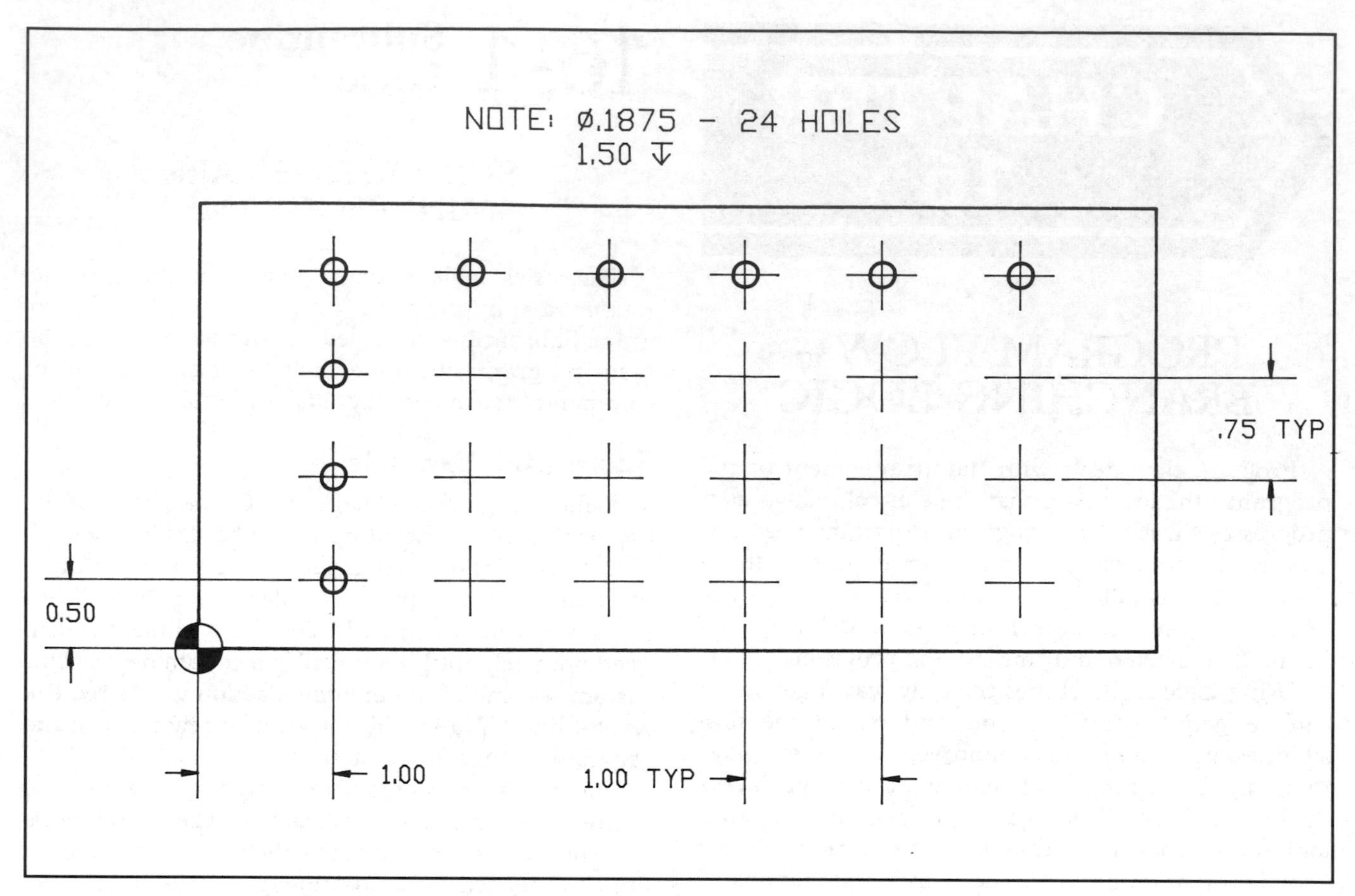

Fig. 16-1. Programming the drilling of this diffuser plate is simplified using programming logic.

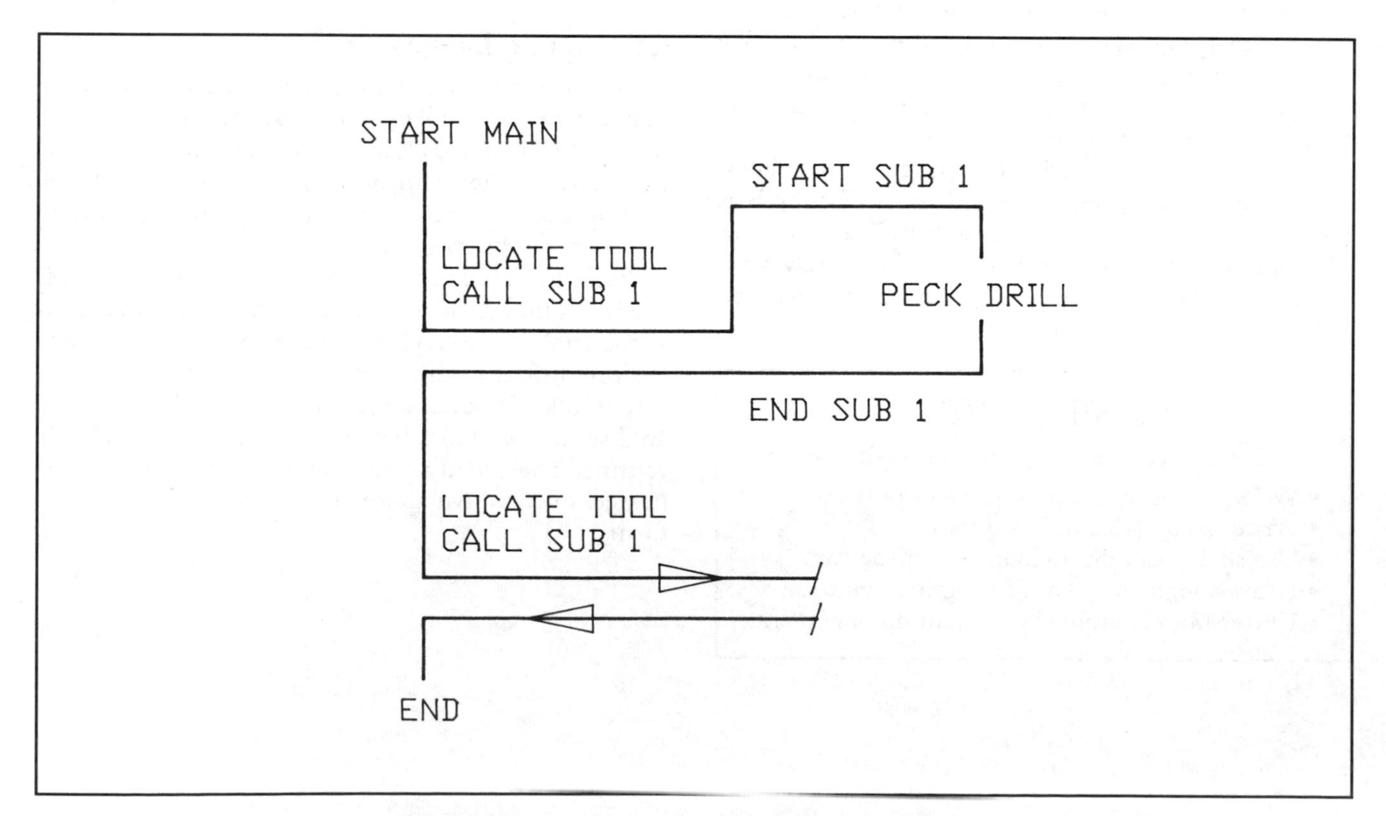

Fig. 16-2. A logic diagram is used to check your program flow.

How Programs Will Be Presented in Chapter 16

The use and arrangement of the logic statements, the syntax, is the same from control to control. However, different controls use different vocabulary words for the logic commands. Thus, all program examples in Chapter 16 will be developed first on the logic diagram then in a conversational program using the same logic statements. You will write a subroutine program in the language of your lab.

You will need to verify the actual program words (coded or conversational) used on your own control to enter the actual statements into the program. This information is on Custom Sheet 14.

You will need to learn three new program words or codes: call subroutine (call sub), subroutine start (sub start), and subroutine end (sub end). Call sub is in the main program. Sub start and sub end are at the beginning and end of the subroutine. Coded controls will have alphanumeric command words that accomplish the above three words.

Using Custom Sheet 14, fill in the blank to indicate the word or code used in your lab.

1. *What code is located in the main program and calls up a particular subroutine?* On coded controls, there may be more than a single word required. One word may be used for the call statement, and another to tell the control the number of the subroutine. List the call subroutine word or codes.

   ____

2. *What code is at the beginning of the subroutine?* The start sub word is used as a file number for the control. When call sub 100 is read by the control, it looks for the subroutine that starts with start sub 100. Some controllers do not require a start sub word because they look for the line number upon which the subroutine is stored rather than at the file number of the subroutine itself. In this type of control, the call sub 100 would cause the control to go to line number 100, execute the program commands found there, and continue on until sub end was read. Does your control require a start sub word?

   ____

   If yes, fill in below. Start sub word or codes:

   ____

3. *What code is at the end of the subroutine?* This word must be in all programs on all controls. Upon reading this word, the control branches back to the main program and executes the command following the call sub word. List the sub end word or code.

   ____

Logic Diagram Example

Refer to Fig. 16-2. By looking at the diagram, you can easily see why this is called "branching logic." The main program begins with a *start main* statement. That symbolizes the beginning of the program. The main program locates the tool and then calls sub 1. The program flow branches away from the main program at the call statement to the start subroutine, where it performs the peck drill program. This subroutine is ended with a sub-end word or code. Then the program flow branches back to the main program—then executes one block beyond the sub-start word. The main program then relocates the tool and calls the sub again. This illustration shows only two runs of the subroutine. Notice that it was not necessary to show the complete branch each time. A lead-out and a lead-in line are sufficient.

This, then, is the form in which we will study program flow, logic diagrams, and conversational statements. Actual program examples in your lab will be on Custom Sheet 14.

16-2 General Characteristics of Subroutines

START AND END STATEMENTS

Start Statement

The type of start statement depends on RAM management. When the Call Subroutine word appears in a program, the control must have a way of knowing where the subroutine lies in RAM. It must know which line in RAM is the first command of the subroutine. This is handled in two different ways. The difference depends on how the MCU of the control deals with RAM. There are two ways: line dependent and line independent.

Read the next section on RAM management. Then answer these questions.

1. Refer to Custom Sheet 14. Is the control you will be programming line dependent or line independent?

   ____

2. Does the control require a start sub word with a file number at the beginning of the subroutine?

   ____

3. Does the control search for a particular line number or a file number?

   ____

You will be more skilled in using syntax in programming a particular control if you understand how the control finds a particular subroutine.

Line-Dependent RAM Data Management

Line numbers are the subroutine address. In line dependent controllers, the line numbers for each line of the RAM serve as file numbers for locating subroutines. Stored subroutines are found by their storage number. Their line number in RAM is their actual address for location by the control.

Consider this example. Assume that you write a subroutine and name it SUB 100. The subroutine must then be stored so that it starts at line number 100 of the controller. The call statement is CALL SUB 100. When the control sees the statement CALL SUB 100, it will go to line number 100 of the RAM and execute the command it finds there. It will continue to follow that group of commands until a SUB END statement is read. Then it will return to the main program and execute the next command statement. The first subroutine command would be at line 100.

No start statement is required. A subroutine in a line-dependent controller does not need a particular statement at the beginning other than the line number on which it is stored. The control looks for the line on which the subroutine is stored. It does not look for an identifying number within the subroutine.

The advantage to this type of data management is that the subroutines may be stored anywhere in the RAM. They may be stored before or after the main program or even inside the main program. The call word tells the control where to look for the subroutine. The control will look forward or backward in the RAM to find a subroutine. It is good planning to store the subroutines beyond the main program by 100 blocks or so. This will leave room for changes. With this type of data management, however, a subroutine can actually be placed inside the main program.

There is a disadvantage to line-dependent controls. During editing, when blocks are added and deleted from the main program and subroutines, the line numbers change. This changes the location where the subroutine is stored. Some line-dependent controllers attempt to track these line number changes and automatically renumber the call statements as lines are added and deleted from the RAM. This usually does not work very well. Due to editing, the program will include odd statements, such as CALL SUB 37 or 198 because of added or deleted blocks.

Line-Independent RAM Data Management

File numbers are used in this type of data management. The MCU finds the subroutine by the file number in the first line of information actually stored *in* the subroutine. A file number must be in the first line of the subroutine. For example, a program calling sub 100 would search through the RAM for the statement sub 100. This sub 100 statement could be anywhere in the RAM *beyond* the main program or beyond the CALL statement. It is standard procedure to store subroutines many lines beyond the main program to allow for editing.

The first line of a subroutine must contain the file number for the RAM to find. Actual machine commands may or may not be on the first line of the subroutine, depending on the control's capacity per line. However, the first piece of information must be a file word.

There are advantages and disadvantages in using line-independent RAMs. Editing does not affect line numbers beyond the main program in this type of data management. The advantage is that no special planning is required to place the subroutines. Just write the program. Then write the subroutines and store them wherever they will fit into the RAM beyond the main program.

There is a dangerous disadvantage. A problem arises when an unerased subroutine in the control matches a call statement in your new program. After a call sub 100 word, the RAM moves forward searching for sub 100. The *first* sub 100 it comes to will be the one it executes. This could lead to a big surprise if the old sub was stored before the new one!

End Subroutine Statements

All subroutines must have an end sub statement. Without the end sub word or code, the controller would not know when to branch back to the main program. There is nothing special about the end sub statement. Each control uses a different statement word or code. They all tell the control that the subroutine is completed and to branch back and continue on with the main program.

One real danger in forgetting to end a subroutine is that the control will continue trying to execute the next lines in RAM. It will skip through blank lines and may find stored information beyond the unended subroutine. It will execute whatever it finds. This can cause the program to execute unwanted commands. Interactive controls will check for this type of syntax error and inform you of logic errors. Passive controls will not.

Activities

Write a Program Containing a Subroutine

1. In the language of your control, write a CNC drill press program that would drill the diffuser plate shown in Fig. 16-1. Here are the planning parameters for the program.
   A) After drilling .75″ deep, the drill will execute six pecks out of the hole in rapid. Then it rapids back in to within .05″ of the drilled depth. There it will feed to the next peck depth. Each peck is .125″ deeper.
   B) Use correct speeds and feeds for 7075 aluminum work material.
   C) The PRZ is the lower left corner, with Z0.0 = 1.0″ from the material.
   D) The PRZ is the departure point.
   E) The gauge height for drill movement is .1″ above the work.
   F) Be sure to include "spindle on" in the program.
   G) Use incremental values for the subroutine. Note, you will use this subroutine again. It will be ready to use again if the commands are incremental.
2. Diagram the program showing the branches the program takes. You do not have to show all twenty four branches. Two at the start and two at the end are sufficient.

16-3 Standard and Parametric Subroutines

There are two kinds of subroutines: standard and parametric.

STANDARD SUBROUTINES

A standard subroutine (sub) is a small program. It has the abilities and limitations of the main program. There can be more than one subroutine in a program and the commands within are the similar to those in any program. A subroutine may be of any length. The format is the same as that of the parent program except the Start and End words. Standard subroutines are written to fit a particular program. Some controls have an upper limit to the number of subroutines allowed in a program.

The subroutine is not stored in a special part of RAM. However, the best practice is to space them beyond the main program end by at least 100 command block numbers. The subroutines are usually downloaded or MDI entered when the main program is entered into RAM.

Some machine shops leave certain standard subroutines in the RAM and never erase them. These subs are frequently used universal operations, such as a center drill and drill routine on a lathe. These can also be stored as a library of subroutines in the permanent storage of the control if it has that ability. They can then be called up to active RAM when the main program is dumped into the control. If variables such as depth of hole and number of pecks occur, then variable parameters are written into the subroutine. The subroutine then becomes a type of parametric subroutine.

A second standard practice is to enter blank spaces in the subroutine. These are to be used for editing, especially on line-dependent RAM controls. These blank lines can also be left in the main program for future editing.

PARAMETRIC SUBROUTINES

There are two uses for parametric subroutines. In each case, the subroutine must cover a broader application than standard subroutines.

Parametric Subroutines Use Variable Parameters

Some or all of the parameters in the subroutine are variables. These variables are filled in. The value is set in the main program as the subroutine is called for use. The variables can be mathematical formulas containing complicated operations. The variables can be conditional statements such as: if A is less than or equal to B, then... The two general applications of parametric subroutines are:

1. *A single subroutine will be used in several programs.* In each program, it must accomplish nearly the same job. The difference is only in the parameters of the commands. These parameters include diameter of cut, depth or length, and how many times an operation is to occur. These are left as variables.
2. *The geometry requires constantly changing parameters but basically the same operation.* An example would be the machining of a spiral scroll plate for a lathe chuck. A scroll plate is the flat grooved plate that opens and closes the jaws when the chuck wrench is turned. This could be machined on a vertical milling machine starting from the center and machining a constantly expanding spiral curve. The cutter would be moving in a circle, but with every increment of arc the radial distance would increase. This could be handled using parametric subroutines. For example, the variables in the parametric subroutine could be set up to increase the radius for each degree of rotation. The radius amount could be incremented for each degree of the spiral cut. Then, given a different spiral for the scroll, the parameters could be re-fixed in the main program. A new program would not be necessary.

16-4 Looping Logic in Programs

GENERAL CHARACTERISTICS

Loops are a second logic manipulation performed on programs. Loops enable the programmer to repeat a group of commands a number of times. For example, use of a loop could shorten the diffuser plate program just written.

In Fig. 16-1, with each horizontal (X) axis move from one hole to the next, the distance was the same. A loop could be programmed using *incremental* values for the locating statements. The loop could call the subroutine. You will write that program after some more syntax skills are shown.

Loops may be in the main program or in subroutines or both. The commands inside a loop are no different than regular program commands. The control may have an upper limit to the number of individual loops permissible in a program. There may also be a limit on the number of times a loop may repeat itself. Where this limitation occurs, we use a technique called factoring, which will be covered later.

There are two categories of loops: *dumb loops* and *smart loops (do over loops).*

Dumb Loops

A dumb loop has little application in programs because, once set in motion, it never stops. The program never reaches the end. One example of a dumb loop is a program used to "exercise" a machine after a long shutdown period. The full range of each axis travel is programmed. Then each is told to go back to the Machine Home. If this is looped, the machine runs each axis through the full range of motion until someone stops the machine manually. Figure 16-3 shows the flow of this exercise loop.

A dumb loop depends upon a programming technique called a jump or skip or goto. A *jump* is an unconditional move to another part of the program. The control simply shifts to another line number and continues to execute commands. There are no conditions on the jump. All the control does is change program line numbers.

The GOTO command is used in computer language as well as in CNC. Those familiar with programming computers will find the logic manipulations in CNC familiar.

If the jump is forward in the program, it is a *bridge*. A bridge allows the programmer to skip commands in the program. If the jump is backwards into the program, it is called a *dumb loop*. Figure 16-4 shows a logic diagram of an unending drill loop. Can you see that the program flow would be endless. It will never reach the end, but what about the halt?

Refer again to Fig. 16-4. The halt word would cause the control to wait until the next part was loaded and the operator touched the start button once again. In coded language, the halt word is M00. NC tapes were literally glued together so that there was no end to the tape when this type of operation was needed. The tape formed an endless "loop."

Smart Loops

The smart loop is commonly called a *do-over loop*, which is shortened to do-loop. It is used to repeat a group of commands a specified number of times. Each time the loop repeats itself, the control counts down until the specific number is reached. When the loop count is satisfied, the control moves to the next block beyond the end of the loop. Figure 16-5 shows a simple do-loop. Every command between lines 100 and 125 will be repeated five times. Once the control cycles through the loop for the fifth time, it moves to line 126 and continues forward in the program.

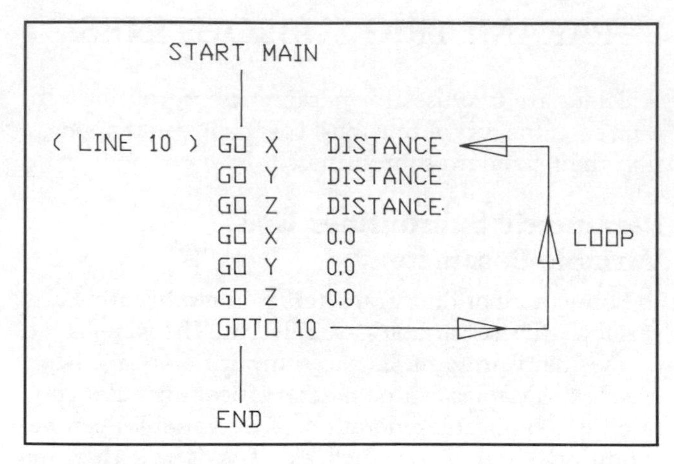

Fig. 16-3. A dumb loop turns back into the program and continues indefinitely.

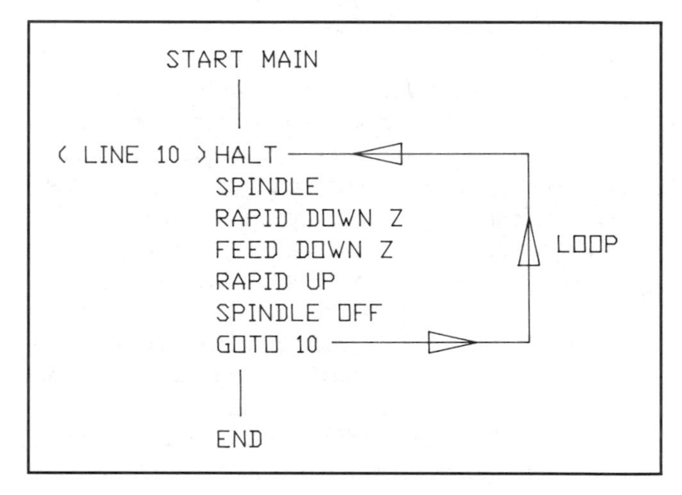

Fig. 16-4. A typical never-ending, drill dumb loop.

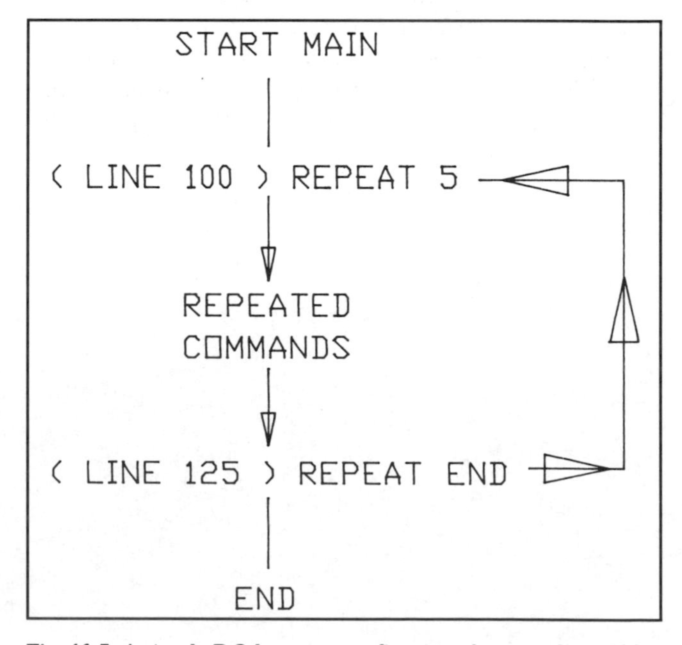

Fig 16-5. A simple DO loop repeats five times between lines 100 to 125.

Code Words for Loops

Evidence of the age of the EIA coded language is found in that there is no command assigned for looping. An NC machine could not perform a do-loop—the tape would have to run backwards through the reader. CNC controls use one of the unassigned EIA command codes for looping. Thus, you may find differences in the codes used from control to control. As an example, general numeric controls use G51 as the assigned word for the do-loop.

Repeat word. A command block to start a loop would look like:

N100 G51 N5

This block would start the loop at block 100 and repeat the loop five times. When the loop end statement was read, the program flow would return to line number 100, four times after the original loop cycle.

Repeat end word. The command block to end the loop would look like:

N150 G50

This command sends the program flow back to the beginning of the loop until the count is satisfied. Then the control would move the program to block number 151. The commands between blocks 100 and 150 would be repeated five times.

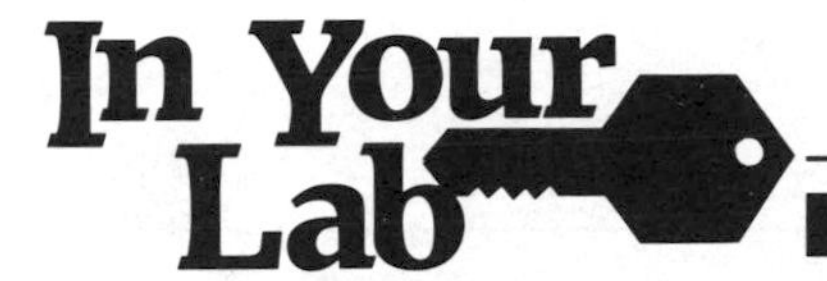

The logic of looping will be the same on every control, coded or conversational. There are three new command words you will need to know for your control. These are given on Custom Sheet 14. Fill in the blanks below.

1. The GOTO command is a word or a code that causes an unconditional jump from one part of the program to another. If a code, it may simply be a letter such as P or N followed by the line to which the jump is to be made.

   The GOTO command word: ____________________

2. The REPEAT command causes a do-loop. It does two things in the program. First, it signals the number of times a command is to be repeated. Secondly, the repeat command is the location to which the program flow returns to start again each time the loop repeats. On coded controls this may require two separate alphanumeric words. One word would be needed for loop start. The second word would designate the number of times to repeat the loop.

   The REPEAT word: ____________________

3. The REPEAT END command ends the loop and signals the control to return to the beginning of the loop.

   The REPEAT END word: ____________________

16-5 Nesting Logic

For the drilling example shown in Fig. 16-1, it was necessary to write twenty four tool-locating statements. That can be avoided using a loop. A loop can be written that will move the tool and call out the drill subroutine. On the drawing, we see that the X axis distance between holes is 1.0″ and the Y distance is .75″. Both the X and Y distances repeat. We can take advantage of that using loops.

LOGIC COMMANDS WITHIN LOGIC COMMANDS

You may use any combination of logic that solves a problem. Loops may be within loops. Subroutines may be inside other subroutines. Loops may be in subroutines and subroutines may be in loops. This is called *nesting*. Figure 16-6 shows a nested subroutine inside a loop. In this example, the tool is moving a repeating distance, then drilling after each move. Nesting can solve many problems for a programmer. In the diffuser plate program, we can write several different programs that would take advantage of the nesting technique. Let's look at three combinations based upon Fig. 16-1. For each, we will call the drill subroutine SUB 1. A logic diagram with the tool path drawing will be shown.

Solution 1. Right Loop—Left Loop

Refer to Fig. 16-7 on page 193. In this solution, we write two loops that each nest the drill subroutine. One loop repeats from left to right along the repeating X axis distance of 1.00″, while the second loop repeats from right to left. Notice that we had to locate the tool in the Y axis before entering the loop and that all values are incremental. Some improvements are possible on this program. Notice that all the commands from line 010 to line 100 are duplicated a second time between lines 110 through 200. Let's see how we can creatively save some work.

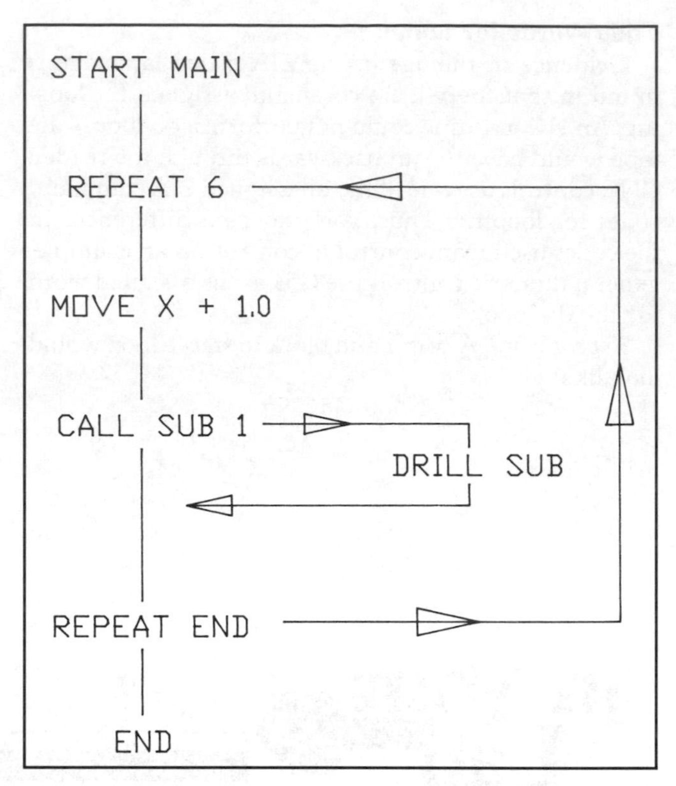

Fig. 16-6. A logic diagram of a loop nesting a subroutine.

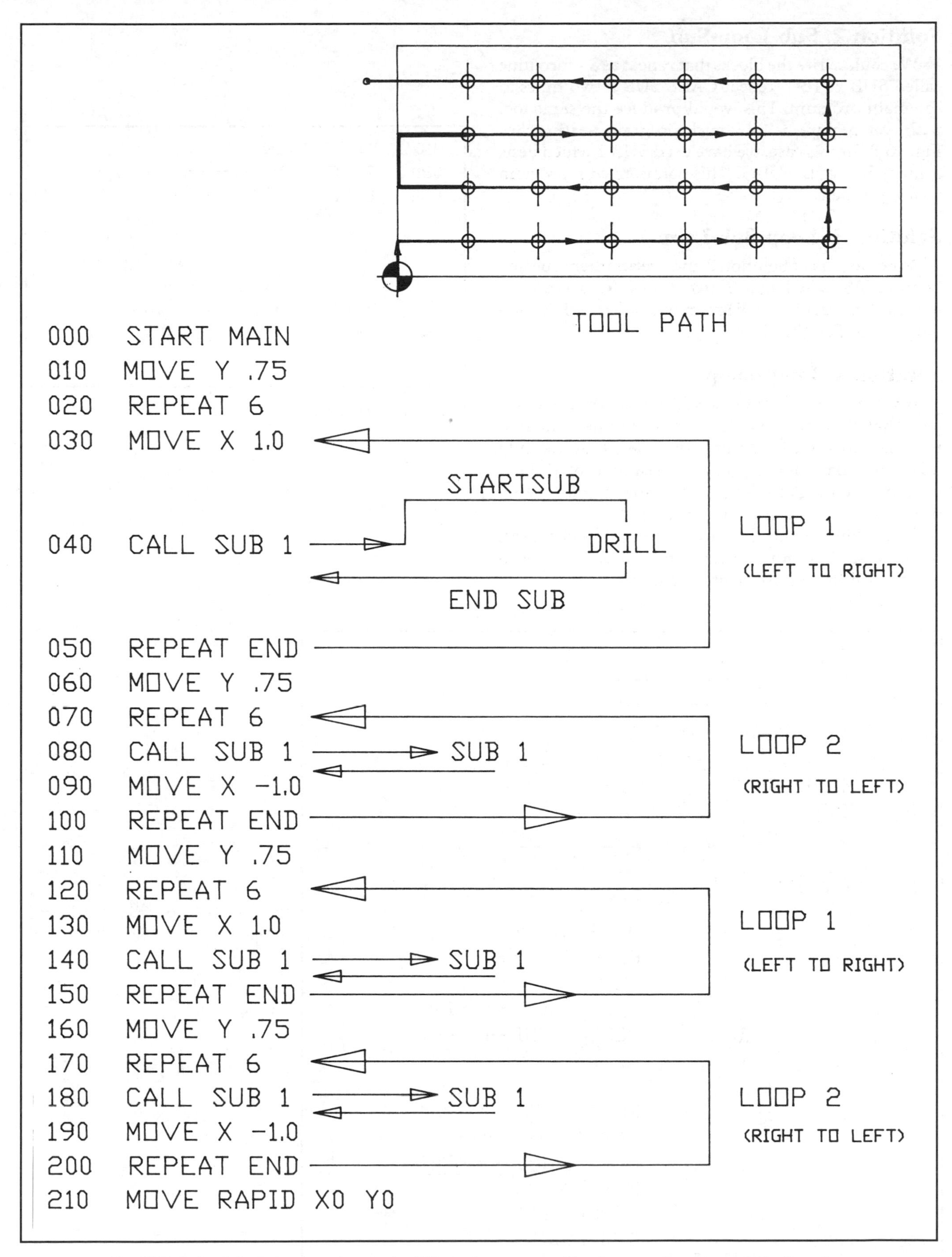

Fig. 16-7. A simple logic improvement to the diffuser plate program. Two loops, each nesting a drill subroutine.

Solution 2. Sub-Loop-Sub

We could write the blocks that repeat as a subroutine called SUB 2. Then repeat CALL SUB 2 two times in the main program. This would produce the same tool path, yet require half as much program writing. See Fig. 16-8. In this case, we have used SUB 2 which nests a loop that calls SUB 1. This solution cuts program writing in half.

Solution 3. Loop-Sub-Loop

Now we look at Solution 2 and see another repeated feature. We called Sub 2 two times. This could be looped. For clarity, the diagram will show SUB 2 without details. See Fig. 16-9.

Solution 4. Loop-Loop

Another simplification would be to write a single loop that steps twelve holes. This would be done first from left to right by drilling six holes, then locating up in the Y axis, then stepping from right to left drilling six more holes. This twelve-hole pattern could be repeated two times using another loop.

If you think about these examples, you will see still other possibilities for nesting loops and subroutines to accomplish the diffuser plate program.

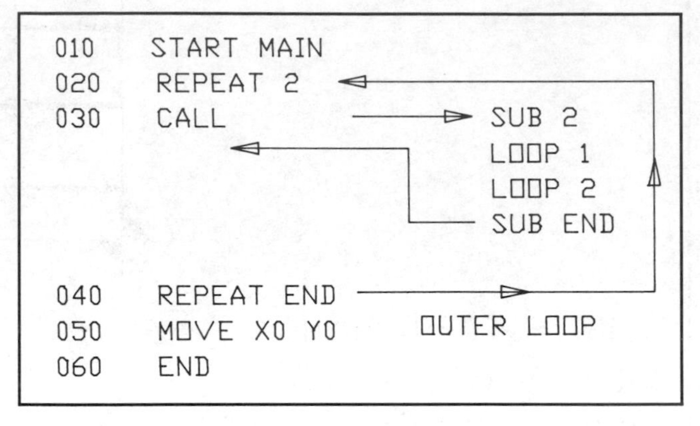

Fig. 16-9. Further refinement, looping the subroutine twice.

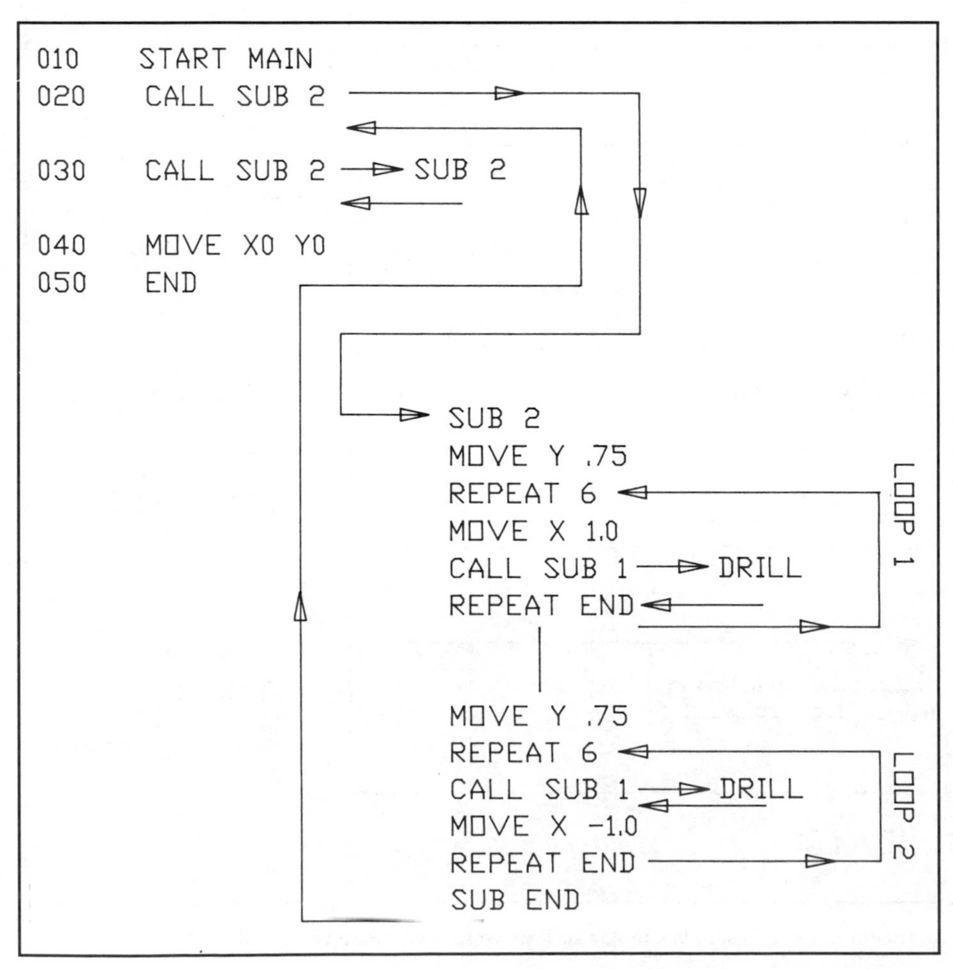

Fig. 16-8. An improvement to the total number of commands to be written: sub, loop, sub.

Write a Logic Program

1. A) Draw a logic diagram of Solution 4 in the previous development, based on Fig. 16-1.
   B) Write the entire inner loop that actually moves and drills.

   ____

   C) You do not have to write Sub 1. It would be the same as the one you wrote in the last activity.

   ____

2. In every solution we have written, notice that upon completing the second row (right to left), the tool moved 1.0″ beyond the last hole to the left. Why was that?

   ____

   ____

   Since this was wasted motion, could a new loop be written that would not move this extra 1.0″? If needed, please write it below.

   ____

   Diagram and write a program that would solve this wasted motion using your shop language. Notice that this would force us to modify the left to right loop also.

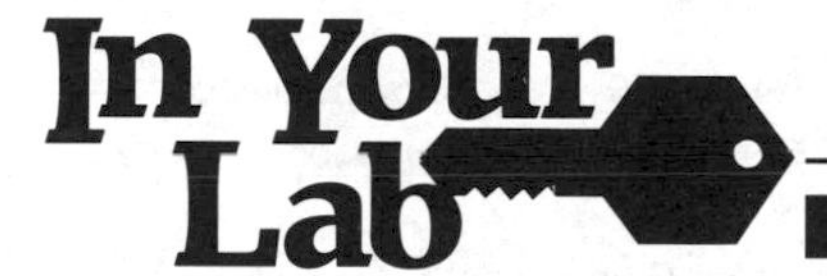

Custom Sheet 15—Logic Exercise

Custom Sheet 15 shows a shape your instructor wishes you to program using logic. Loops and subroutines may not be as common in basic turning as they are in milling.

Write a logic program for Custom Sheet 15. Follow these guides:

A) Diagram the solution you intend to program.
B) See if any improvements are possible.
C) Diagram the improvements.
D) Write the program using the language of your shop.

Immediately to the right is the answer to the improved loop that would not contain the extra move to the right of 1.0 after drilling the right-to-left row.

Six Holes—Five Spaces

This solution is based upon the fact that in a row of six holes, there are actually only five spaces between them. Refer to Fig. 16-1. Count the spaces between holes. The extra use of Sub 1 was to drill the end hole. If you were writing this program to drill 2000 plates, eliminating the extra motion would be a great timesaver.

```
010  START MAIN
020  MOVE X .75
025  REPEAT 2 ---------------------------------------------|
030  MOVE Y .75                                            |
040  REPEAT 5 ----------------|                            |
050  CALL SUB 1 --> SUB 1     |                            |
060  MOVE X 1.0               | LEFT TO RIGHT LOOP         |
080  REPEAT END --------------|                            |
090  CALL SUB 1 --> SUB 1                             REPEAT
                                                      PATTERN
100  MOVE Y .75                                        LOOP
110  REPEAT 5 ----------------|                            |
120  CALL SUB 1 --> SUB 1     | RIGHT TO LEFT LOOP         |
140  REPEAT END --------------|                            |
150  CALL SUB 1 --> SUB 1                                  |
160  REPEAT END -------------------------------------------|
```

16-6 Logic Planning and Errors and Factoring Loops

Creatively using logic in a program can save much programming, but it can also create problems. As you have seen, syntax gets complicated and errors occur. Here are some suggestions.

PLANNING LOGIC

1. Always start with the simplest form of the program. Diagram the syntax. Then improve by adding further steps, such as repeating loops.
2. Watch out for which mode value you use for coordinates. Absolute and incremental coordinates both have their place in logic programs. But be sure that the two are compatible. For example, a main program may be in absolute and the subprograms or loops may require incremental. Switch to incremental at the beginning of the sub-program, but be sure to switch back to absolute at the end of the subprogram.
3. Often, the simplest form of the program, in terms of writing, may not be the most efficient tool path—as shown in the last activity. Inspect programs to see if they could be more efficient. The number of parts to be run determines whether you will improve beyond a certain point.
4. Once a program is written using logic, it is often difficult to test using graphics and plotting. A cautious dry run may be the only way to finally verify the program.

ERRORS

Nesting Errors

Often a control will return a nesting error. This can be caused by one of two things.

1. You have exceeded the maximum number of nestings possible by the control. For example, one control may accept only three deep nestings, while others will accept more.
2. You have called up a branch that does not return correctly. The controller cannot work its way through the path you have designed. Review your diagram.

Cycle and Mode Errors

Many cycles and modes are permanent until cancelled. This can cause difficulty if a mode is in effect in the main program or subprograms but not needed elsewhere. Be sure to add cancellations of modes when they are used in syntax planning.

UPPER LIMITS ON REPEATS—FACTORING SOLUTION

Some controls impose an upper limit on the number of times a loop may repeat. This is easily overcome using a technique called *factoring*. For example, suppose you need 400 repeats of a loop and the upper limit for a single loop is 50 repeats. Simply write a double loop as shown in Fig. 16-10. Program lines 030 through 060 will be repeated 400 times. Each loop is broken down into a factor of the total repeats needed. You can see how any number could be produced, even if three or four factors were needed.

Refer to Fig. 16-10. The inner loop repeats eight times each time the outer loop is cycled.

$$50 \times 8 = 400$$

Sequence Number	Command	Comment
N030	Repeat 50	First outer loop
N031	Repeat 8	Inner loop
N032		
	(Program Commands)	
N058		
N059	End Repeat	Ends inner loop
N060	End Repeat	Ends outer loop

Fig. 16-10. A factored loop will overcome control limits on the number of repeats possible.

16-7 Propeller Example

This section presents a program that produces a propeller. If you are confident of your understanding of logic and polar values, this program will enable you to test your knowledge. This example is offered for enrichment and entertainment. While you will not be quizzed on it, studying it is a very good self-check of your understanding. This program is for one side of the propeller. It would have to be turned over and precisely coordinated, then run a second time for the back side.

```
N001 Rem:Propeller program - produces three blades roughed out
     one side only. PRZ is center of part - Z departure .1"
N002 START INS 99
N003 SPINDLE ON
N004 FR XY=5.0       Rem: Blocks 2-6 are prep info about feed
N005 FR Z =1.0            rates and tool diameters.
N006 TD   =.125
N007 SETUP ZCXYU

N008 REPEAT 3        Rem: Blocks 8-22 are the outer loop.  This
N009 GO   R .20           loop positions the cutter at an absolute
N010 GO   Z 0.0           radius of .2" from the center. Then brings
                          the cutter to the material. This loop
                          nests the two inner loops to make the
                          blades.  See also #20, 21 & 22
N011 REPEAT 25       Rem: Blocks 11-14  - a nested inner loop that
                          produces a single pass down one
N012 GR Z-.008            blade of the propeller - concentric to PRZ.
N013 GR A- 1.0            The cutter will move relative (incremental)
N014 REPEAT END           Z - .008 then a relative angle of 1 degree.
                          It will do this 25 times thus a sweep of 25
                          degrees (blade width) with a prop pitch of
                          .200 deep (25x.008=.2)
N015 GR R .010       Rem: Block 15 moves the cutter out relative .010"
                          along the radius.

N016 REPEAT 25       Rem: Blocks 16-19 a second inner loop to sweep up
N017 GR Z .008            the blade parallel to the first pass but out
N018 GR A  1.0            .010 along the radius.
N019 REPEAT END

N020 GO Z .010       Rem: Block 20 raises the cutter to gauge height
N021 GR A 120.0           Block 21 indexes the cutter 120 degrees
N022 REPEAT END           Block 22 returns the outer loop to block 8
                          where the cutter repositions for the next
                          blade.

N023 END
```

Fig. 16-11. A propeller program.

CHAPTER 17

PROGRAMMING WITH CANNED CYCLES

Another laborsaving programming technique is the *canned cycle*, sometimes called a *fixed cycle*. These two terms will be used interchangeably in this book, although there is disagreement in the machine tool industry about these definitions.

Canned cycles are not part of a logic syntax. They are preprogrammed groups of commands that can simplify the programming of long operations. Canned cycles simplify operations that are commonly done. For example, they are used to cut a thread on the lathe, turn a straight shaft, bore a hole, mill a pocket, peck drill, center drill, and drill a bolt circle. Each of the above operations could be programmed faster using a canned cycle.

OBJECTIVES

After studying this chapter, you will be able to:

- Using Custom Sheet 16, list the available canned cycles on your machine.
- Write a canned cycle into a program.
- Define the difference between canned and user-designed cycles.

17-1 Canned Cycles in Programs

All CNC controls feature some canned cycles. A *canned cycle* is a group of commands stored for permanent use. A canned cycle is "hard wired" into the control. This means that it cannot be erased, but it can be customized for a particular operation when needed. Many new controls feature amazing amounts and varieties of canned cycles. On some controls, purchasers may even buy extra canned cycles from a list of options.

A canned cycle shortens standard operation programming because the programmer need only enter the cycle identification and parameters of the desired result. For example, assume that a straight turn cycle is to be used on a lathe. We know that all straight turning cuts are similar. The program must rough to a given diameter and length, rapid back for a second pass, feed in, semifinish to a given distance allowing material for a finish pass, rapid back, feed in, and feed to finish dimensions.

Many new lathes feature a turning cycle that will program all of the above moves by simply entering the parameters of the part to be turned. The factors that would change from shaft to shaft are the parameters. For the shaft-turning canned cycle, they would be:

1. The raw stock size.
2. The amount of material to take per pass or the number of passes to take.
3. The distance to the end of the cut (Z and X).
4. Where to start the cuts (Z and X).
5. Amount of material for a finish cut.

Once the parameters for the desired shape are entered, a canned cycle will complete a particular operation, such as rough, semifinish, and finish-turn a shaft.

17-2 Programming Canned Cycles

SELECT THE CYCLE FIRST

The programmer would first select the cycle from a library that the controller offers. Once the cycle code or command is entered in the program, the parameters required by the cycle would be entered. *These parameters must be entered in a particular order and form.*

Canned cycles save time and program lines. The advantage to canned cycles is twofold.

1. They allow the programming of often long operations into a single line or small group of lines.
2. Canned cycles simplify and speed up programming. They save program lines and time.

If you were to purchase a C/NC controller, you would compare the number and kind of canned cycles the control features.

STANDARD CANNED CYCLES

Coded Cycles

Several fixed cycle codes are used in industry. Each of these has a definite order in which the parameters must be entered. A good example of a simple cycle in the coded language is the drilling cycle.

As an example, refer to the *G81 drill cycle* shown in Fig. 17-1. The line number would be followed by the G code for the cycle, then by the parameters in the correct order. (Note that some controls would require the cycle code G81 at the end of the entry line.)

1. G81 select drill cycle.
2. Rapid-travel to X,Y location.
3. Rapid-travel Z spindle to gauge height.
4. Feed to Z depth.
5. Rapid-back to gauge height = R.
6. Feed rate = 2 IPM.

N100 G81 X4.0 Y3.0 K–1.0 Z–2.50 R–1.0 F2.0

Coded cycles are modal. If a second hole was desired, much of the above information would remain acceptable. The G81 code is *modal*. This means that it remains in effect until cancelled with a G80 code. Also, the parameters that do not change do not need to be entered a second time. The command to drill the second hole (Fig. 17-1) would look like:

N101 X7.5 Y1.0

Note that the only parameters required were the X and Y locations. The other parameters were carried forward. If a large number of holes were to be drilled, this could save time, as well as space in RAM. Every hole after that would need only the location. Refer to Fig. 17-1, hole 3. If a new depth was required, then the new Z parameter would be added to the line.

N102 X8.0 Y.50 Z–3.0

Every following entry would be the same, changing only the parameters that changed from hole to hole. Once all holes were drilled, the cycle would be cancelled by a G80 code, which cancels fixed cycles.

N103 G80

In the EIA codes, there are several standard coded cycles for milling and fewer for turning. Manufacturers add custom-canned cycles to enhance the machine.

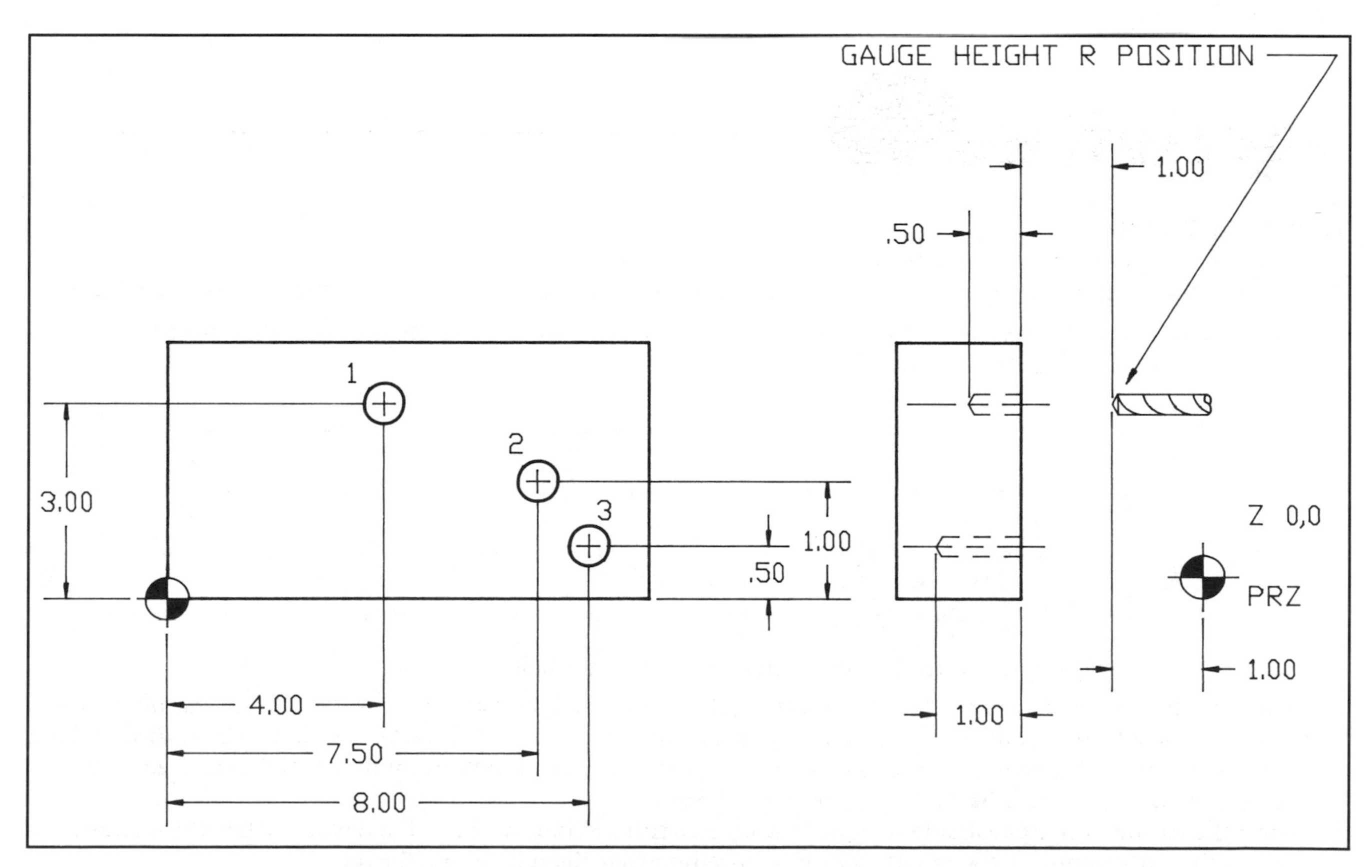

Fig. 17-1. A typical program problem solved by canned cycles. This part could be programmed using a drill cycle.

Conversational Canned Cycles

Program entry of canned cycles into conversational controls is very easy. The user selects the cycle needed from a menu or keypad. The control then asks for the parameters, one at a time, in the correct order. If the control is interactive, it evaluates each entry for certain errors, such as exceeding the machine limit or incorrect form.

The advantage here is that you do not have to memorize the order of the parameters in the command. Canned cycles work very well on conversational controls. There are so many options that it is possible to write an entire program using mostly canned cycles. The Mazak and Okuma lathe controls, for example, which are menu driven, are capable of writing programs entirely from the cycles offered. Many conversational controls consider their canned cycles a major feature and try to make them as useful as possible.

How to Use Canned Cycles

Unless your control has only a few canned cycles, you will not be able to memorize their entry formats. You should know what cycles the machine offers. You should know where to find the correct order in which to enter the parameters. When a geometry arises that could be simplified using canned cycles, be prepared by knowing where to find the correct entry format. Do not try to memorize the parameter format at this time.

CUSTOM SHEET 16

Microprocessors are becoming less expensive. Competition is causing manufacturers to produce more canned cycles. Nonetheless, there is little standardization among coded controls. Coded controls use the unassigned codes for canned cycles. Conversational controls feature a wide range of canned cycles. The same control on two different machine types will have different canned cycles.

1. Scan Custom Sheet 16, and note the shapes and operations that may be programmed using canned cycles. Notice the order in which the parameters are entered for each cycle.
2. Staple Custom Sheet 16 in the back of this book or put it in a binder.

Fixed Cycle Program

Safety Note: Don't forget to cancel the canned cycle once it is completed.

The objective of this activity is to familiarize you with the cycles and their entry format on your machine.

1. Referring to Custom Sheet 16 or the programming manual for your machine, list the number of canned cycles possible. Label the basic operations or cycles.

____

____

____

____

Using the language of your control, choose from activity 2 or 3 below.

2. Write the diffuser plate program using a fixed cycle that is used on your machine. To do this simply rewrite the drill subroutine part of the program. If your machine has a peck drill cycle, use that. If not, drill without pecks. This canned cycle will be set aside as Subroutine 1 and used in the way that Sub 1 was used in the previous examples. It will be in Sub 1, the body of Sub 1.
3. Using the canned cycles available on your machine, write a program for a shape your instructor assigns. For a lathe, this might be a straight shaft or a bored pocket. Keep it simple for now.

17-3 User-Designed Fixed Cycles

CUSTOM CYCLES

A user-designed fixed cycle, or canned cycle, is one not built into the machine at the factory. A user-designed cycle would be the same as any canned cycle once it was written by the user. It would generate a certain group of commands once the parameters were entered. The difference is that the user writes the basic shape or operation to be performed. The user also specifies the parametric entries within the cycle. This offers an advantage to shops that specialize in certain types of repetitive work. Many of their programs are similar, but the parameters differ from program to program.

For example, a shop may turn chain sprockets on the lathe. The sprockets may differ only in size. They would each have a center hole, a hub, and an outside diameter. All of these could be written into a fixed cycle, with parameters for each feature.

On the mill, the cutting of the sprocket teeth is similar, yet each size of sprocket requires different parameters. These include items such as the number and pitch of the teeth.

This user-designed cycle is stored in the permanent program storage component of the control. It is called into use by the main program. When the special cycle is called for use in a program, the parameters are called out. This technique is also called a parametric subroutine.

In Your Lab

1. Does your control feature user-designed cycles?

___

2. How many user-designed cycles are possible? Most controls have a limit on the number of custom cycles that can be stored in permanent storage.

___

3. Briefly, what is the difference between a canned cycle and a user-designed fixed cycle?

___

___

___

___

CHAPTER 18

THREADING ON A C/NC LATHE

One useful application of a canned cycle is the lathe threading cycle. Most manufacturers offer internal, external, and scroll or face thread cycles. Other thread options include multiple and tapered threads, plus special pitches.

The production of threads using a canned C/NC cycle is exciting to watch. The speed with which a C/NC lathe is able to start and pull out of the thread is amazing. Threads are better on a C/NC lathe because the RPM may be at or near the correct cutting speed for the material. This is something standard lathes cannot achieve. At the spindle speeds possible in C/NC threading, qualified carbide-insert thread tools can be used near the correct surface speed. This produces excellent finish and prolongs tool life.

OBJECTIVES

After studying this chapter, you will be able to:

- Write a program using threading cycles.
- Calculate starting and stopping distances for threading on a lathe.
- Define acceleration and deceleration of axes.
- Chart and compare the threading cycles available on the lathe in your shop.

18-1 Thread Programming Concepts

THREADING PARAMETERS

Threading is a complicated process that is made simpler by canned cycles. The following parameters are needed by the control to thread correctly.

1. Initial departure point (X axis and Z) axis. This must include an acceleration distance.
2. Thread depth.
3. Lead of thread.
4. Thread form angle.
5. Nose angle of cutting tool.
6. Full length of threads. This must include a deceleration-pullout distance.
7. Number of passes (not on all controls; automatic on others).

Safety Rule. *To thread correctly, you must read and practice the individual instructions given in the manual for the machine. I can think of no other area where this advice is more important than in C/NC threading. There are differences between controls. Although the parameters are similar, their usage may not be. The order is which the threading parameters are entered may differ from one controller to the next. Due to the fast axis movements, a crash occurs quickly when parameters are reversed in the program.*

Some controls will calculate all programmed moves from the basic facts about the thread and material position. Others require you to calculate much of the above information. In general, the newer controls require fewer calculations.

LEAD AND ACCELERATION/ DECELERATION

The are two concepts of lead and acceleration/deceleration must be understood to thread successfully. C/NC machines are not like standard threadcutting lathes. On a C/NC lathe, there is no direct mechanical connection between the saddle feed and the spindle rotation.

C/NC lathes use drive motors. Standard lathes have a gear drive. On a standard lathe, if the half-nut clutch is correctly engaged, the tool will trace the same thread each time a pass is taken. A C/NC lathe has no mechanical connection. It has no gear drive. The saddle is powered by a drive motor. A separate motor drives the spindle. The controller is responsible for the exact coordination of the two drives for threading.

The control senses the spindle position and starts the saddle feed at a coordinated time. Each successive pass must start in exactly the same spindle/saddle relationship. This means that some details must be seen to by the programmer or the control. To understand this, we need to look at the two basic C/NC concepts of lead and acceleration/deceleration.

Lead

In threading, the ratio between the *spindle RPM* and the *tool advancement* is crucial. The tool must move exactly the right pitch distance for one revolution of the spindle. If you expect to turn 16 threads per inch, then the saddle Z axis must move 1⁄16″ for each revolution of the spindle. This ratio is called *the lead of the thread.*

To calculate the lead, divide 1″ by the number of threads to be cut:

$$\text{Lead} = \frac{1}{\text{Threads per inch}}$$

As an example, program a thread with 13 threads per inch.

$$\text{Lead} = \frac{1}{13} = .0769231$$

The lead is not an even number. This could be a problem. If the standard feed rate on your lathe will accept information to four decimal places, you will enter a lead of .0769 per spindle revolution. You can see that for each thread on the shaft, the extra .0000231 will be left off. A long thread will become inaccurate. In one inch, there are thirteen threads. The error will be 13 × .0000231 = .0003. On most threads that would not be enough to make any correction. But for a 10″ thread it could be a problem. Some lathes handle this problem by accepting thread lead information to six decimal places. Others will accept the threads per inch required and automatically compensate for the round off. Some lathes, however, make no provision for this problem. Usually, these are lathes with older controls. You must be aware of these possible problems. Make your decisions based upon the tolerances for the finished part.

Acceleration/Deceleration

A second potential problem caused by the lack of a direct mechanical gearing between the spindle and Z axis saddle is the fact that the spindle is already turning when the saddle is commanded to move. The Z axis drive motor must go from zero RPM to a specific RPM that produces the correct lead. Secondly, when threading to a shoulder, the saddle cannot simply stop. Neither function can be instantaneous. There is a short time and distance in which the saddle must accelerate or decelerate.

Acceleration

Refer to Fig. 18-1. The short distance required to get the saddle to speed up to the correct ratio is called the acceleration, or start, distance. During this period, the lead will be finer than programmed. This acceleration must take place off the workpiece to produce a correct thread. Tool contact can occur only after the acceleration of the tool Z axis.

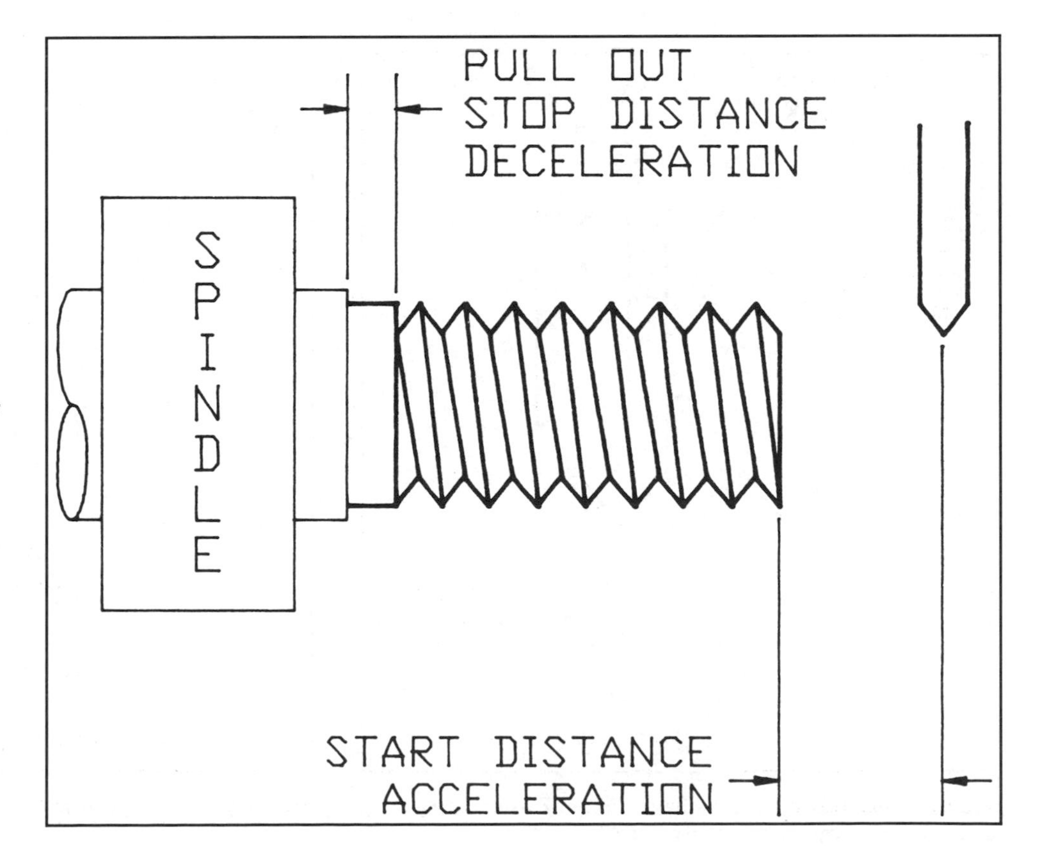

Fig. 18-1. A minimum distance must be programmed to allow for acceleration.

In Fig. 18-1, the departure point for the thread cycle is away from the part in the Z axis distance. This start distance is a minimum. The tool could be positioned further away, but it then becomes less efficient. There are two factors in calculating the start distance:

1. The calculated acceleration position.
2. The offset distance.

The offset distance is the amount of lateral tool (Z axis) movement with each change in the infeed X axis. See Fig. 18-2. As the tool is fed into the work at 29° for each succeeding pass, the tool moves in, as well as sideways.

The start distance is found using a formula based on the mechanical properties of the machine. The faster the RPM, the longer the time required. The start distance formula for each machine will be found in the programming manual unless this is automatically calculated by the control.

A typical formula is:

Start Distance = RPM × lead × .015 + Offset Distance

Refer to Fig. 18-3. The offset distance is found through the formula:

Offset Distance = Tan 29 × Infeed Amount

If the tool is fed in directly with no compound angle infeed, or if there is only a single pass required to complete the thread, then the offset distance is equal to zero. The angle in the above formula changes if the thread is a different form. An acme thread would use 14°, the same as setting the compound on a standard lathe. Again, remember, that this start distance is a minimum distance required to get the saddle going.

As an example, assume that a lathe is making 500 RPM and turning a 20 pitch thread. The start distance would be:

S.D. = 500 × 1/20 × .015 + O.D.
S.D. = .375 + O.D. (if used)

Deceleration

If the saddle has to stop the tool bit against a shoulder, then a slow-down distance will be needed. This distance must also be calculated. This distance must be subtracted from the total Z axis length of the thread. The machine must pull out in time to stop.

A typical stop distance formula is:

Stop Distance = RPM × Lead × .015

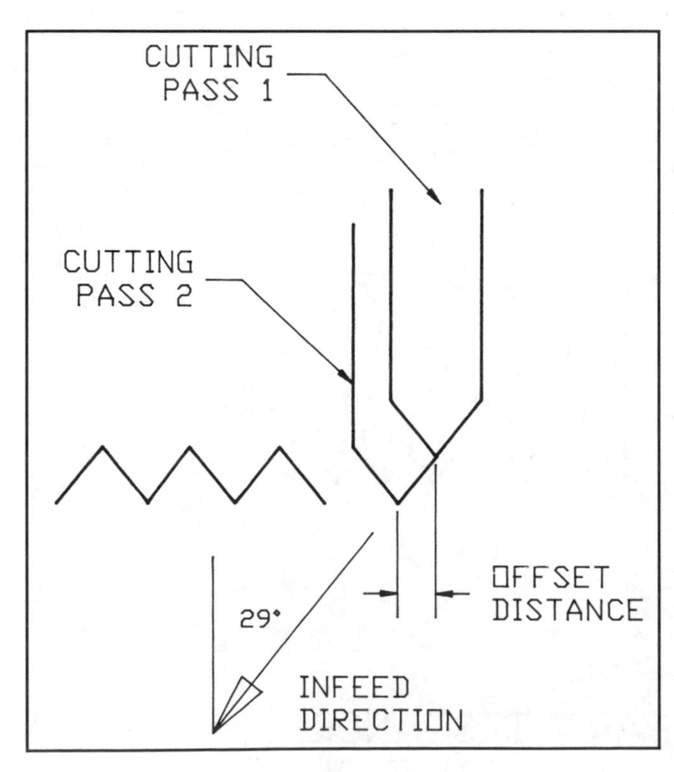

Fig. 18-2. The offset distance is the amount of Z axis movement (sideways) with a given X distance in-feed.

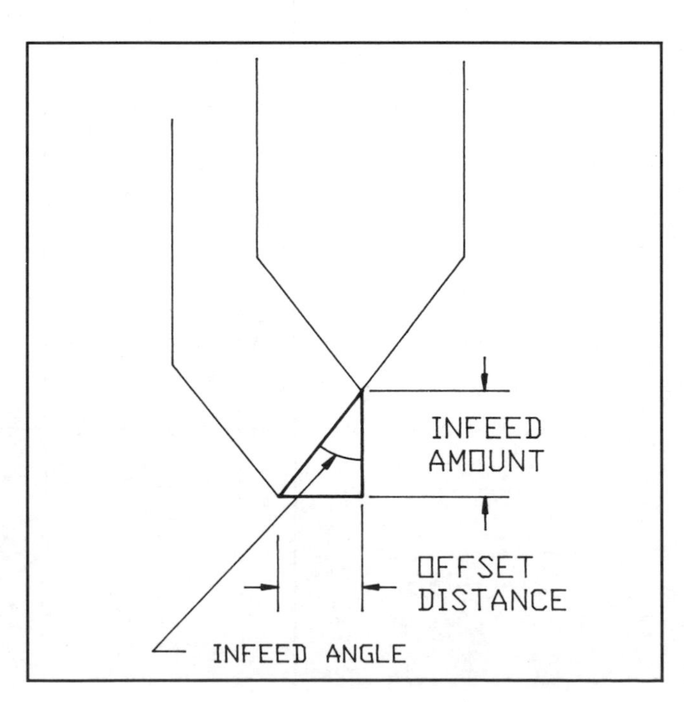

Fig. 18-3. The triangle is solved to determine offset distance.

18-2 Limiting Factors in Thread Programming

There are five limits imposed upon the programming of threads.

LIMIT 1. SPINDLE RPM

The programming manual will clearly define the maximum RPM for threading for two reasons.

A) Lathe chucks are dangerous beyond a given RPM.
B) The Z axis drive can move only so fast. Beyond a given spindle RPM, the saddle can not move fast enough to produce the correct lead.

LIMIT 2. FEED AND RPM CANNOT BE OVERRIDDEN

Since the ratio of RPM to feed rate is crucial, most C/NC lathes disallow any manual override from the control during a thread cycle. RPMs should be calculated very carefully.

LIMIT 3. MACHINE RESPONSE

Where the programmer must calculate the start and stop distances, the formula is different from lathe to lathe. The .015 multiplier in the two formulas presented is for a specific machine. Depending on the responsiveness of the machine—how quickly the axis drive can speed up—this number will change. The basic formula could even change.

LIMIT 4. MAXIMUM AND MINIMUM POSSIBLE LEADS

Each machine will have a limit on the coarsest and finest pitches possible. Programming beyond this limit will result in an error code.

LIMIT 5. AMOUNT PER PASS

The depth taken for each pass of the tool, or the number of passes needed to rough and then finish the thread, is controlled either by you or by parameters hardwired into the control. These same limits are imposed during manual threading to prevent tool breakage, torn threads, and overstressing of the machine.

Program Commands for Threading—Custom Sheet 17

No attempt has been made to display the variety of thread cycles or the entry formats for the program parameters. There is a great deal of difference in controls in threading. In general, as controls become "smarter," threading becomes less difficult to program. Procedure. A threading cycle proceeds as follows:

1. Rapid to departure point. (This is based on the calculated acceleration.)
2. Feed to first pass depth. (The X axis distance is based upon rough stock size.)
3. Accelerate and cut thread. (The Z axis distance is based on deceleration and print dimensions.)
4. Rapid pullout to X departure distance.
5. Rapid return to Z departure.
6. Feed in to next X axis depth.

Repeat this sequence until finished depth, based on print, is reached.

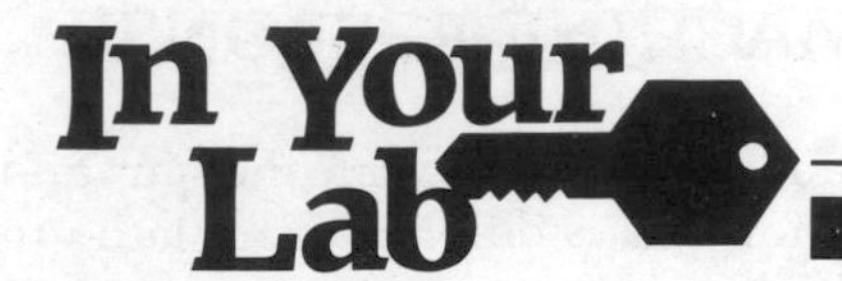

If you have a C/NC lathe, please look over Custom Sheet 17 or the programming manual for the control. Then answer the following questions:

1. Check the threading cycles your control has.
   - A) Single pass threading ☐
   - B) Multiple pass ☐
   - C) Tapered threads ☐
   - D) Internal threading ☐
   - E) Multiple start threads ☐
   - F) Scroll threads ☐
   - G) Variable lead threads ☐
2. Does your control require you to calculate the acceleration and deceleration distances?

   Yes ☐ No ☐
3. If Yes, what is the acceleration formula?

   ____
4. What is the deceleration formula?

   ____
5. List the order of parameter entries below.

Parameter	Symbol or Code Letter

6. Refer to Fig. 18-4. Write a program command block for the following thread:

   ⅝-11 2A UNC × 1.5 lg
   Material = Mild steel
   Cutter = Carbide

   Recommended surface speed = 200 FPM.(Note: This may violate an RPM limit on your machine if you were using a chuck.)

   Holding method: Collet.
   Two passes. No compound angle feed. Feed straight in.
7. What was the acceleration distance for the above thread?

   ____
8. What was the deceleration distance?

   ____

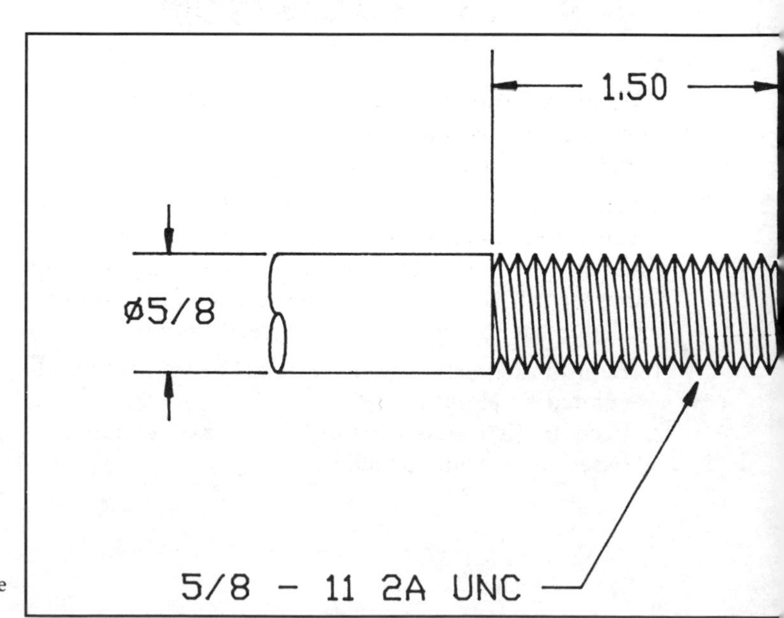

Fig. 18-4. Write a program canned cycle command to produce the following thread on your control.

CHAPTER 19

ADVANCED PROGRAMMING CONCEPTS

The subjects in this chapter are grouped together because each is found only on advanced CNC equipment. Not all machines offer each feature. Thus, your instructor will want to set some priorities for study. Have your instructor set priorities for this material by checking off one category of involvement for each subject. Choose either *scan, read,* or choose *study and do exercises.*

- *Scan.* Scan the material to identify the important points.
- *Read.* Make sure that you thoroughly understand the material. Read all material.
- *Study and complete the exercises.* Study the information and techniques presented. Then do the assigned exercises.

OBJECTIVES

	Involvement Level		
	Scan	Read	Study and Do Exercises
• Write a program using mirroring.	☐	☐	☐
• Use constant velocity programming for a lathe.	☐	☐	☐
• Write a safety work envelope program to avoid crashes.	☐	☐	☐
• Enter program scaling to change part size.	☐	☐	☐
• Plan a 2½ axis cut and a multiple-axis program.	☐	☐	☐

19-1 Mirror Imaging Programs

ADVANTAGES OF MIRRORING

If you hold a blueprint up to a mirror, the image you will see is the exact opposite of the print. The image will be shown as a left hand, or mirror image, part. You can also see this if you hold a blueprint up to a strong light and look from the back side.

Not all CNC controls offer mirror imaging programs. Others offer it as an option. There are three advantages to mirroring:

1. Once a program is written, a second need not be written to produce a left-hand version of the part. This sounds great but causes a problem, as explained later.
2. Mirroring can reverse the position of a part on a fixture. The part is not actually then a left-hand part. Instead, it is a flopped-over right-hand part.
3. Mirroring can be used in a program to redefine a symmetrical pattern without writing the program over.

MIRRORING REVERSES AXIS DIRECTION

The program applies mirroring to one axis at a time. The mirror axis code or command is entered before the axis commands in the program. The result is easy to understand: the action of the axis mirrored is reversed. If the program originally called out a +Y movement, a mirrored version would produce a –Y movement.

CODES FOR MIRRORING

The EIA language has no code for this mirroring, since NC machines were incapable of mirroring. The code will be from the unassigned codes list. A machine, for example might use G60 as the mirror code. The G60 is followed by the axes to be mirrored.

CANCEL MIRRORING

You must remember to cancel mirroring because it is a modal command. There are two ways in which to cancel mirroring. Your control will use either or both.

1. **Cancel code or word.** A cancel mirroring code or word that will end mirroring is put into the program.
2. **Flip flop commands.** This is a second entry of the exact same command that will double-reverse the axis sign—this brings you back to a right-hand version. Example:

   G60 X—Turns on mirroring X axis
   G60 X—Turns off mirroring X axis (reverses the reverse)

This is called a *flip-flop command.* One command turns it on. A second turns it off. This is similar to a flip-flop switch in electronics.

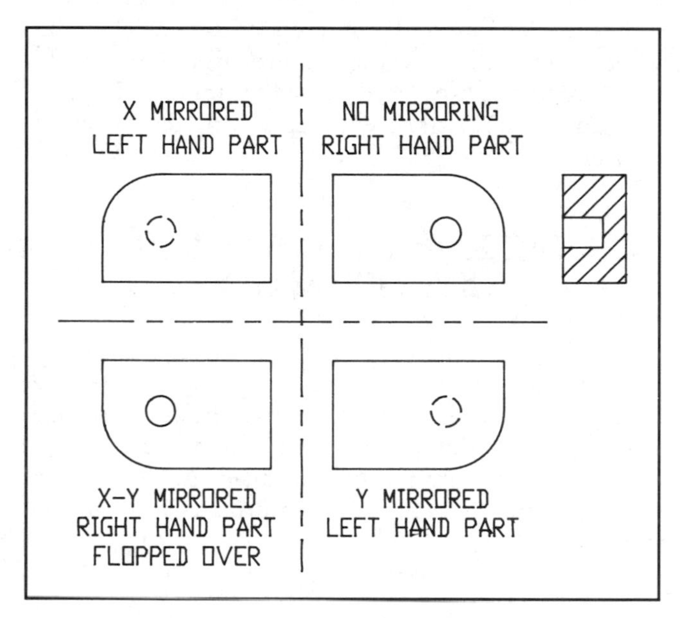

Fig. 19-1. Individual axes can be mirrored to cause different effects.

QUADRANT MIRRORED

Refer to Fig. 19-1, which shows the result of mirroring different axes. The program as written produces the part shown in the first quadrant. To distinguish right- from left-hand parts, visualize the round pocket milled halfway through the part.

Note the effect of mirroring different axes.

- Mirroring only X produces the left-hand part in quadrant 2.
- Mirroring X and Y produces the flopped-over right-hand part in quadrant 3.
- Mirroring only Y produces the left-hand part in quadrant 4.

EXAMPLES

The following two examples show how mirroring can be used as a timesaving program manipulation.

Example 1. Make a left-hand part from a right-hand program. Refer to Fig. 19-2. A problem arises using this technique. Notice that the cutter is climb milling the right-hand part, but conventional milling the left-hand part. Shops where the programs are long enough to justify mirroring axes reverse the spindle rotation and use special left-hand cutters that have reversed helix and cutting action. The result is climb milling on both parts. The special cutters are costly, however.

Example 2. This shows how to run four slots in a part program. This example shows how a symmetrical feature may be reversed or repositioned on a single part. Refer to Fig. 19-3 on page 209. The slot in the part in quadrant 1 is as programmed. By mirroring various axes, we may run the slot three more times on the same program commands. Absolute dimensioning would be best for this application. The PRZ would need to be in the center to facilitate positioning of the cutter.

Fig. 19-2. A left-hand part is produced from a right-hand program using a mirrored X axis.

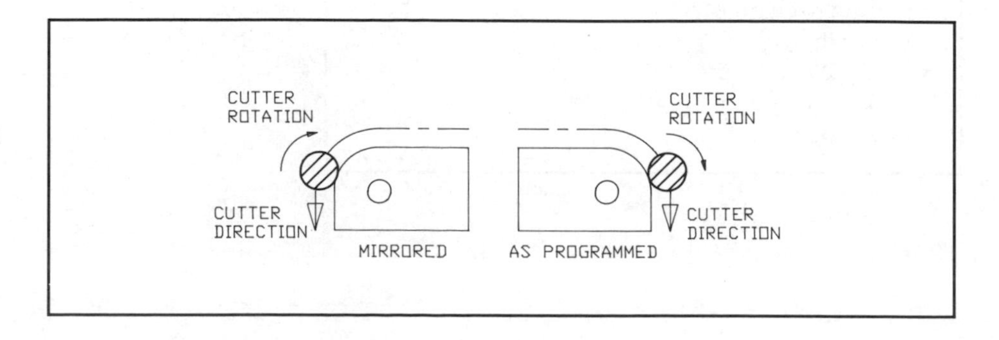

MIRROR X

MIRROR X-Y

MIRROR Y

Fig. 19-3. A single feature may be reproduced by writing a subroutine and using mirroring.

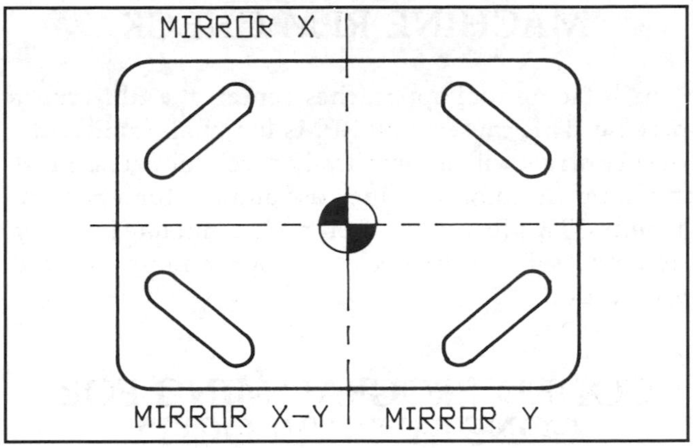

In Your Lab

CUSTOM SHEET 18

1. Refer to Custom Sheet 18-A or your programming manual. Does your control feature mirroring?

____

2. How is mirroring entered on your control? Write a program command line:

   To mirror the X axis: ____________________

   To mirror the Y or Z axis: ____________________

   To mirror the X and Y/Z axis simultaneously:

   ____

   To turn off mirroring: ____________________

19-2 Constant Velocity Programming

CONSTANT VELOCITY

Constant velocity, also known as constant surface speed, is another feature found on CNC lathes. The RPM of a particular cut depends upon the radius distance from the center and the recommended surface speed for the cutter/work combination. On a facing cut where the radius is constantly changing, the RPM should change, as well.

Constant velocity programming solves this problem. Any change in X axis distance from center will result in a proportionate increase or decrease in the spindle RPM. Once the code or word for constant velocity is entered in the program, the distance of the tool from the centerline, along the X axis, is considered the model for surface speed. Any subsequent change in the X distance (radius), will cause a change in the RPM.

MACHINE RPM CHECK

As a facing tool approaches center, the RPM must increase. This causes high RPMs for small radius cuts. Most controls will automatically check for excess RPM and limit the amount of increase allowed for a particular setup. This feature is often programmable in a way similar to safe work envelope programming. See the next section.

CODED PROGRAMMING FOR CONSTANT VELOCITY

G96 is the assigned preparatory code to initiate the CV mode. G97 cancels the mode on many controls. Your control might require a different code.

CONSTANT VELOCITY IN MILLING

Some advanced milling machine controls offer CV as an option to solve a minor problem. As can be seen in Fig. 19-4, an end mill progressing around a part will actually produce different feed rates at the tool contact point with the work. In standard velocity programming, the centerline of the cutter is where the feed rate applies. Refer to Fig. 19-4. If the feed rate is 20 IPM, then the actual contact velocity will be 20 IPM going straight. But it will be less on the outside radius and greater than 20 IPM through the inside radius. If the control has CV capability, the tool contact point, rather than the cutter centerline, will be maintained at a constant feed rate.

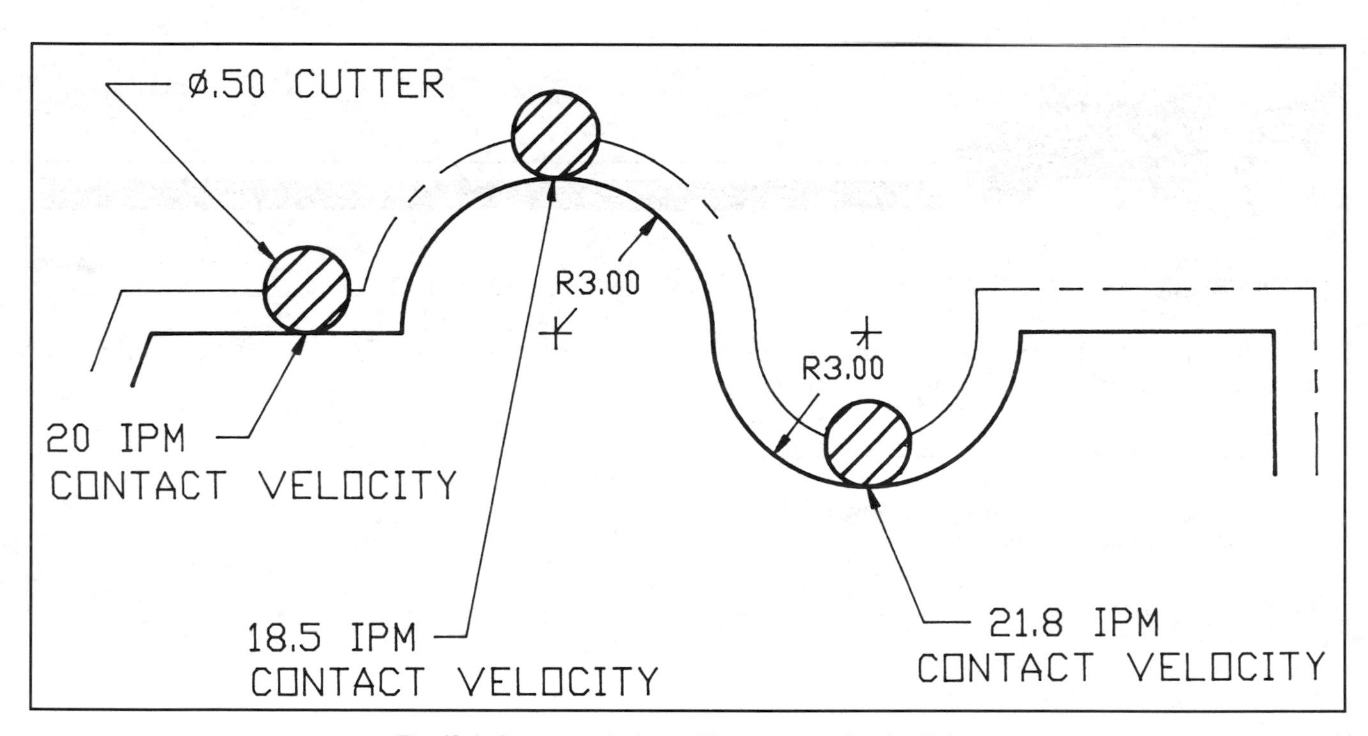

Fig. 19-4. Constant velocity problem on a peripheral mill cut.

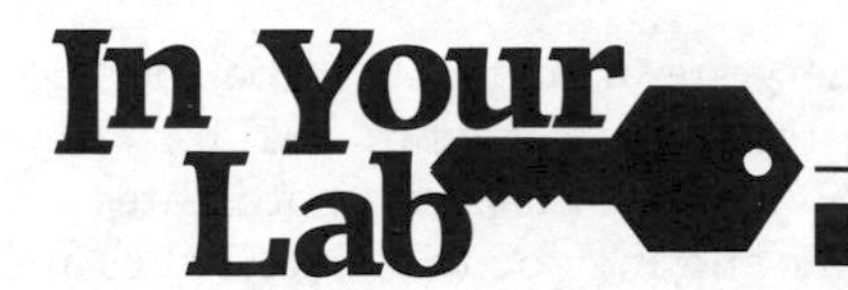

Referring to Custom Sheet 18-B or your programming manual, answer the following questions.

1. Does your control have constant velocity?

____

2. What is the code or command word to initiate CV?

____

3. Write a program line that would initiate constant velocity.

____

19-3 Programming A Safety Work Envelope

As mentioned in Chapter 7, many controls have a safety feature that allows the programmer to define a "sacred zone" into which the machine will be prevented from moving. Two other names for a *safety work envelope* are: *soft limits* or *crash zone programming.* All three define a way of preventing accidents by writing limits to the travel allowed away from the *machine home position.* This feature finds equal use on lathes and mills. The safe zone is defined by coordinates of the corners of the rectangular zone on lathes and the extreme diagonal corners of a rectangular box on mills.

SAFETY ZONE FOR LATHES

The safety zone for a lathe is a rectangle in the X-Z plane. After entering the code or command word for safety zone, the corner nearer the machine home position is identified. Then the corner farthest away is identified. See Fig. 19-5.

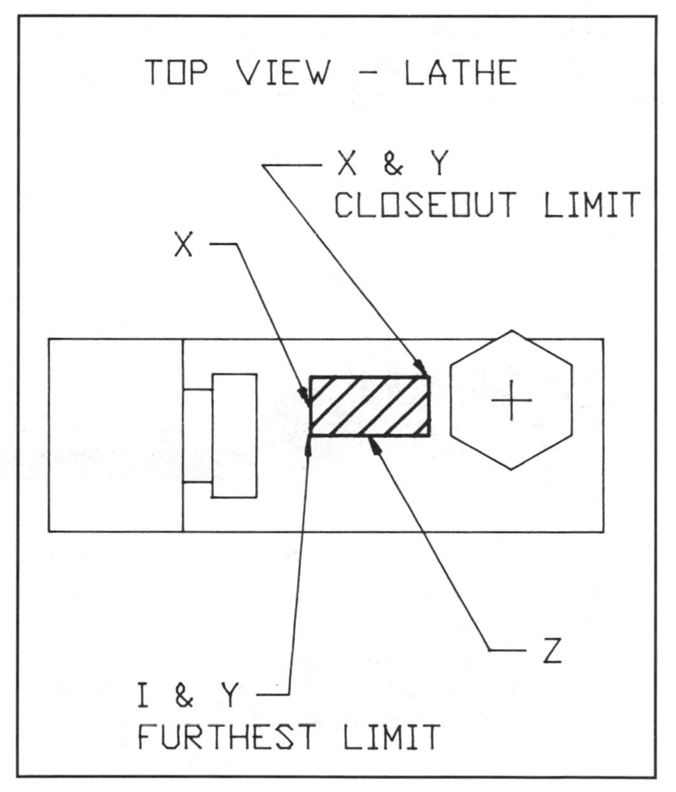

Fig. 19-5. A "safe zone" is defined on a lathe as a rectangle in the X-Z plane.

SAFETY ZONE FOR MILLING MACHINES

The safety zone for a milling machine is three dimensional, but is primarily concerned with the Z axis. The zone is in the shape of a rectangular box. The corner of the box closest to the machine home position or to an extreme limit of each axis travel is the first coordinate of the zone. The second coordinate is the extreme diagonal corner. See Fig. 19-6 on page 212.

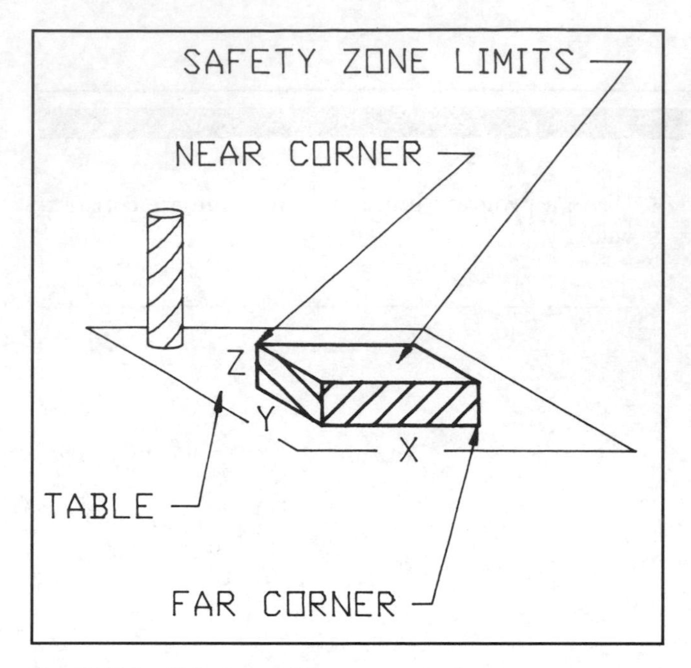

Fig. 19-6. A mill "safe zone" is a rectangular box.

CODED PROGRAMMING OF SAFETY ZONES

The alphanumeric code word to initiate safety zone programming will be taken from the unassigned codes in your machine library of commands. If your control has this ability, the general format will be the code word followed by the near corner in X-Z for lathes and X-Y-Z for mills. The second set of coordinates will be for the far corner and may be in I-K for lathes and I-J-K for mills. There are some differences in safety zone programming because each manufacturer has used a different code number and written their own standard procedure. Refer to Custom Sheet 17 or your programming manual.

You may want your program to cross the safe zone on purpose. This is possible, but the program must include a temporary unlock command. This command will not be modal. It must precede any crossing or movement within the defined zone.

CANCELLING A SAFETY ZONE

Because cancelling a safety zone would be unusual, most controls do not provide a simple code that will cancel the zone. The way to end safety zone programming is to redefine a new safety zone that is at the far limits of the machine travel. In other words, expand the safety zone to the maximum travel of the machine and allow the full travel of the machine.

Remember that the safety zone feature may be called several different names. Find this information in the programming manual.

1. Does your control feature a safety zone limit?
2. What is the command word or alphanumeric code to initiate this safety zone?
3. What is the axis or coordinate format? Write a program command that will define a safety zone.
4. Will the control cancel a safety zone, or must you redefine the zone at zero?

19-4 Programmable Scaling

Programmable scaling is defined as a percentage of original programmed moves. This interesting feature allows the programmer to expand or contract the entire part being programmed. The scaling factor is applied to programmed axis moves based on a percentage of the original size. For example, if an original program move was 1″, a scaling of 85% would produce a movement of .850″. A scaled move of 1.25% would produce a movement of 1.25″.

USES OF SCALING

Example 1. Moldmaking

One very good application of scaling is in plastic moldmaking. Hot plastic injected into a mold shrinks after removal from the mold. This shrink factor must be built into a precision mold. The mold must be bigger than the finished part to allow for shrinkage. Expanding all dimensions of the mold cavity can be mathematically difficult. The programmer writes the cavity program to the print specifications for the milling machine. He or she then inserts the scaling factor in the program to increase the dimensions by the inverse of the shrink factor of the plastic.

Example 2. Tool Wear Compensation

A program not written with tool radius offset provisions can still be compensated. The part is machined and measured. The difference is entered as a scaling factor to bring the part to size or to compensate for wear.

Individual Axis Scaling

Some controls allow scaling of individual axes. This can cause unusual distortions. On a milling machine, if only the X axis was scaled, a circle would become an ellipse.

Cancelling Scaling

There is a problem when cancelling scaling on some controls. Once the cutter leaves the PRZ in scaling mode, all moves are not what the programmed distance said. *The machine is not where the control has it positioned on the absolute registers.* The tool is in a different position than the control thinks it is. The tool is at a scaled version of the position.

If you cancel scaling at some position other than PRZ and then command the tool back to PRZ, the tool will not be able to return to the original PRZ because it will move the now unscaled distances to return. To overcome this problem, *return to PRZ inside the scaling mode, then cancel. This insures that you have not lost the PRZ.*

For example, a program calls for a first move of 3.0″ from PRZ in the X axis. The programmer inserts an upscale of 1.50%. This means that the first move actually moves 4.5″ (1.50 × 3.0″). The machine position registers have the tool located 3.0″ from PRZ, though it is really at 4.5″. If the operator cancels scaling at that point and then commands the tool back to PRZ, can you see what will happen? Will it over- or underrun the PRZ? (The machine will return only 3.0″ and will underrun the PRZ by 50%.)

Codes or Words

There is no EIA code word assigned for scaling, it will be one of the custom words on your machine. There are two ways to cancel the scale mode:

1. A cancel code or a conversational word.
2. A return to 100% scaling. This would simply be a program command that causes scaling, but at a ratio of 100% of the original moves.

Refer to Custom Sheet 19C or your programming manual. Answer the following questions.

1. Does your control offer part scaling?

____

Be careful of your decimal point on the next two exercises.

2. What is the program command to upscale by 103.7%?

____

3. What is the program command to downscale by 97%?

____

4. Can your control scale individual axes?

____

19-5 Multiaxis Machining

SCOPE OF THIS SECTION

The purpose of this section is to familiarize you with some of the terms and procedures in multiaxis machining and the production of three-dimensional shapes. There are multiaxis lathes, mills, and a variety of sophisticated machinery available. Actual multiaxis programs are beyond the scope of this book. Some machines are programmable for a limited variety of geometries at the keypad MDI. Others require sophisticated graphic or offline languages, which we will cover in Unit 7. We have touched on some of the ways these machines are programmed. At this point, though, the programming becomes very complicated and application specific. We will first look at milling machine three-dimensional work. Then we will look at lathes.

MULTIPLE AXIS MILLING MACHINES

There are four classifications of three-dimensional machining on mills. In order of increasing complexity, they are three-axis linear, two and one-half axis machines, true three-axis machines, and multiaxis machines.

Three-Axis Linear Machining

Many, but not all, modern CNC milling machines will perform a three-axis linear interpolation. You will want to ascertain if your machine will perform this movement.

Three-axis linear machining is not difficult to envision. It is a straight-line move involving three axes: X, Y, and Z. For interpolation, consider the example of a manual Bridgeport. If you and a friend could crank three handles and keep them turning at the correct rates in comparison to each other, you could machine a three-axis line. The control calculates these rates.

The *feed rate* is a vector calculation for each individual axis. The resultant tool path is at the programmed feed rate.

Refer to Fig. 19-7 on page 215. In that example, the programmed feed rate is 40 IPM. We wish to move from point A to point B—3.0 in X, while moving –2.0 in Y and –2.5 in Z. The calculation made by the machine is shown.

The actual calculation is based upon solid geometry and a vector solution shown in the solid figure. The lengths of the sides are proportionate to the actual feed rates. Assume that we solve for the distance A-B and then compare that to the 40 IPM feed rate. This will give a ratio that we can apply to the other sides, which represent the feed rates for the individual axes.

For example, if the length of side A-B is found by labeling points C and D as shown, we know there are three right triangles involved. You can make these calculations yourself from what you already know. We

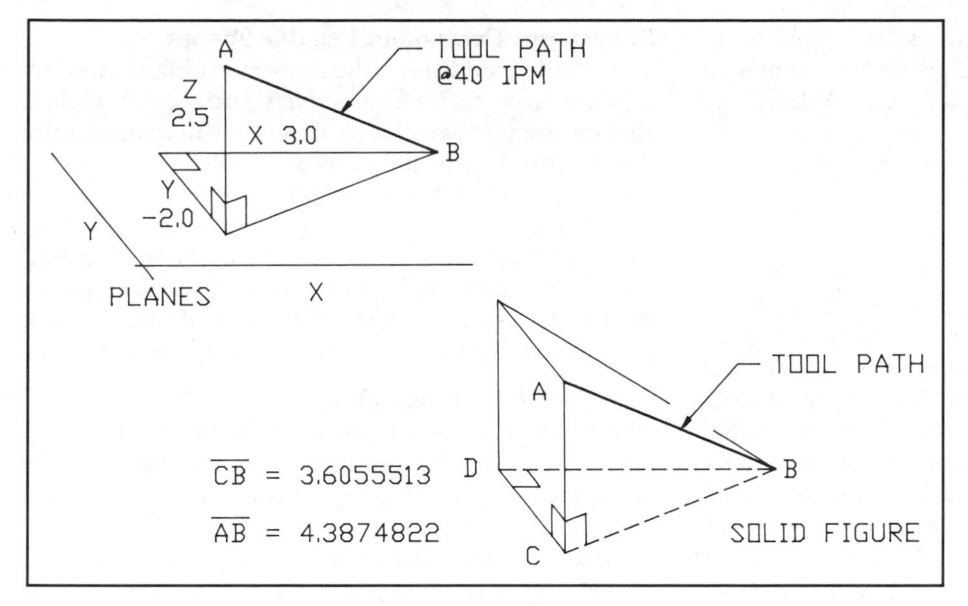

Fig. 19-7. A three-axis linear interpolation at 40 IPM.

need to find the distance traveled from A to B. (Note that all distances are positive for calculations.)

First, using the Pythagorean theorem, find side C-B, which equals 3.60655513.

Second, using the Pythagorean theorem, find side A-B, which equals 4.3874822.

With this we have the three sides to form a proportion. Each axis move will be compared to the ratio of 40 compared to side A-B:

$$\frac{40}{4.3874822} = \frac{X\,\text{rate}}{3.0} = \frac{Y\,\text{rate}}{2.0} = \frac{Z\,\text{rate}}{2.5}$$

X portion of 40 = 27.3505 IPM
Y portion = 18.2337 IPM
Z portion = 22.7921 IPM

In other words, if the X axis moves at 27.3505 IPM while the Y axis moves at 18.2337 IPM and the Z axis moves at 22.7921 IPM, the resultant tool path will be 40 IPM.

Two and One-Half Axis Machining

A two and one-half axis machine will perform a flat circular interpolation while performing a linear interpolation in the third axis.

Figure 19-8 shows the machining of a helix. A milling machine could cut this spiral groove in this cylindrical part by circling around the outside while the Z axis simultaneously drives down in a straight line. The program would have to make sure that the Z axis had dropped the pitch distance at exactly 360° of rotation. In general, polar controls are much easier to program for this type of movement.

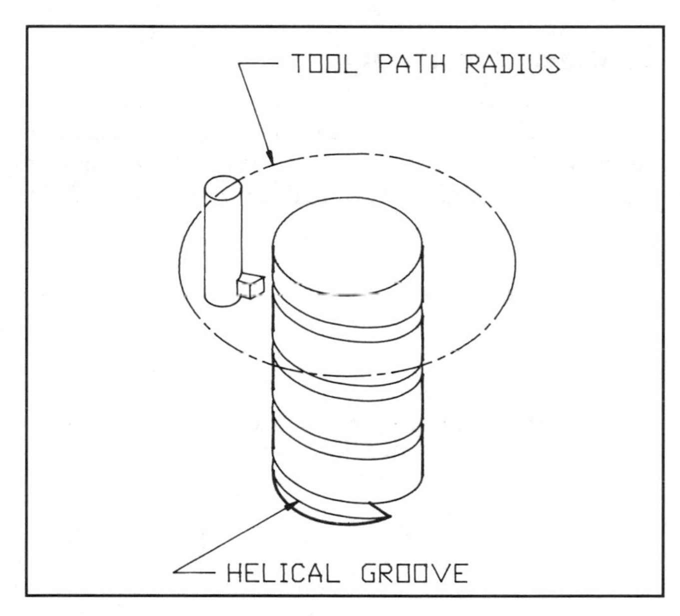

Fig. 19-8. Two and one-half axis movement allows two axis to perform circular interpolation while the third axis performs linear movement. In this example, a mill produces a helix.

Three-Axis Circular Interpolation

The next higher level of geometry possible on CNC machines is true three-axis movement. Many CNC machines will perform two-axis circular machining in all three standard planes—X-Y, X-Z, and Y-Z. Only a few, however, will machine a true circular path in X-Y-Z simultaneously. Let's refer again to the Bridgeport example to get some feel for how complicated that might be. First, you and your friend would have to crank all three handles at three different rates so that the tool path would be the programmed feed rate. Next, at every instant, as the cut progressed, you would change the ratio between all three feed rates.

In Figure 19-9, the cutter moves from point A to point B. The plane formed by A-B and the center of the sphere is not parallel with any axis. This is true three-axis circular interpolation.

The feed rates are calculated with the spherical formula:

$$X^2 + Y^2 + Z^2 = R^2$$

This formula yields the distance relationship for each axis for any instant in time. A ratio is then formed that is similar to that in the three-axis linear example. Each axis is assigned its portion of the programmed feed rate. This ratio is constantly changing as the tool progresses. The point C shown has its own set of X,Y, and Z distances for which there is a correct feed ratio.

CNC machines must slow down the tool path feed rate to a maximum speed. This is necessary because of the speed required within the M.C.U. to output the calculations to drive all three-axes of circular interpolation. This is called the *3-D default feed rate.* Again, in general, polar controls are simpler to program in this area.

Multiaxis Machining

There are two types of multiaxis machining. The first type of multiaxis machining can be programmed by hand. The second type is harder to program and requires a computer designed to program difficult shapes.

Simple and Compound Regular Shapes

To understand the difficulties in multiaxis machining, we need to look at regular surfaces. A regular surface is one that can be described mathematically. Regular surfaces include cones, cylinders, spheres, flat planes not parallel to any machine plane.

Surfaces are simple or compound. A surface is said to be *simple* if a sheet of paper could be wrapped around it without wrinkling. A cylinder and a cone are simple shapes. A surface is *compound* if a sheet of paper cannot be conformed to it. A sphere is a compound shape.

Easy Multiaxis Machining

An example of a multiaxis program that would be possible by MDI would use a programmable rotary table mounted on a milling machine table in a vertical position. Fig. 19-10.

Machine a helical, tapered, conical screw. An example of such a screw would be a tapered drill or the auger that forces meat through a meat grinder. In this case, the program would involve only the X and Z axes and the A axis. As the A axis rotates, the Z axis must move down, while the X axis travels also. This would not be difficult to program. One rotation of the A axis would have to occur simultaneously with one length of the pitch in the X-Z plane line.

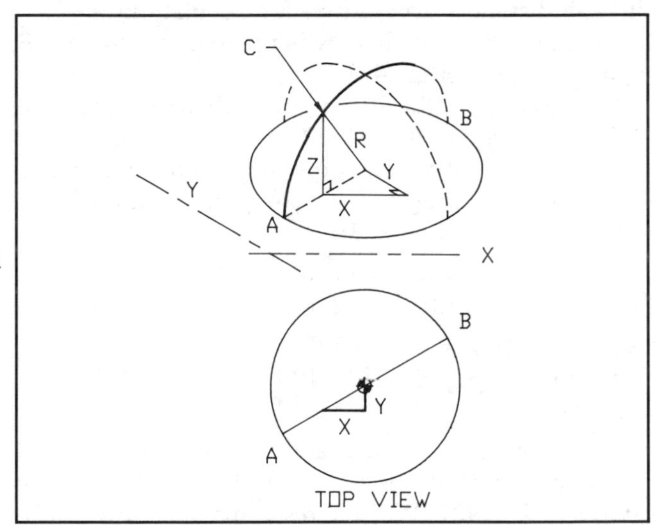

Fig. 19-9. True three-axis circular movement requires the control to continuously interpolate each axis.

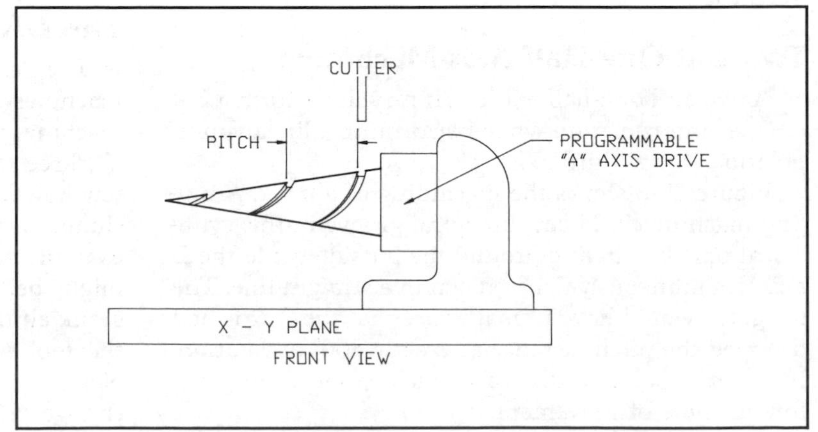

Fig. 19-10. A multi-axis cut on a milling machine.

Further Study

We will not explore advanced multiaxis machining on mills. However in Unit 7, you will be exposed to the programming techniques used for programming these machines. Before you consider learning to program advanced shapes, you should study some higher mathematics. A good math background will give you a solid understanding of what is going on within the computer and how to get the most out of it. Here are some math courses and the order in which they are offered at your local college or tech school.

1. Algebra (intermediate and advanced).
2. Trigonometry (plane and intermediate).
3. Descriptive geometry (solving 3-D problems graphically).
4. Analytic geometry (the equations of shapes).
5. Computer programming (BASIC, FORTRAN, or PASCAL).
6. Calculus differential (the mathematics of changing equations).
7. Calculus integral (the mathematics of changing equations).

MULTIPLE-AXIS TURNING MACHINES

There are three classifications of multiple-axis lathes. Here we will discuss the first two classifications.

1. Secondary operations: drilling, light milling.
2. Auxiliary axes: extra tool slides for faster production.
3. Specialized machines: custom turning machines for special purposes.

All turning machines can be equipped with automatic bar feeding devices. These allow the stock to be fed forward for the next part without operator intervention. Many can feed the material without stopping the spindle.

Class 1 Turning Centers—Secondary Operations

These machines are more than a lathe. They perform small secondary operations as the turning takes place. This gives them an obvious advantage. Some have optional devices that bolt on. Others are factory equipped with special-function devices such as a cross-drilling head or a milling attachment on a tool turret. These machines are more difficult to program due to the timing of the operations and the planning of certain secondary operations. Secondary operation turning centers often complete in one station parts that might otherwise have required several setups.

Tooling and Setup Advantage

Secondary machines gain more than just the time saved in completing secondary operations at a single station. With the part held exactly on center, in a reliable position many operations are simplified. The lathe becomes the setup tooling for secondary operations. This saves more than just machine time. Also, these lathes have an orientible spindle. This means that the spindle can be programmed to work like an indexing head. It can rotate the part a given amount for drilling and milling at specific spacings around the part.

Class 2 Auxiliary Axis Turning Centers

These machines have extra functions, but they are basically lathe functions. These machines can be further broken down to common X-axis machines, independent X-Z and U-W axes machines, and multiple spindle machines. None of these concepts are new to machining. Automatic lathes and screw machines also have these features.

Common X-Axis

The most common extra functioning turning center has a secondary tool turret. This second tool turret is mounted opposite the primary turret and works with it on a common X-axis. This type of machine has twice the toolholding capacity. It is more complicated to set up, but it can accomplish a great deal more in terms of tool shapes available. A slight timesaving is realized because the tools can be ready for the next operation by switching from back to front tool turrets. However, the machine cannot turn out the part significantly faster. Under normal circumstances, only one tool is used at a time.

Independent X-Z and U-W Axis

This machine has two complete tool stations that are programmed independently of each other. This machine turns out parts in significantly less time. Since the second tool station can move in both the X and Z direction, it could be easy to get confused in programming. These secondary axes are called out as U and W.

Multispindle Turning Centers

These machines can accomplish amazing part-cycle time. The machine features three or more spindles, each with its own bar stock. The entire spindle head, carrying all spindles, rotates to each machining station. A six-spindle machine turns out parts in one-sixth the time.

Combinations of the above features, plus automatic part loading and robotic part transfer, are found in industry.

UNIT 6

Compensating Programs for Tool Differences

Mystery surrounds the compensation of programs for tool length, radius, and shape. Many technicians have little understanding of the basics. This subject, like so many others, is simple when studied one fact at a time. For two reasons, we will study manual compensation of programs first.

1. While some excellent controls will plan automatic tool compensation for you, others will not. On these controls, the compensations must be made by you.
2. You are handicapped when requesting computer compensation if you do not understand the basics of manual compensation. If you simply jump to automatic compensation you will never quite understand what is going on.

CONTENTS

CHAPTER 20

WRITING A MANUAL COMPENSATION OF A PROGRAM

This chapter discusses one of the least understood skills in CNC programming. This chapter will make geometry live for you because every manually compensated program is a series of solved geometry problems. After you can solve these problems, you will learn how to use the microprocessor to assist in problem solving.

OBJECTIVES

After studying this chapter, you will be able to:

- Sketch a cutter path for a part geometry.
- Solve for unknown points in the part path.
- Solve for new point coordinates in the cutter path.
- Correctly apply the geometry rules to solve the trigonometry problems involved in point solutions.

20-1 Why Use Cutter Compensation?

If you enter a *part path* program into a control and plot it with a pen, the result will be an exact drawing of the profile of the part. Then mount a tool (lathe or mill) in the machine, and run a part, the shape will be incorrect because of the extra material the cutter takes. Compensating a program corrects for the cutter radius and thus allows the correct shape to be produced. The compensated cutter path line is an expanded version of the part. The expansion amount per side is the cutter radius.

Manual compensation (MComp) is a skill all programmers should possess. When writing programs for any shape on any C/NC machine, the path produced must be changed (compensated) to adjust for the cutter used. These adjustments to the program—compensations—are discussed in this section.

A REMINDER

MComp programs have no flexibility. The cutter size and shape for which they were compensated is the only cutter you can use. Do you remember the difference between part path and cutter path? If not, turn back to Chapter 1, Section 3 (Fig. 1-9) and Chapter 8, Section 1.

After the planning decisions have been made and the general tool path and cutter size have been selected, you are ready to prepare a program. There are four skills in manual program compensation:

1. Making the sketch.
2. Locating the significant points.
3. Calculating the part point coordinates.
4. Calculating the compensated coordinates.

We will study each skill in turn.

20-2 Sketching the Cutter Path

THE IMPORTANCE OF A SKETCH

When writing an MComp program, always draw a sketch. There are hundreds of mathematical traps in writing an MComp program. A good sketch will prevent many of them. A good sketch will:

1. Provide a permanent record of your work for future reference.
2. Give you a clear understanding of the problems and thus prevent incorrect assumptions.
3. Provide a quick way to verify results. Measuring a sketch is a simple way to double-check your calculations.

Special Concession to Programmers

In machine shops, there is a rule that states "**never write on a part print.**" There is an exception to this rule for programmers. Often, a company will allow a programmer to sketch on a print. Once marked upon, this print never leaves the programming area. Drawing on an accurate print saves much time. Always ask before drawing on a part print.

Use color to draw the cutter path. Figure 20-1 on page 220, shows a part outline. We will refer to the part in Fig. 20-1 in this chapter and in Chapter 21. It could be a mill part or one-half of a lathe part. The concept of compensation is the same, with the exception of the tool flank shape in turning.

The radius of the tool is sketched in several locations, using a circle template if possible. Then the centerline of these circles is connected using a colored pencil. Suppose a .25″ radius tool is chosen to machine the part shown in Fig. 20-2 on page 221. The circle will have a .50″ diameter. This colored line will be the line you actually program. It will be the cutter path line.

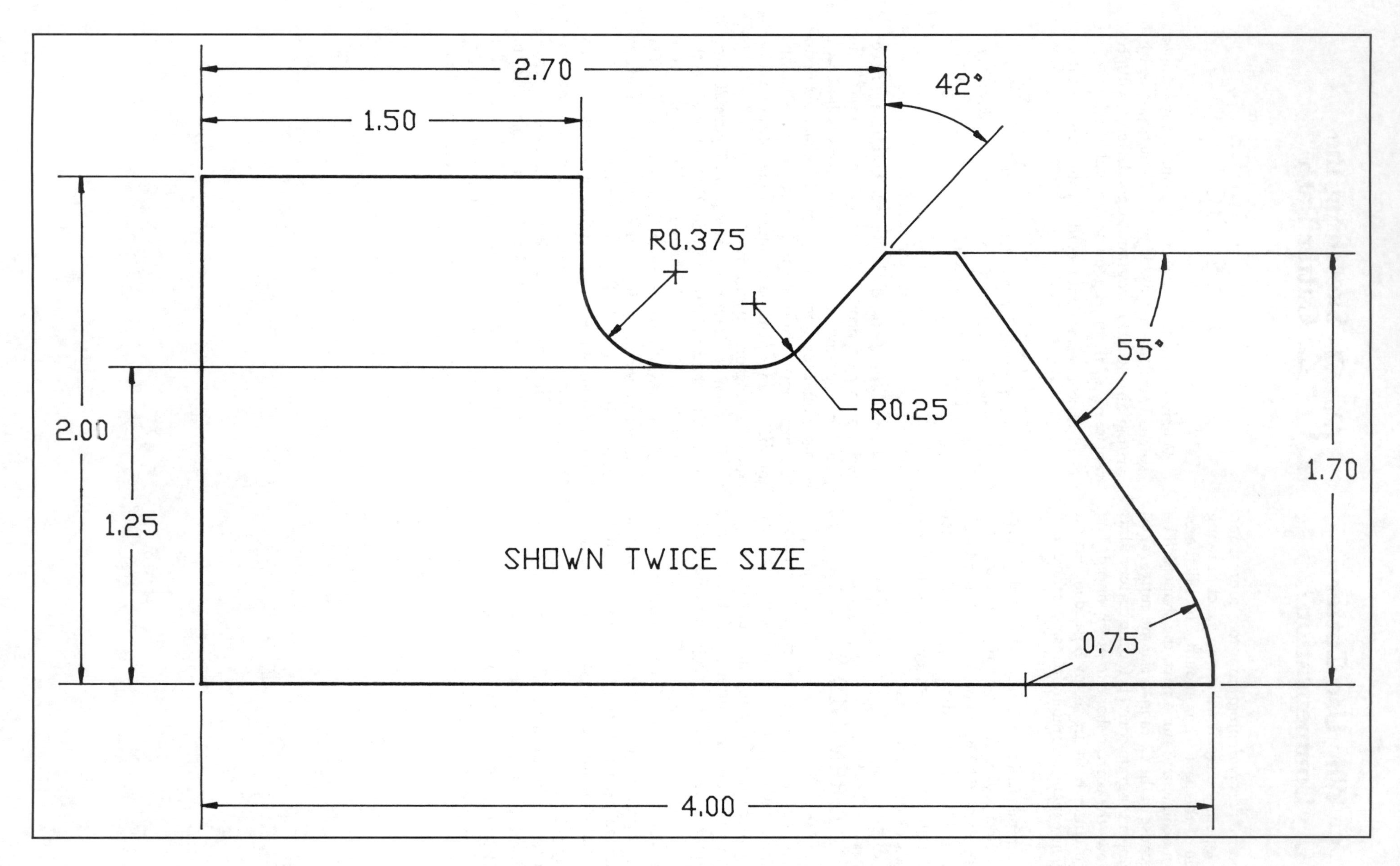

Fig. 20-1. A typical part print to be programmed by compensation.

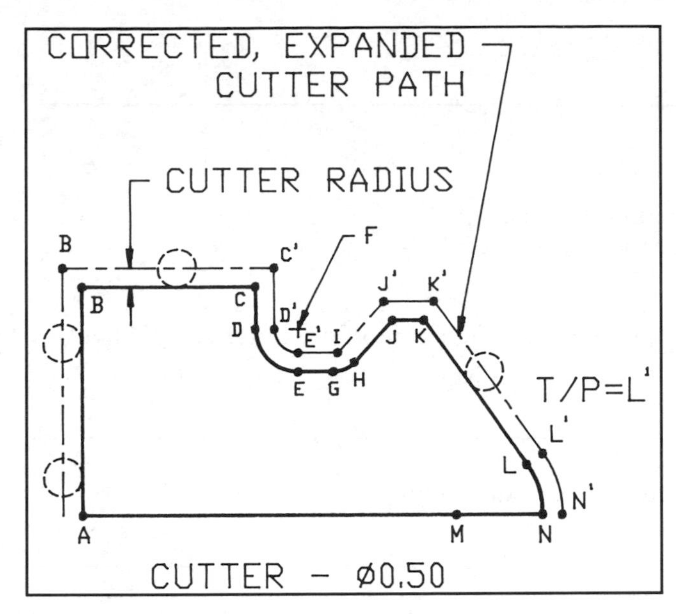

Fig. 20-2. A sketch of the cutter path.

LATHE NOTE

While sketching the tool path, consider the shape of the tool. On most part shapes any tool shape problems should be obvious. They will be situations where the tool simply cannot fit into the intended part shape. Either the tool radius is too big or the tool flanks and shank are the wrong angle or size to produce the part shape. If you are in doubt, draw several possible tool shapes on tracing paper. Hold these tool shapes over the part outline to see if they fit the intended application.

Activities

Sketching Cutter Centerline

Using a sharp, pointed colored pencil, machinist's rule, and circle template, draw a cutter path sketch on Fig. 20-1. Use a cutter *radius* of .25″ and a diameter of .50″. Be accurate. This sketch will be used again in future activities. (Note that the drawing is twice the size of the part. Therefore, the cutter circles will also need to be scaled up to twice size. Actual tool radius will be .50″.)

20-3 Locating the Significant Points

In general, each significant point on the part path has a corresponding point on the new cutter path.

Refer to Fig. 20-2. Points B, C, D and E on the part path have corresponding points B', C', D', and E' on the compensated cutter path. These are the two following exceptions:

1. *Centers of arcs remain the same.* Point F at the center of the circle is the same for both paths. No new coordinate must be calculated.
2. *Formed radii do not require tangent points for the arc.* The cutter moves to point I and forms the .25″ radius. There is no arc to program. Thus, there are no tangent point coordinates. They are eliminated. Points G and H in the part path have no corresponding point in the cutter path.

1. Drawing on Fig. 20-1, make a dot at each significant point on the part path.
2. Label these points the same as those in Fig. 20-2, where they are labeled A through M.
3. Make a colored dot on each significant point on the tool path.
4. Label these A' through N'.
5. There were two significant points on the part that had no corresponding point on the cutter path. What were these points?

____

6. There were three points that remained the same for each path. List these points.

____

____

20-4 Calculating the Significant Points on the Part Path

AN EXAMPLE OF A PROBLEM

Refer to Fig. 20-1. We will assume that Fig. 20-1 shows a lathe job. Line A-N is the centerline of the part. Point A is the PRZ.

Often, before the new significant points can be calculated, the part points must be given a coordinate value. Notice on your prepared Fig. 20-1 that there is no Z dimension from PRZ to point K. Before calculating point K', point K must be solved. A Cartesian value must be found. Point A is the PRZ on the centerline of the part. We know that point K has an X axis distance of 1.7". We do not know the Z value.

We need a Cartesian coordinate value for the Z distance of point K from the PRZ. A polar coordinate machine would simplify the calculation, but for this example we will assume a control limited to Cartesian only. The following solution is offered as an example. After studying this example, we will look at the nine geometry rules that are used most often in these programming geometry problems. The actual trigonometry is omitted.

Two triangles must be solved to calculate the Z value of point K. The solution is shown in Fig. 20-3. Note that the X distances are given as radii. This is to simplify the concept. Remember, many lathes are programmed using diameter values for the X axis. The radii values in the following example would be doubled to represent diameters then entered into the control.

The Z axis distance is found by subtracting distances [MO] and [MN] from 4.0"; [MN] = .75" (the part radius).

Therefore: ZK = 4.0 –.75 –[MO]

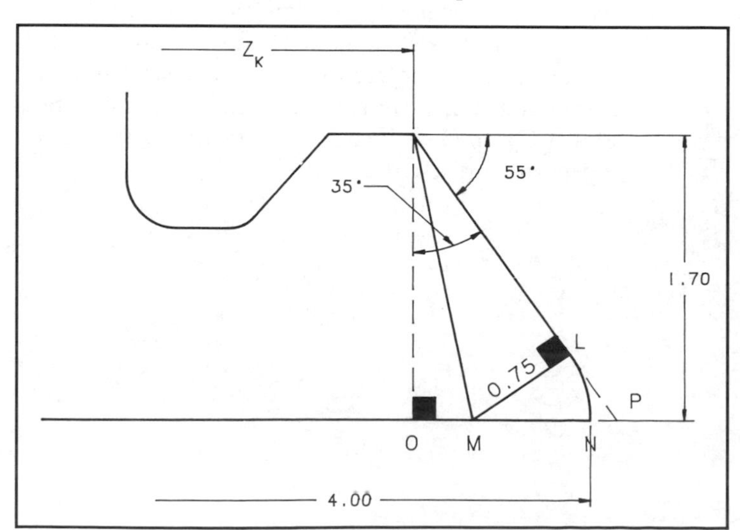

Fig. 20-3. Solution triangles to find point K.

TO FIND DISTANCE [MO]

It can be proven that angle LKO = 35°.
(Complementary angles 90-55 = 35.)

Side [KL] is extended to point P.
(We already know two facts in this triangle: side [OK] and angle PKO = 35.)

Solve triangle KOP for side [OP] = 1.19035.

Line [ML] is drawn to the tangent point L. This creates a new triangle MLP.
(We use the tangent point rule to draw [ML]—discussed later.)

To find [MO], we now solve subtriangle MLP for side [MP] and subtract this from [OP].
[OP] – [MP] = [MO]

In triangle MLP, side [ML] = .75″.
Angle PML = 35°.
(Rule 5 shown later.)

Size [MP] = .91558″.

[OM] = ([OP] – [MP])
[OM] = (1.19035″ – .91558″) = 0.27477″.

Distance ZK = 4″ –.75″ –.27477″ = 2.97523″

This type of math is necessary for the programmer. This proof is typical of those encountered in general shop math. If you have difficulty with this type of geometry or trigonometry, ask your instructor for supplementary problems and suggestions for extra study.

20-5 The Rules of Geometry for Finding Points

The following rules of geometry are used in the majority of problems requiring solutions such as that above. Put a cello-tab on this rules page and refer to it often.

- Rule 1. When any two lines intersect, opposite angles are equal and the sum of any two adjacent angles is equal to 180°. Fig. 20-4.
- Rule 2. The sum of the three angles in a triangle equals 180°. The sum of the two acute angles in a right triangle equals 90°. The two acute angles are complementary angles. Fig. 20-5.
- Rule 3. When any line cuts across two parallel lines, corresponding angles are equal. Alternate exterior and interior angles are equal. Fig. 20-6 on page 224.

  The sum of any two noncorresponding angles is equal to 180°. These two noncorresponding angles are supplementary angles.

$\angle A = \angle C$
$\angle B = \angle D$

$\angle A + \angle D = 180°$
$\angle D + \angle C = 180°$

A B C D

Fig. 20-4. Intersecting lines. Rule 1.

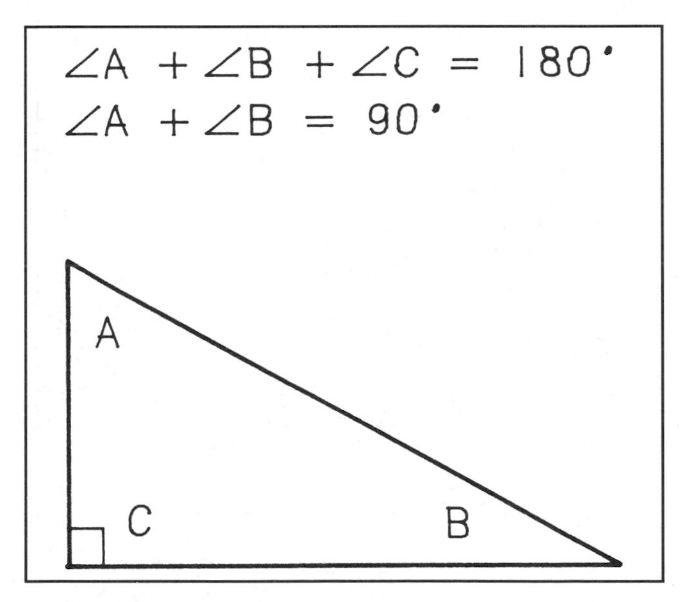

Fig. 20-5. Angle sums in triangles. Rule 2.

- Rule 4. If a straight line is tangent to an arc or circle, a line drawn from the center of the arc to the point of tangency will form a 90° angle with the straight line. Fig. 20-7 on page 224.

  This constructed line then becomes useful as a side of a right triangle, as in the previous development. See Fig. 20-3. Side ML became a side of a solution triangle.

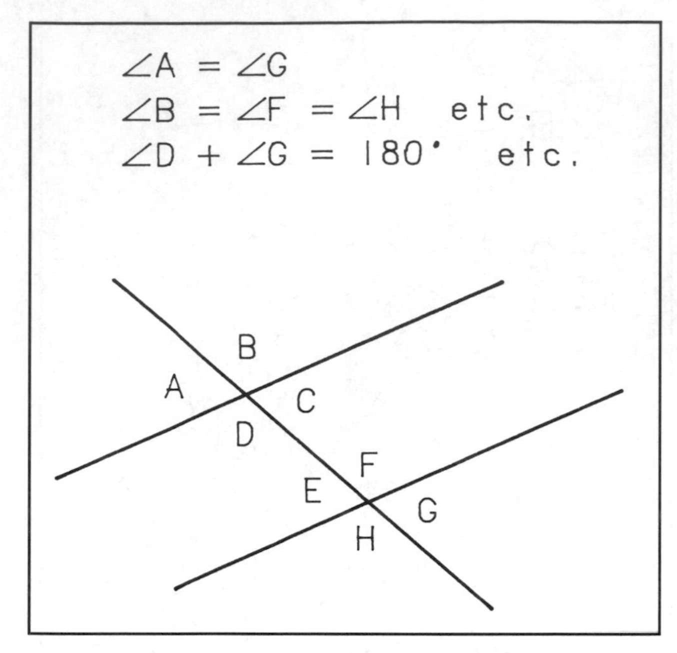

Fig. 20-6. Parallel lines cut by a transversal line. Rule 3.

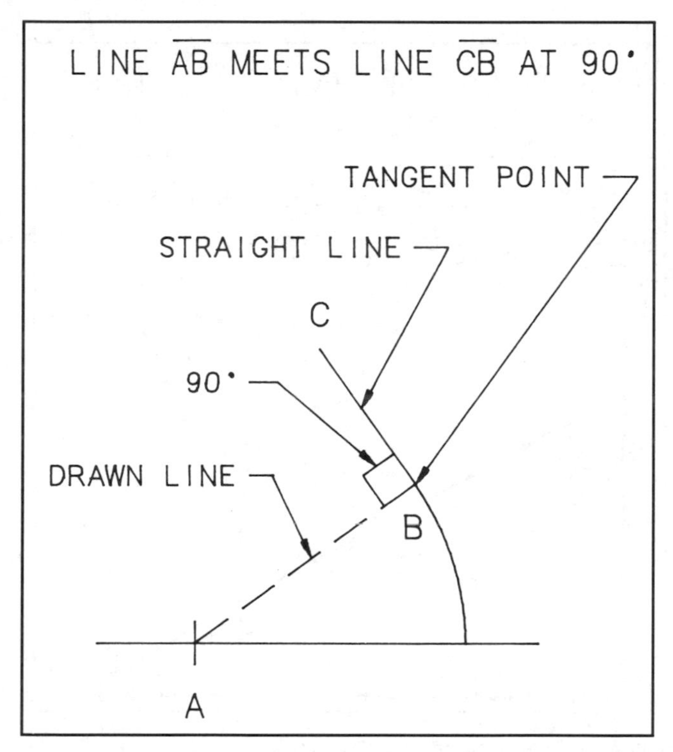

Fig. 20-7. Tangent point. Rule 4.

- Rule 5. Two angles may be proven equal if the relationship between corresponding legs is proven equal. Fig. 20-8.

  In Fig. 20-8, the left leg of angle A is at 90° to the left leg of angle B. Also the right leg of angle A is at 90° to the right leg of angle B. To understand Rule 5, refer again to Fig. 20-3. Notice that angles OKP and LMP were proven equal. Find the left leg of each angle (sides KP and ML). Notice that they are at 90° to each other. Now find the right legs of each angle (sides OK and MP). Notice that they, too, were at 90° to each other. Right legs and left legs have equal relationships—*the angles are equal.*

- Rule 6. When an arc is tangent to both legs of an angle, a line drawn from the vertex of the angle to the center of the arc will bisect the angle. Fig. 20-9 on page 225.
- Rule 7. If an arc is tangent to both legs of an angle, the degree spanned by the arc will equal 180° minus the angle between the legs. Fig. 20-10 on page 225.

Rule 7 is most useful when programming arcs in polar coordinates or using the radius method on a Cartesian controller. If you know a certain arc is tangent to both legs of an angle, the degrees spanned by the arc will be equal to 180 minus the number of degrees between the legs of the angle. For example, in programming arc B in Fig. 20-10, we find that the included angle between the legs of angle A is 40°. Then arc B is 140°. (180° – 40° = 140°.)

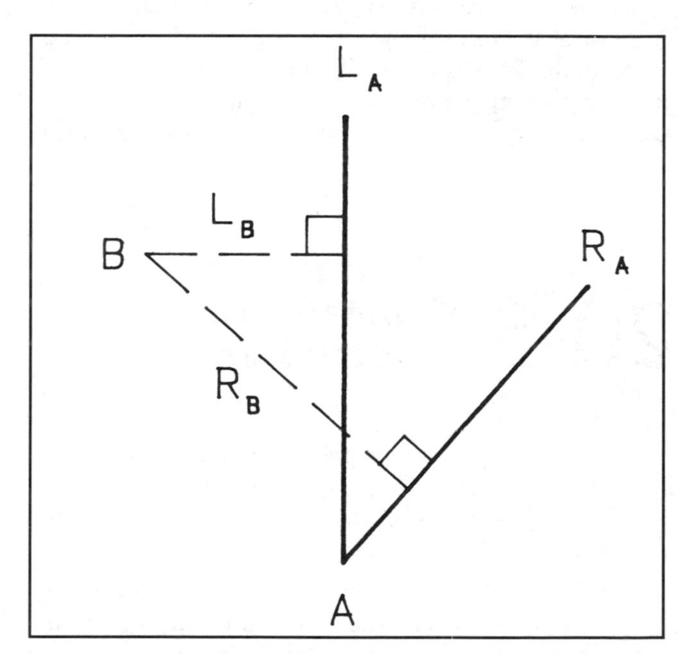

Fig. 20-8. Right leg-left leg. Rule 5.

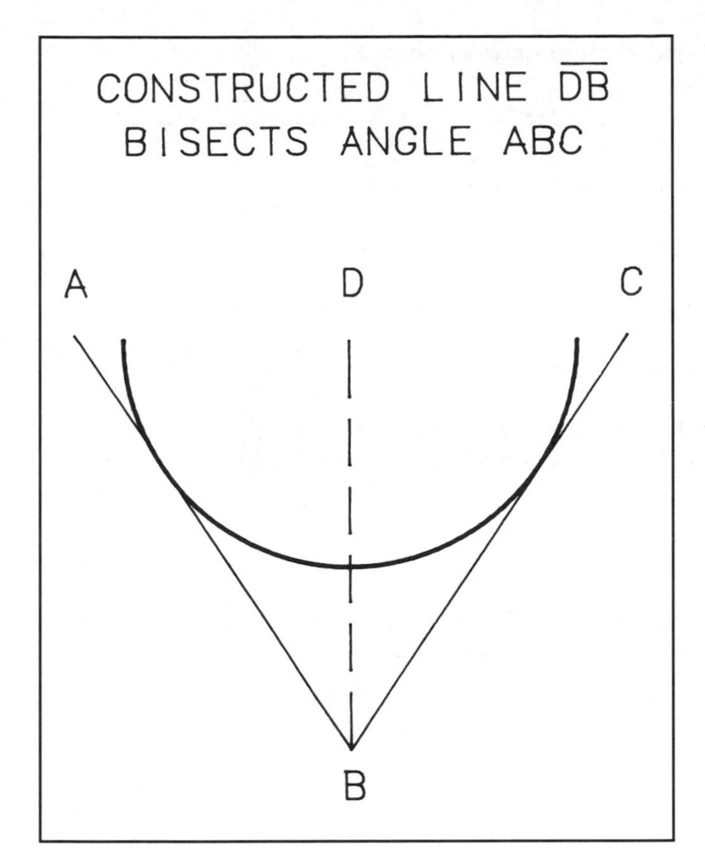

Fig. 20-9. Angle tangent to circle. Rule 6.

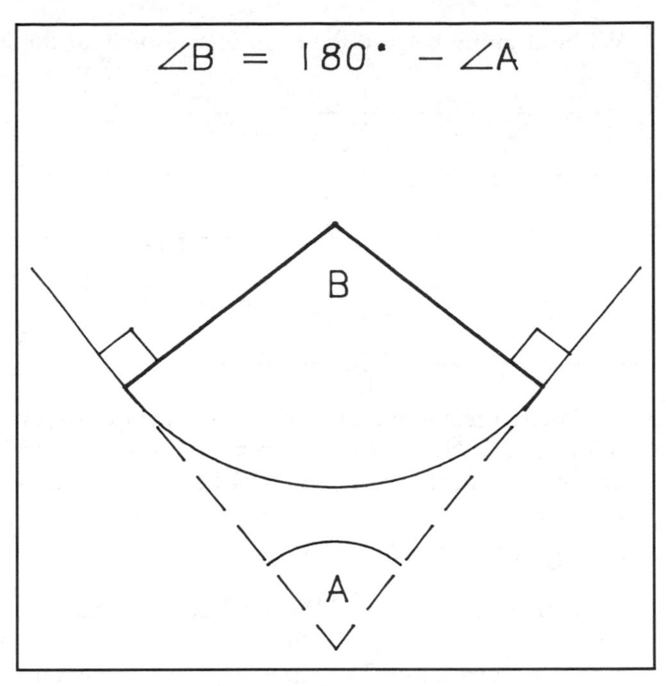

Fig. 20-10. Complimentary arcs. Rule 7.

Find the Cartesian X and Y distance of each of the following points on the part shown in Figs. 20-1 and 20-2.

Each of the following significant point problems can be solved using the rules of geometry given previously. For this set, each problem point is on the part path. It is not on the compensated cutter path. Measure your sketches or the drawing as a check for your work. This is important. All X point coordinates are radius as shown and vertical on the page. If entered into a control, they would probably be converted to diameters. The PRZ is point A; Z distances are to the right from point A.

Answers are in the Reference Point following this activity.

1. What is the X distance for point I? Refer to Figs. 20-1 and 20-2.

____

2. Using Fig. 20-1 as a worksheet, calculate the Z distance from PRZ for point I.

3. Without going back to the example, resolve for the X and Z coordinates of point K.

4. Complete the command block below for a Cartesian I-K curve method command for the curve between points L and N. L is the start point. This command block is in absolute. Calculate the absolute value for X and Z. Calculate the incremental values for I and K. Line MN is the centerline of this part. The radius of this curve is .75″.

G90 G02

Remember, the X and Z coordinates are the end point identifiers. These may be absolute or incremental depending on the mode in which the program is operating. The I and K values are the center identifiers from the start point. They are always incremental.

5. What is the X value for point F?

Draw your solution sketches on Fig. 20-11 for the next two problems.

6. Refer to Fig. 20-11. Solve for the X and Z absolute value for points N and O.

7. Refer to Fig. 20-11. Solve for the incremental X and Z distance between points P, Q—starting at P.

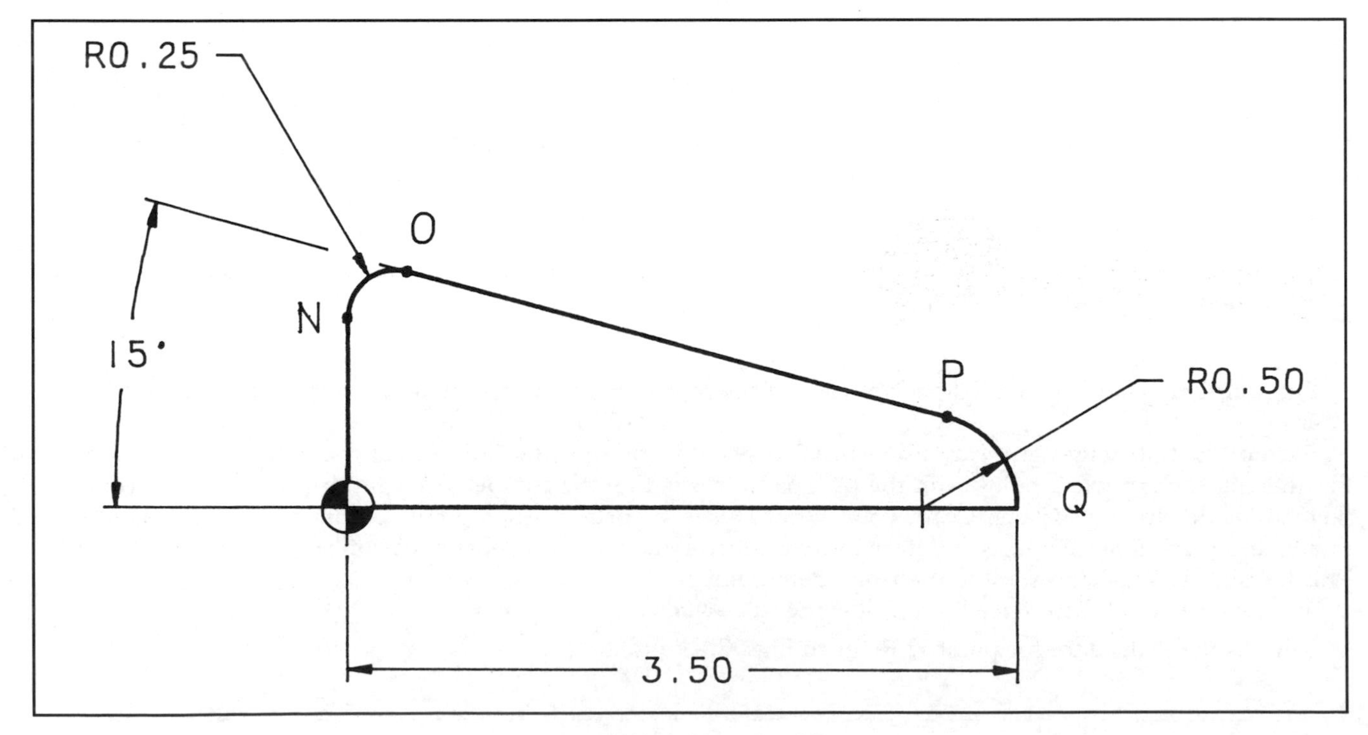

Fig. 20-11. Problems 6 and 7.

Here are the answers to the questions in the previous activity. How many of your answers were correct?

1. Add 1.25″ + .25″ = 1.50″.
2. Refer to Fig. 20-12 on page 228. Calculate the Z distance for point I:
   Z = 2.7 – distance J-R.
   2.7 – .5165 = 2.1835
   To find distance JR:
   Construct line IS (Rule 4).
   Construct triangle TRJ.
   Angle RJT = 48° (Rule 1).
   Angle RTJ = 42° (Rule 2).

   Solve for side IT in triangle IST:
   IT = CoSecant 42 × .25″ = .3736″.

   Solve for JR in triangle JRT.
   Distance IR = .20″ (1.7″ – 1.25″ – .25″ = .20″).
   Side TR = .5736″ (IT + IR).
   JR = Tan 42 × .5736″ = .5165″.
3. See the example solution for details.
   X = 1.70″ Z = 2.9752″
4. G90 G02 X0.0 Z4.0 I–.4302 K = – .6144
   Refer to Fig. 20-13 on page 228.
   Angle LMN = 35° (Rule 5 and complementary angles).
   Distance I= Sine 35 × .75″ = .4302.
   Distance K= Cosine 35 × .75″ = .6144.
   Place minus signs on I and K because of direction from start point to center point.
5. X for point F= 1.625 (1.25 + .375).
6. Solving for absolute values of point N: X = .9957 Z= 0.0. Refer to Fig. 20-14 on page 228.
   Construct triangle ADC.
   Subtract DN from AD.
   To find AD, we need to complete triangle ACD by finding AC.

   To solve for the length AC:
   Construct line BP (Rule 4)
   Solve for line BC= CoSecant 15 × .5 = 1.9319.
   Subtract BQ from BC: 1.9319 – .5 = 1.4319.
   Add QC + AQ for the side AC: 1.4319 + 3.5 = 4.9319.

   Now solve for the height AD:
   AD = Tan 15 × AC
   .2679492 × 4.9319 = 1.3215.

   Solve for ND:
   Angle NDO = 75° (Rule 2).
   Angle NDF = 75°/2 = 37.5° (Rule 6).
   Solve for the height ND:
   ND = CoTan 37.5 × .25 = .3258

   To find the X distance, subtract ND from AD:
   AD-ND = 1.3215 –.3258 = .9957

   Solving for absolute value point O:
   X = 1.2372 Z = .3147
   Construct triangle ODG.
   Angle ODG = 15° (Rule 3)

   Solve for DG = CoSine 15 × ND.
   Where line DO = ND (Rule 6).
   = .9659 × .3258 = .3147
   Solve for OG:
   OG = Sine 15 × ND.
   .2588 × .3258 = .0843
   AD - OG = 1.3215 – .0843 = 1.2372
   (OG could also have been found using the Pythagorean theorem.)
7. Refer to Fig. 20-14. Solving for the incremental value from point P to Q:
   X distance = –.4830 (minus sign due to direction).
   Z distance = +.3706

   Construct triangle PHB.
   Solve for sides HQ and BH.
   X solution = PH.
   Z solution = .500 R – BH.
   Angle BPH = 15° (Rule 5).

   X Solution
   Side PH = CoSine 15 × .5
   = .9659 × .5 = .4830

   Z Solution
   BH = Sine 15 × .5 = .1294
   .2588 × .5 = .1294
   .5 – BH = .50 – .1294 = .3706

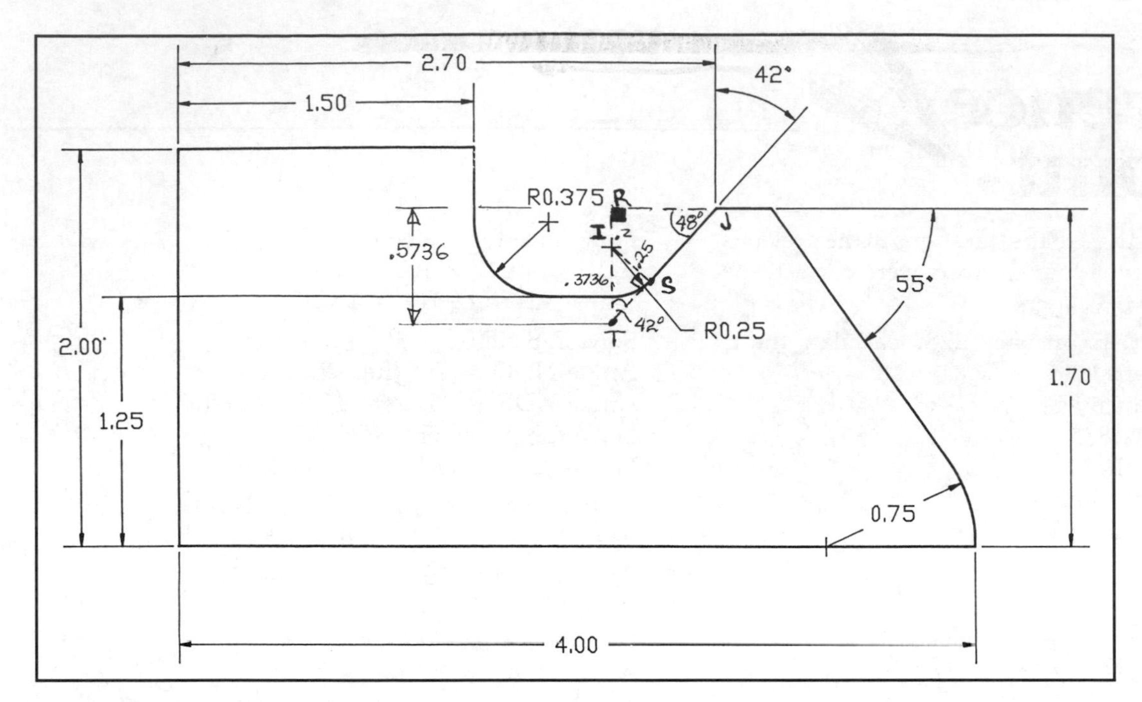

Fig. 20-12. Sample solution.

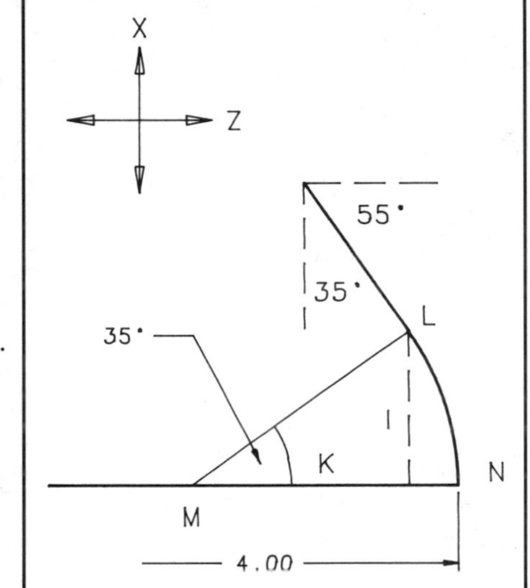

Fig. 20-13. Sample solution.

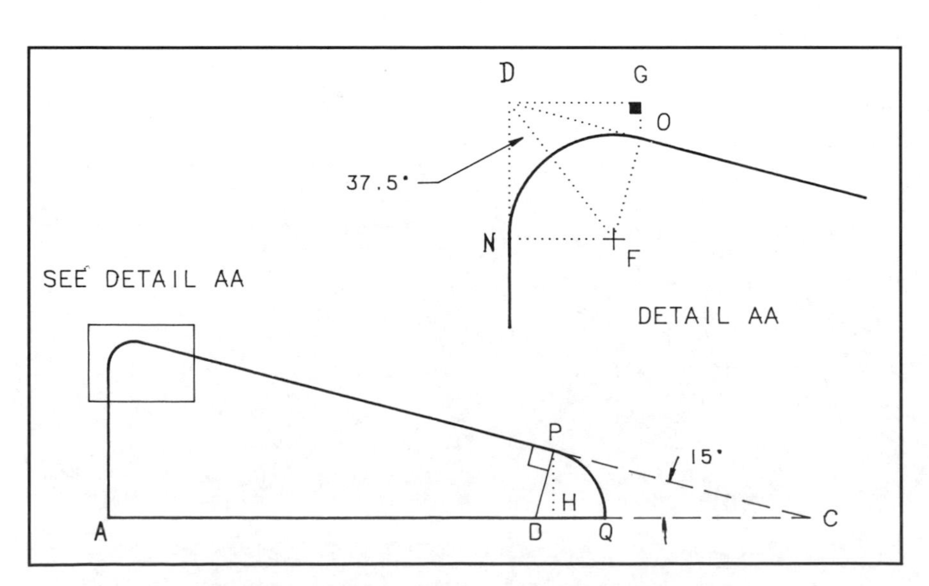

Fig. 20-14. Solutions using geometry and trigonometry.

20-6 Calculating the Compensation Adjustments

COMPENSATION ADJUSTMENTS

After finding the values of the significant points on the part path, you are ready to calculate the compensating distances. These compensating distances are caused when we apply a tool radius to the geometry of the part path. The significant points along the cutter path will be shifted. To write a manual compensation program, we must calculate the shifts.

The following compensation adjustments apply equally to lathes and mills. They occur when a cutter centerline (cutter path) is drawn parallel to a part geometry. With lathes, the additional obligation is to make sure that the cutter flanks do not interfere or rub on the work.

STRAIGHT LINE INTERSECTIONS—OUTSIDE CORNER

Refer to Fig. 20-15. A cutter with radius R must go around any outside corner intersection of two straight lines, where the included angle A is less than 180° and greater than 0°.

The compensating amount (L) is *added* to the part path. It is found with the following formula:

$$L = R \times \text{CoTangent}\ (A/2)$$
$$R = \text{Cutter Radius}$$
$$L = \text{Compensating Amount}$$
$$A = \text{Included Angle } 180 < A > 0$$

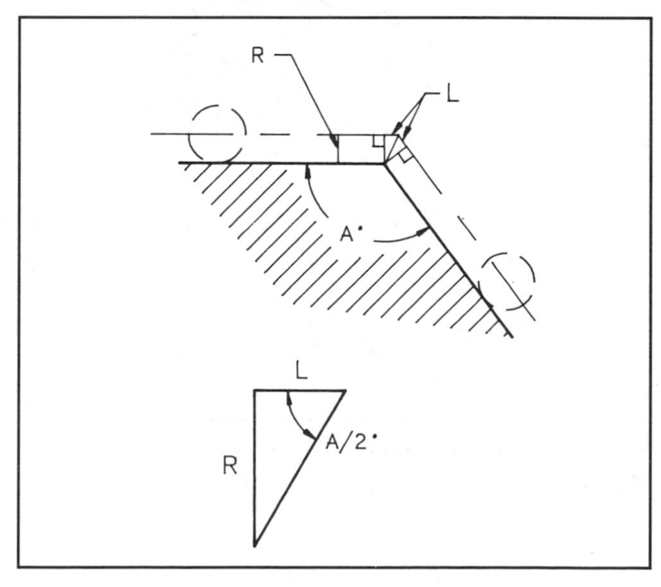

Fig. 20-15. A cutter must go around an outside corner.

COORDINATE COMPENSATION

There will be an adjustment to the coordinate value of the new point. The amount of shift in the X and Y or X and Z axis depends on the relationship of the corner to the basic axis of the machine. See Fig. 20-15A. Determine the amount of shift as follows:

First, determine the length of line P-P′, using the following formula:

$$P\text{–}P'' = \text{CoSecant}\ (A/2)\ X\ R$$

Second, draw triangle PP′B, which has its lesser sides parallel to the machine axis system. Put another way, the two nonhypotenuse sides of this triangle will be the compensation amounts. They should be parallel to the axis system.

Third, determine the slope of P-P′ as compared with the X,Y, or Z axes of the machine (shown as angle Q). Solve triangle PP′B for the horizontal and vertical legs.

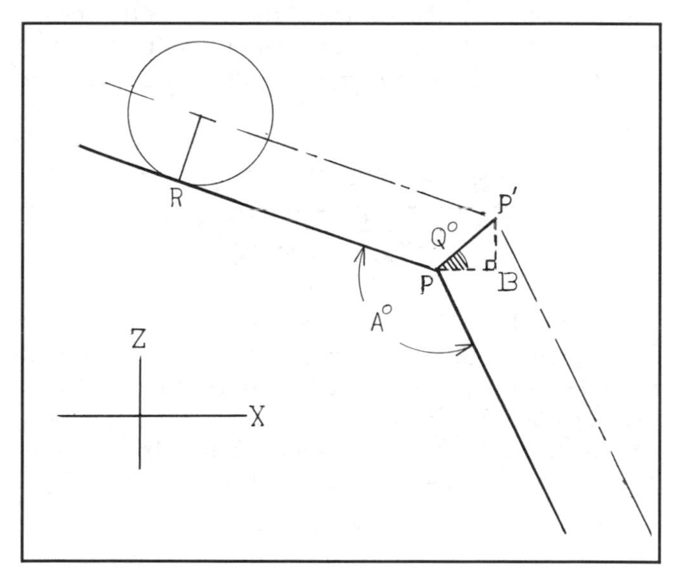

Fig. 20-15A. Coordinate adjustment as a cutter with radius R goes around an outside corner.

These will be incremental compensation adjustments to the part points. They will be added to or subtracted from the coordinate position of the old part point to create the new cutter path point.

Using the following formulas:

$$GB = \text{Cosine}(Q) \times PP'$$
$$G'B = \text{Sine}(Q) \times PP'$$

POLAR SHORTCUT—OUTSIDE CORNER

Refer to Fig. 20-16. If you have a lathe or mill control that is capable of polar arcs, there is a shortcut to compensating around an outside corner. You might program a circular arc that does not cut away the corner but instead takes the shortest path around. Program the cutter path to the tangent point T/P. Then program an arc with radius R to the next T/P. The number of degrees this arc will travel is determined by the following formula:

$$\text{Arc Degrees} = 180 - A$$

This is a direct application of Geometry Rule 7, stated previously. The center of the arc is at the corner intersection of the part. If the center is shifted slightly inward, the arc will also remove the sharp edges for you—a bonus.

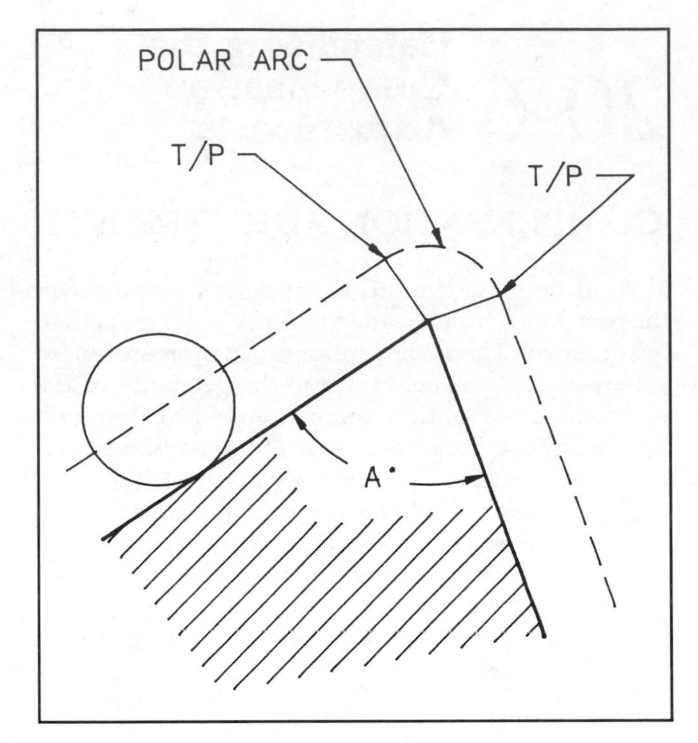

Fig. 20-16. Polar shortcut for outside corners.

STRAIGHT LINE INTERSECTIONS—INSIDE CORNER

Refer to Fig. 20-17. A cutter with radius R must go around an inside corner intersection of two straight lines. Here the included angle A is less than 180° and greater than 0°. The compensating amount L is subtracted from the part path and is found by the same formula for outside corners:

$$L = R \times \text{CoTangent}(A/2)$$

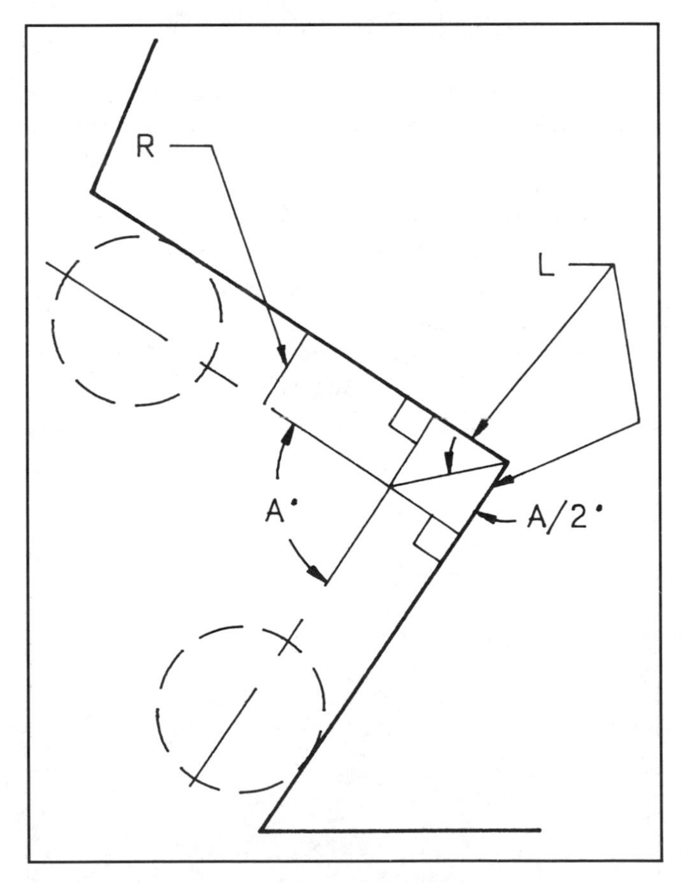

Fig. 20-17. The distance L is subtracted from the tool path.

COORDINATE COMPENSATION

Inside corner compensations require a point shift in the same fashion as outside corners. The solution is as follows.

1. Find the length of the line connecting the points. This is shown as PP′ in Fig 20-15A.
2. Find the slope of this line as compared to the normal axis system of the machine.
3. Draw the solution triangle with the perpendicular legs parallel to the machine axis system.
4. Calculate the horizontal and vertical legs of this triangle.

These are the incremental adjustments to the original point to create the new cutter path point.

Straight Line Intersecting an Arc—Outside Compensation

Refer to Fig. 20-18. A cutter with radius r must go around an outside tangency. The basic radius of the part geometry R must be increased by adding the cutter radius r. The new cutter path arc radius equals:

$$R + r = \text{Cutter Path Radius}$$

Tangent Point Shift

Refer to Fig. 20-19. When the curved cutter path is tangent to a straight line that lies at an angle A to a reference axis of the machine, the tangent points will be shifted in both the X and Y or X and Z axes. The amount of shift can be determined by solving the small triangle shown. There will be a shift for the parallel and perpendicular sides. The angle in the compensation Triangle is A and proven by Rule 5.

The side parallel to the reference axis is found by the following formula:

$$\text{Parallel Shift} = \text{Sine } A \times R$$

The perpendicular shift is found by the following formula:

$$\text{Perpendicular Shift} = \text{CoSine } A \times R$$

Straight Line Intersecting an Arc—Inside Compensation

Refer to Fig. 20-20. When the control is compensating for a cutter inside a part where a straight line is tangent to an arc and the straight line is not parallel to one of the reference axes, the tangent points will be shifted.

1. The cutter path radius is the difference in the part radius and the cutter radius: R – r.
2. The solution triangle for tangent point shift is solved using the formulas shown above.

Parallel Shift = Sine A × r
Perpendicular Shift = CoSine A × r

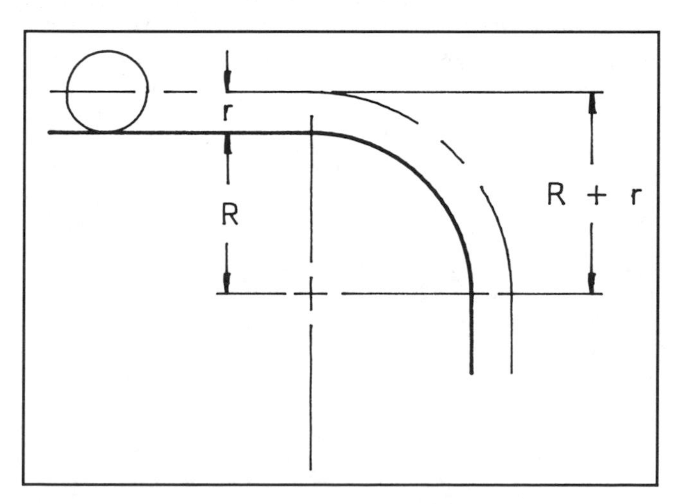

Fig. 20-18. A compensated outside radius is equal to R + r.

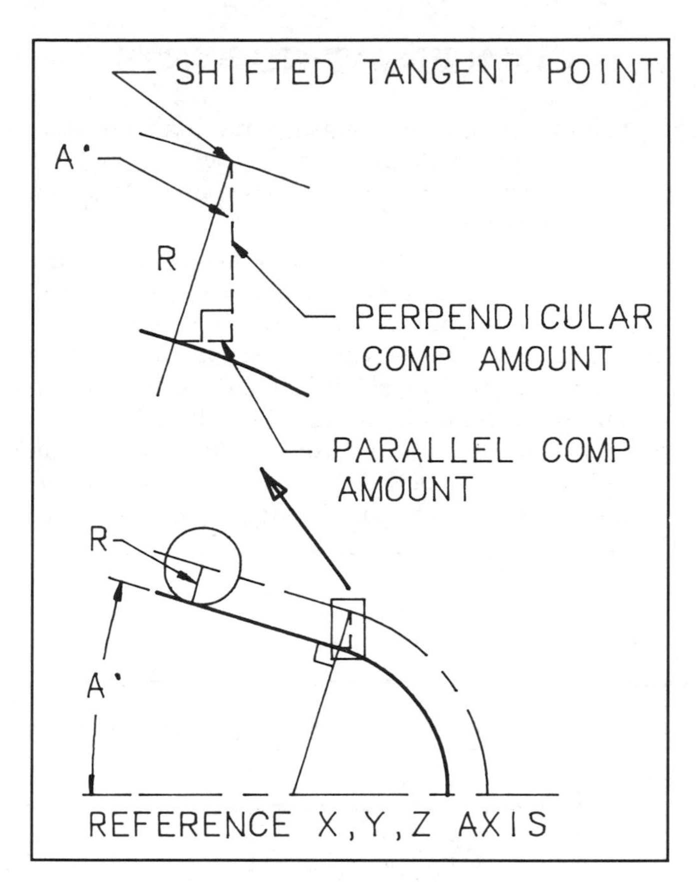

Fig. 20-19. Tangent point shifts may be calculated using these formulas.

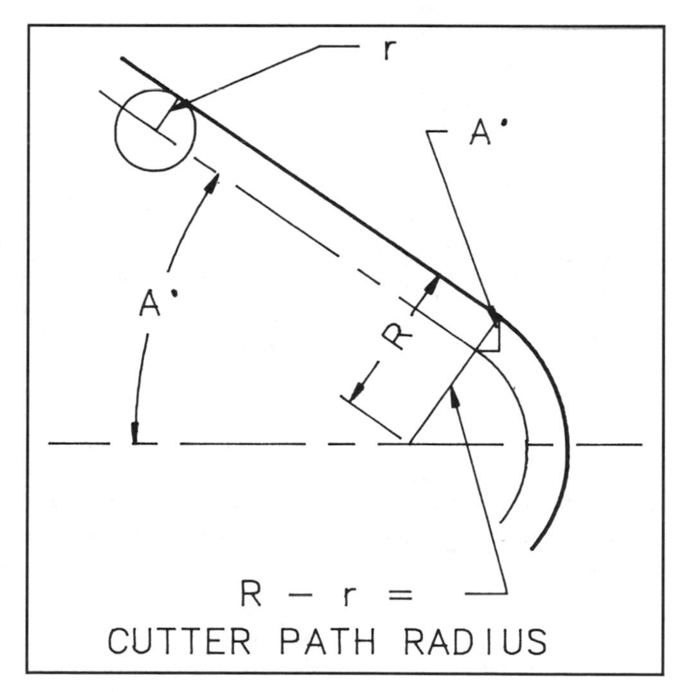

Fig. 20-20. Calculating a point shift as a sloping straight line becomes tangent to a curved line.

COMPENSATING HINTS

The following three compensating hints are often overlooked. Although simple, they can evade you unless you have drawn a sketch of the path to check as you write the program.

1. *Compensating outside features.* When compensating around the outside of part features, remember to add the cutter radius to both sides of the dimension.
2. *Compensating inside features.* When compensating inside features, correct the dimensions by subtracting the radius on both sides. Fig. 20-21.
3. *Compensating stepped features.* When going from an outside to an outside line or inside to inside line, do not change the part distance. Shift the cutter path by a distance equal to the radius. Fig. 20-22.

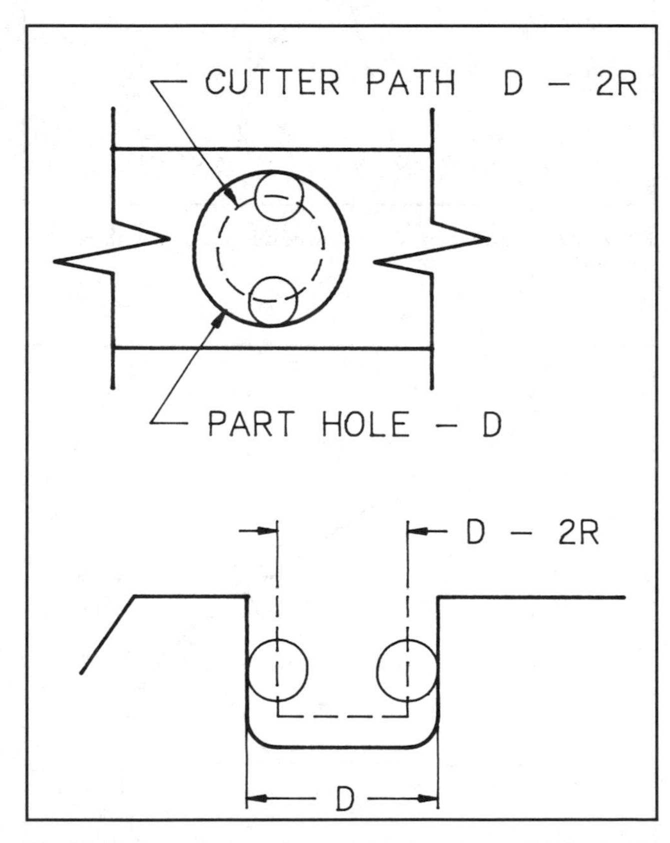

Fig. 20-21. Remember to subtract twice the cutter radius for inside features.

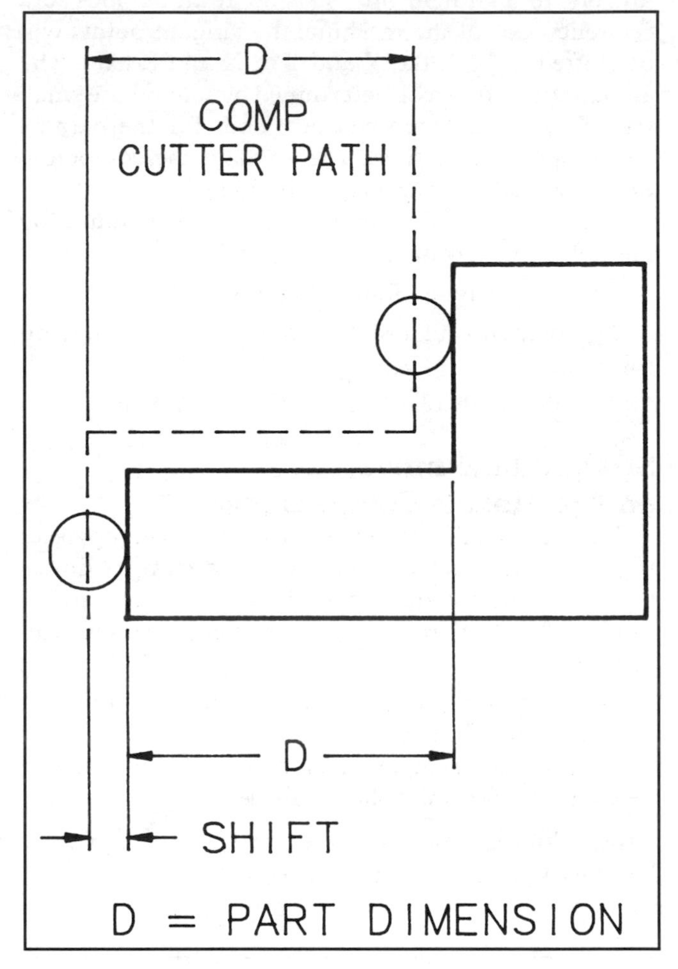

Fig. 20-22. There is a side shift when compensating stepped features.

Activities

Writing a Compensation Program

Write a compensated program for one of the three suggestions below in the language of your lab. Follow the general instructions. To check your program, you will need to plot or run the part or use the graphic display on the control. The compensated program choices are:

1. A part shape assigned by your instructor that has angular surfaces and radii.
2. One of the drawings in Appendix A.
3. Refer to Fig. 20-23 and the following instructions for an advanced challenge.

General Instructions

A) *Identify the significant points on the part.* Draw a dot on the drawing for each significant point on the part geometry. The part shown in Fig 20-23 has twelve such points if your control will program arcs through a quadrant line. Otherwise, it has thirteen. There is an extra point at the top of the upper .375″ radius.

B) *Calculate absolute values for each significant point.* To write a program, you will need to calculate certain significant points on the part geometry first. List each point on the chart provided—accurate to four decimal places.

Measure the drawing as you make a calculation—it is to scale and will be a quick check of your work. If you cannot, with some effort, calculate the significant points the answers are on the next page—Reference Point.

C) *Draw the cutter path centerline.* Make a colored dot for each significant point on the cutter path. Use a photo copy of Fig. 20-23.

D) *Solve for the incremental compensation adjustments to each significant point on the cutter path.* Retain your calculations. They will be needed to program each curve. Use the chart provided.

Lathe Instructions

A) Start at point A. Turn to point C.
B) Line A-B is the centerline.
C) Assume point B is the PRZ.
D) Use a tool nose radius of .20″. This is far too big. It is chosen to coincide with the mill problem so the cutter path will agree in the Reference Point.
E) Remember to turn the radius X axis dimensions into diameters if your lathe accepts diametrical entries.
F) Label the chart axes as X and Z.

Mill Instructions

A) Assume point B is the PRZ.
B) Write a program that proceeds from point A to C using climb milling.
C) The cutter has a radius of .20″.
D) Label the chart axes as X and Y.

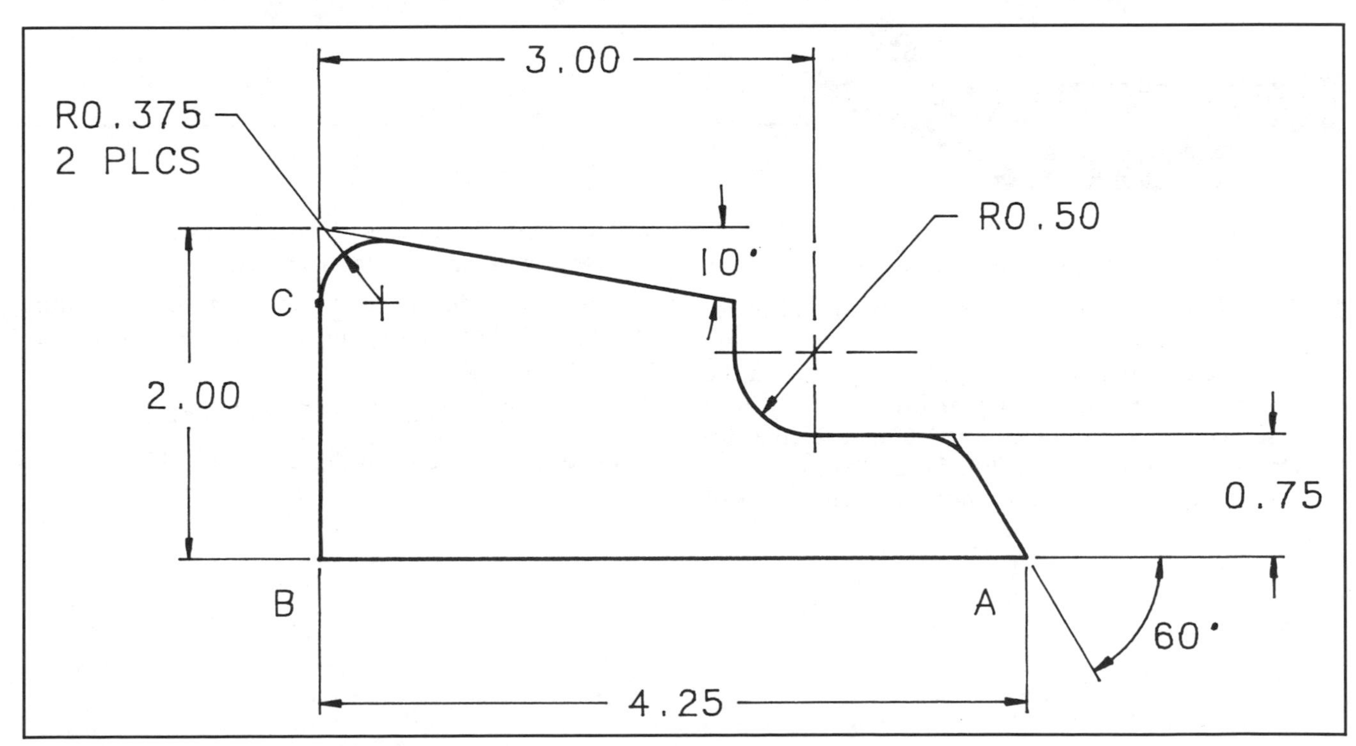

Fig. 20-23. Manual program compensation student activity drawing.

Absolute Point Value			
Coordinates (Mill or Lathe)			
Point	___Axis	___Axis	
	(X or Z)	(Y or X)	
B	0.0	0.0	PRZ
C	0.0		Start Point
D	.375		Center of .375R
E			
F			
G			
H			Center of .500R
I			
J			
K			Center of lower .375R
L			
A		0.0	End Point

Incremental Compensations			
Coordinates (Mill or Lathe)			
Point	___Axis	___Axis	
	(X or Z)	(Y or X)	
C'			
D'			*No change to center*
E			
F'			
G'			
H			
I'			
J'			
K			
L'			
A'			

Significant Points

This Reference Point deals with four topics:

1. The significant points on the part path and cutter path. Fig. 20-24 on page 235.
2. Absolute distances for the significant points on the part path. These are in tabular form.
3. A sketch of the triangles needed to find the distances. This drawing, Fig 20-25 on page 235, shows only the lines used. It does not show the results of any work with trigonometry.
4. Incremental distances for the points on the cutter path. These are in a tabular form.

You can use this information to write your program. Programs will differ, depending upon the type of controller in your lab.

Part Path Point Coordinates

Absolute Values. Refer to Fig. 20-23 and Fig. 20-24.

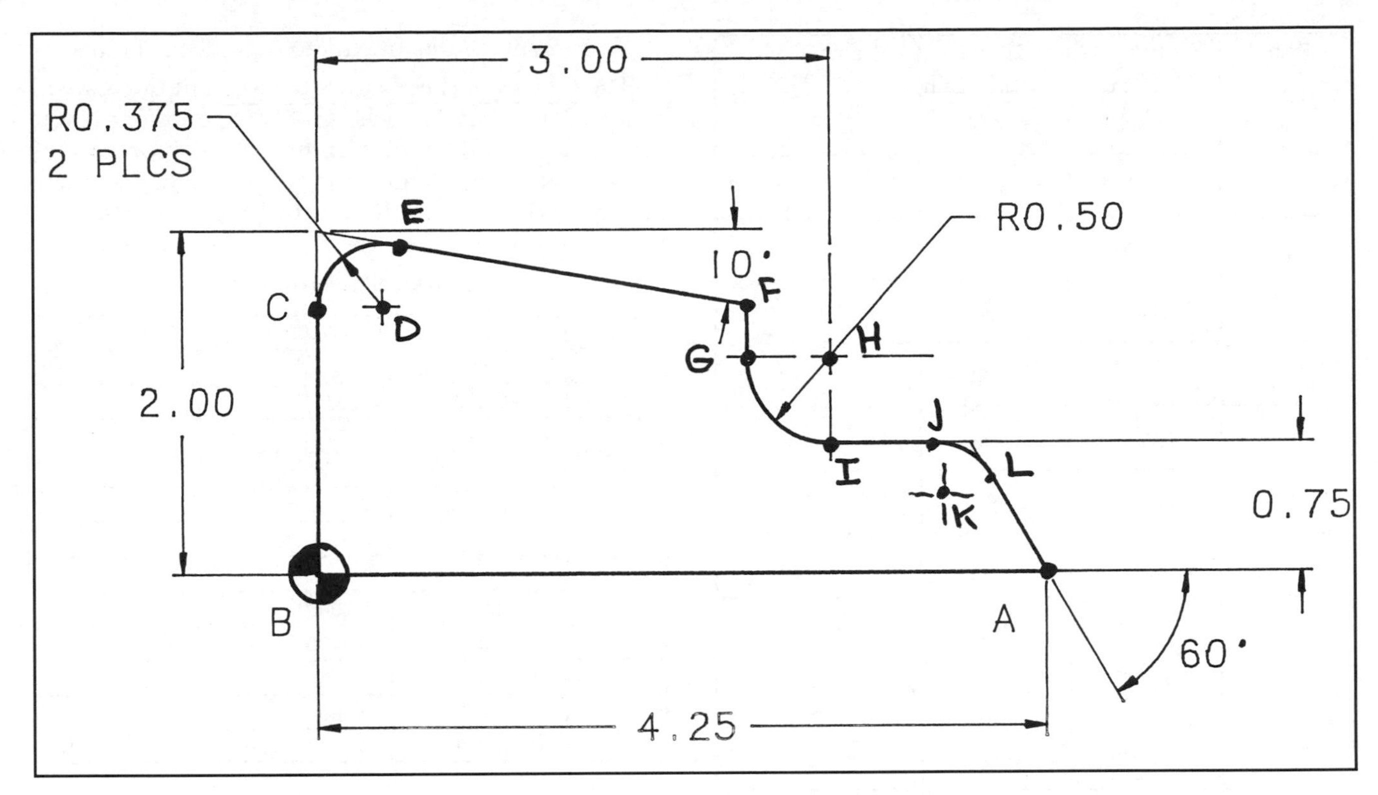

Fig. 20-24. The significant points in this problem.

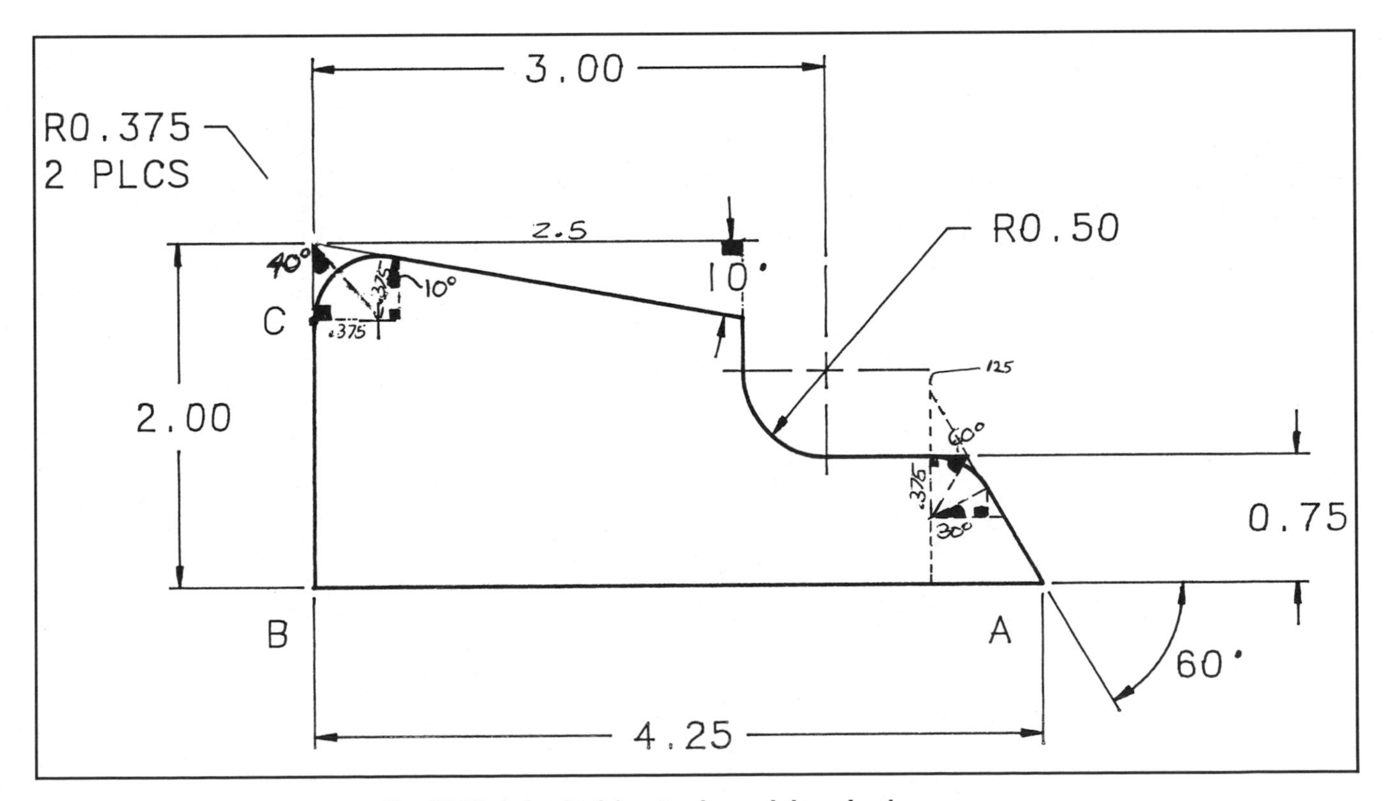

Fig. 20-25. A sketch of the triangles needed to solve the program.

Point	X Value X Value	Z Value Y Value	(Lathe Frame) (Mill Frame)
B	0.0000	0.0000	
C	0.0	1.5531	
D	0.3750	1.5531	
E	0.4401	1.9224	
F	2.500	1.5592	
G	2.500	1.2500	
H	3.000	1.2500	
I	3.00	0.750	
J	3.6005	0.750	
K	3.6005	0.375	
L	3.9253	0.5625	
A	4.2500	0.0000	

See Fig. 20-25 for sketches.

Incremental Adjustments to Significant Points

The following chart shows the amount of compensation adjustment for each significant point for a cutter path using a .200″ radius cutter. The compensations will be given as horizontal and vertical amounts. These amounts will be in the X and then the Y direction for mills or the Z and then the X direction for lathes

Incremental Compensations

Point	Horizontal	Vertical Amount
B	Not affected	
C'	–0.2000	0.000
D	Not affected	
E'	0.0347	0.1970
F'	0.0695	0.1908
G'	0.2000	0.0
H	Not affected	
I'	0.00	0.2000
J'	0.0	0.2000
K	Not affected	
L'	0.1732	0.1000
A'	0.1732	0.1000 (stopping at tangent point)

AUTOMATIC TOOL COMPENSATION

This chapter shows you how to let the controller compensate the part path for the tool radius and shape, which is the tool geometry. With a few commands, many CNC controls can take over the work of compensating. There are differences as to how the program must be prepared to suit a particular control, but the basic parameters remain the same.

In general, the newer controls have more sophisticated compensation ability; they are easier to use and have more flexibility in the part and tool shapes for which they can compensate. Interactive controls can analyze your program and show you where the part-tool combination will not work.

OBJECTIVES

After studying this chapter, you will be able to:

- List the compensation parameters needed by any control.
- List the possible compensation planes and associated codes.
- Make a diagram of compensation, tool-left, tool-right, and the associated codes.
- Write a computer-compensated program for the following:

  Fig. 20-1.
  The drill gauge (Appendix A Drawing C).
  A custom sheet.
- Select and enter a tool shape into a lathe program control.
- Safely enter and exit automatic compensation in a program.
- Define the difference in tool radius and length compensation.

21-1 Compensation Types Radius and Length

This section examines two aspects of tool compensation in CNC work: Radius compensation and length compensation. Both types are used on lathes and mills.

RADIUS COMPENSATION

Radius compensation is more complicated than length compensation. Controls must have special internal programming to handle the math required to adjust the significant points in the program. The control can solve the compensation amounts discussed in Chapter 20.

To understand CComp and see what the control actually does we will look at the information all controls need to initiate radius CComp.

The Radius CComp Parameters

To initiate CComp in a program, the programmer must set up the program by adding three parameters in the correct order. The setup person then must also enter the final parameters into the offset registers. With these parameters in the program and in the tool offset registers, the control can adjust for the cutter shape and lock onto the shape of the part. The control "sees" the part path as a line to be followed. The control knows what tool shape it is using.

Four Program Entry Parameters

1. *The part path.* This is the shape of the part in program form. This is no more than the definition of the part — a plot of the part path. The control will see this as a line to be followed. *This information is the program.*
2. *The cutter geometry.* This is the size and shape of the tool for which the control must compensate. With this parameter, the control can determine how far to stay away from the line. It can also calculate the amount of shift in the significant points. *This information is in the tool offset registers, sometimes called the tool information page or memory.* These two facts — the cutter radius and shape — are entered at machine setup. They are not in the actual program.
3. *The compensation plane.* This tells the control in which two-axis plane the compensation is to be accomplished. Some machines are not capable of compensation in more than one plane. A lathe is a good example of a single-compensation plane machine. Thus, this parameter is not always needed. If needed, this information is in the actual program.

4. *The direction sense.* This is the direction in which the cutter is to proceed around the part path. This will define to which side of the line to adjust the cutter. This parameter, which is found in the program, has two parts.
   A) *The side of the line to which the control is to move.* This is either "tool right" or "tool left" of the work when the work is viewed from above the comp plane. This is in reference to the direction of cut. See Fig. 21-1. Think of this as "pushing a lawn mower along a fence." The fence is the part and your view from behind the mower (tool) is either to the right or left of the fence.
   B) *The actual tool move direction.* Given all the above parameters, the control cannot actually move the cutter to tangency with the work until a first move of the cutter is programmed. The control must know the direction in which the tool is to move. The following two methods are used to put the control on track by giving a move command.
      1. An actual program move.
      2. A dummy move that is not actually on the part path but is parallel with the first move intended when actually cutting or is an extension of the part path beyond the actual part.

Either of these moves gives the control the final sense of the direction in which the cut is going. This tells the control which direction to compensate.

Both parts of Parameter 4 work together to show the controller which direction to shift the cutter so it is tangent to the line.

Refer to Fig. 21-2. Pretend you are the controller using the cutter shown. You are located on Line A-C at Point B. You are looking down on a mill table. First, you are to be *cutter right* of the part. Since all you actually see is a line — not the part — to which side of the line could you move? You have two choices. Fig. 21-3 shows the choices. If you are moving left on the page, the cutter will shift above the line by the radius amount. If you are moving right, the cutter must shift down on the page by the radius amount.

To get a sense of the first compensation, the control needed to know both the side of the line and the direction of movement. After obtaining that sense of direction, the control simply follows the line as described in the program. It maintains the tangent relationship established at the start of the program by the four parameters. It then moves around the part.

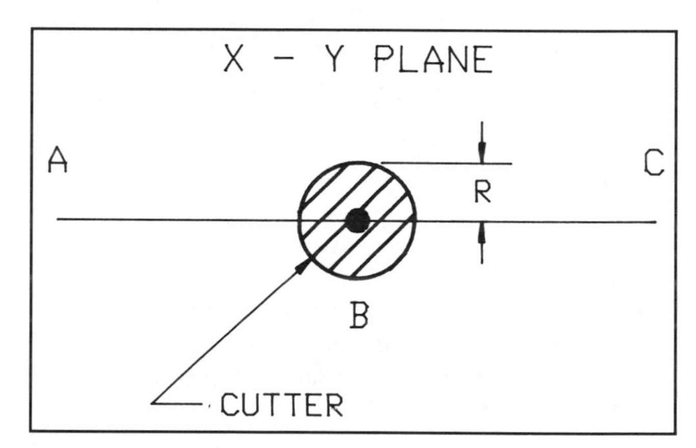

Fig. 21-2. To be "cutter right" of this line, you need more information.

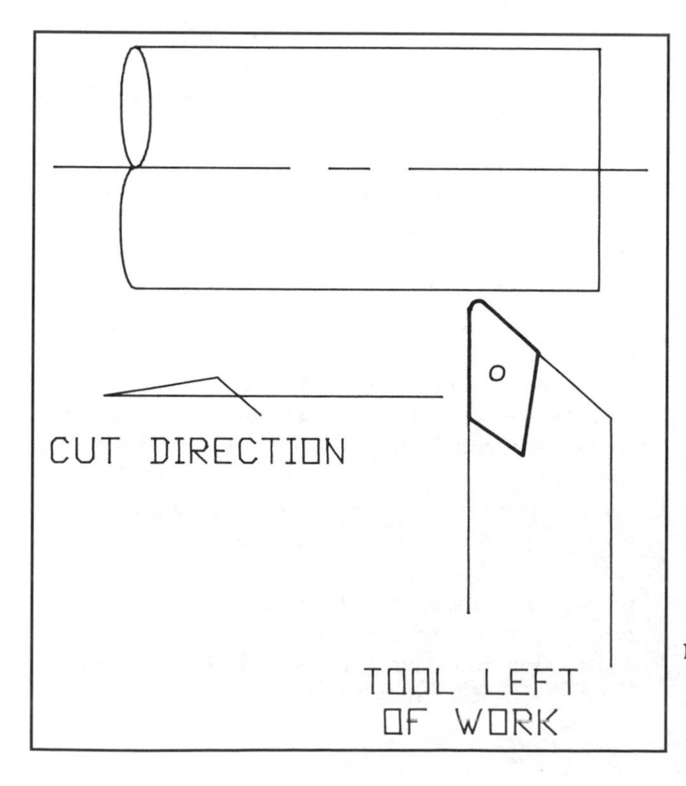

Fig. 21-1. On this lathe job, the tool is "left of the work."

Normal Compensation

Normal compensation means that the compensation is at right angles (90°). Notice that no matter how you tilt the paper and line shown in Fig. 21-3, the compensation move would have to be at 90° to the slope of the part line. That is a normal compensation. The cutter moved at 90° away from the line.

LOOKING AHEAD

The ability to "see" the entire part shape and to "see" how the tool shape interacts with the part shape is a big part of the controller's overall intelligence. Consider a low-level compensation control. It does not see any shape considerations beyond the segment of the line it is following at the time. This is called a *non-look-ahead control.*

Examples of Seeing the Entire Part Shape

Refer to Fig. 21-4. In this example, the cutter is following an inside shape and is on line A-B. It is proceeding cutter right as it goes from A to B. The program command is X 3.0″. The cutter blindly goes 3″ to the right. Without the ability to "see" line B-C coming up, the cutter will bump into the corner and cut into the part by the radius amount.

Look-Ahead Example

To correct this problem, most modern C/NC controls read the program from three to fifteen blocks ahead of the current command. Some controls actually "know" the entire part shape after scanning all program lines. This gives the machine the ability to adjust the program command lines. Consider Fig. 21-5. In this drawing, although the program command was X 3.0″, the control saw line B-C coming up and held back by the correct radius amount.

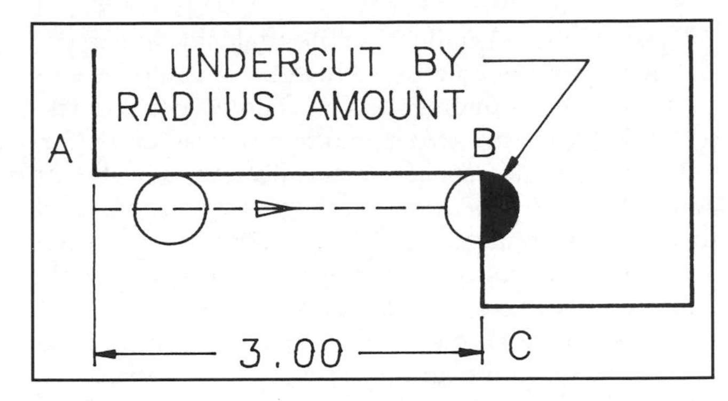

Fig. 21-4. A control must "look ahead" to avoid this problem.

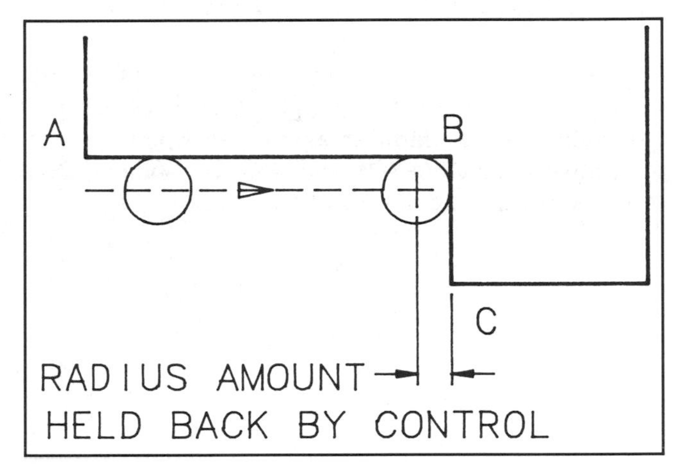

Fig. 21-5. The control knows the radius amount to hold back to avoid undercutting.

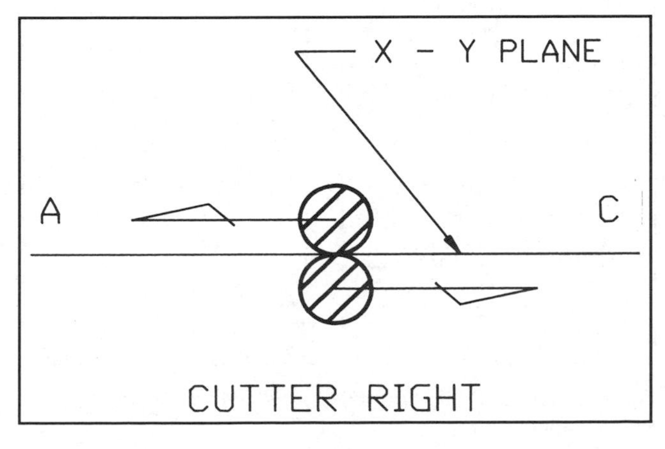

Fig. 21-3. Now you know the direction and can compensate for the cutter.

Refer to Fig. 21-6. Because the angle is less than 90° in the corner, the amount to hold back would have to be solved by trigonometry. The control sees this and solves the problem based upon the cutter radius. Different cutter radii would cause different adjustment amounts to the axis move. A bigger cutter would need to stay away from the corner a greater distance, while a smaller cutter could go further into the corner without undercutting the next part line.

If the program is correctly set up for CComp, changing the cutter radius in the control's tool information memory will automatically adjust the points with no further work from you.

TWO LATHE FACTS IN RADIUS COMPENSATION

A lathe control must consider two additional factors when told to compensate to a part line. These are the tool shape and the lathe imaginary tool point. If the program and machine setup are correct, modern CNC controls will handle these extra factors.

Tool Shape Affects the Moves

As you can see from Fig. 21-7, the lathe tool shown could not fit into the groove because of the tool flanks. Either you or the control must have the ability to analyze this situation. The shape of the tool can affect the shape of the part. Lathe compensation is not just a round-nose tool maintaining a tangent path to a line. The total tool must be considered. If the control lacks the intelligence to analyze this problem, then the programmer or setup person must assume this task.

As mentioned earlier, a library of standard tool shapes will be available with the control. Certain advanced controls can create and store user-designed special tool shapes in this library.

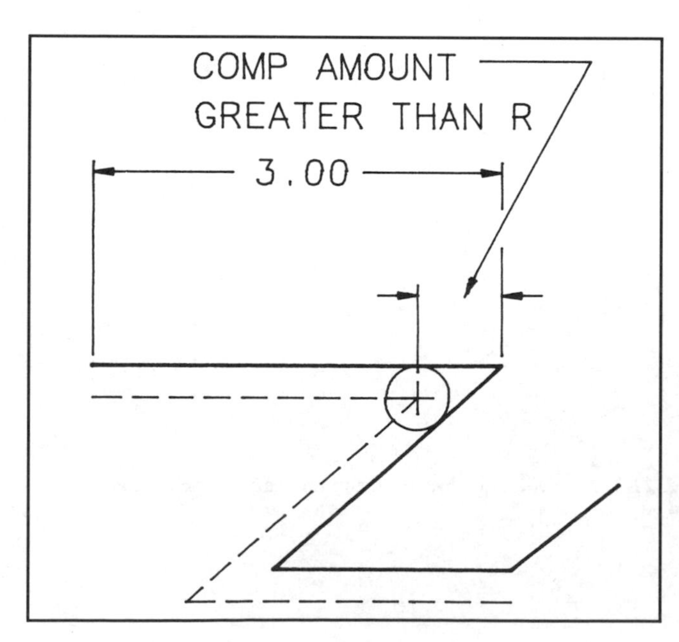

Fig. 21-6. Because the corner angle is acute, the compensation amount held back is greater than R.

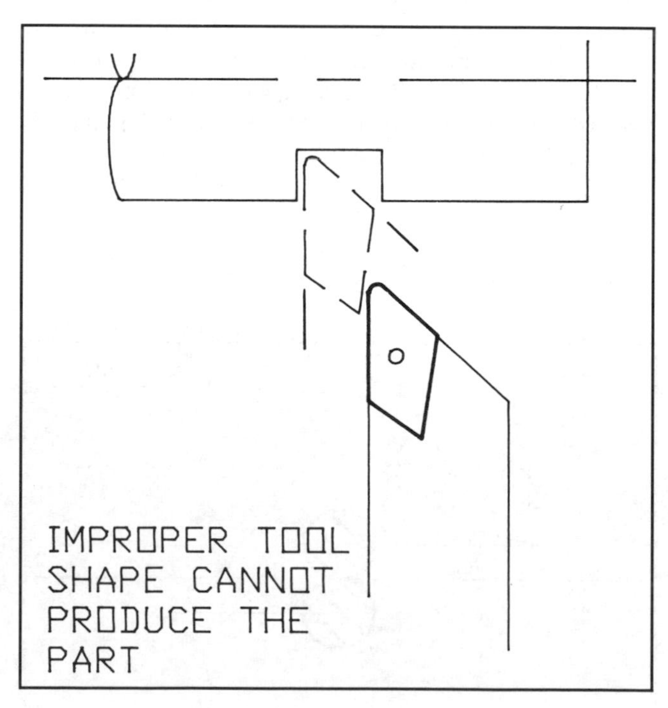

Fig. 21-7. Total tool shape must be considered for turning compensation.

Lathe Imaginary Tool Point

The second unique complication in lathe CComp is the imaginary tool point. As shown in Fig. 21-8, a tool with a nose radius would cut the same diameter and machine to the same shoulder as a tool with no nose radius (a pointed tool). When this tool with a nose radius is coordinated to the workpiece, the control assumes that it has a pointed tool with the point at B. This is the case until we enter the tool radius and standard shape into the tool offset register memory.

Without the tool shape and nose radius, compensation adjustments would be made as though the tool had no radius and Point B was real. The control is operating as though Point B is the cutter centerline. Once the control is given the tool shape data, it then "sees" the center of the *tool nose radius* as the cutter centerline Point A. Any compensations are then made to this real point on the actual tool. Most modern CNC lathe controls can handle this problem as long as the four parameters outlined above are in the program and tool offset registers. Also, the tool must be correctly coordinated to the workpiece.

21-2 Program Entries to Initiate Tool Radius Compensation

PROGRAM EXAMPLE

Whether a conversational or coded control, each needs the parametric information outlined above to initiate CComp. For this development, we are going to compensate a part path program for Figure 20-1 (page 220). We will assume that this is a mill problem, and that the PRZ is at the lower left corner of the part. Since the left side and bottom edge are not contoured, they will not be machined. There are table clamps on these surfaces.

We are using a .500″ diameter cutter and will be machining from the upper left corner in a climb milling path to the lower right corner. The cutter is positioned over the PRZ with a 2.0″ departure distance in the Z axis. Will this be tool left or tool right? If you answered "tool left," you were correct!

COMPENSATED PROGRAM EXAMPLE

In this example, we will look at a standard coded program that is written to machine the part shown in Fig. 20-1. Conversational programs will require similar entries. These basic parameters must be in the program or control regardless of the format. Underlined command blocks are those that directly initiate compensation. These will be dicussed directly after the program example.

```
N001 G40 G90 F25.0 S1500  (G40 CANCELS ANY PREVIOUS
                          COMP)
N002 T0103 M06            (T0103 TOOL CODE ENTRY,
                          M06 TOOL CHANGE)
N003 M14 G00              (SPINDLE AND COOLANT ON,
                          RAPID MODE)
N004 X– .75 Y2.0          (POSITION TO DROP CUTTER,
                          CCOMP NOT IN EFFECT YET,
                          EXTENSION OF PART LINE)
N005 Z–1.75
N006 G01 Z-2.5            (AT FEED RATE MODE,
                          DROP THE SPINDLE)
N007 G41 G17 X1.5         (G41 START COMP. TOOL LEFT
                          G17 XY PLANE—MOVE
                          CUTTER RIGHT)
N008 Y 1.625
N009 G03 X1.875 Y1.625 R.375
```

Most of the remainder of the program is simply the part path program. If there were program commands that you did not understand, look them up in Appendix B at the back of this book.

```
N016                      G00 Z0.0
N017                      G40 X0.0 Y0.0 M02
```

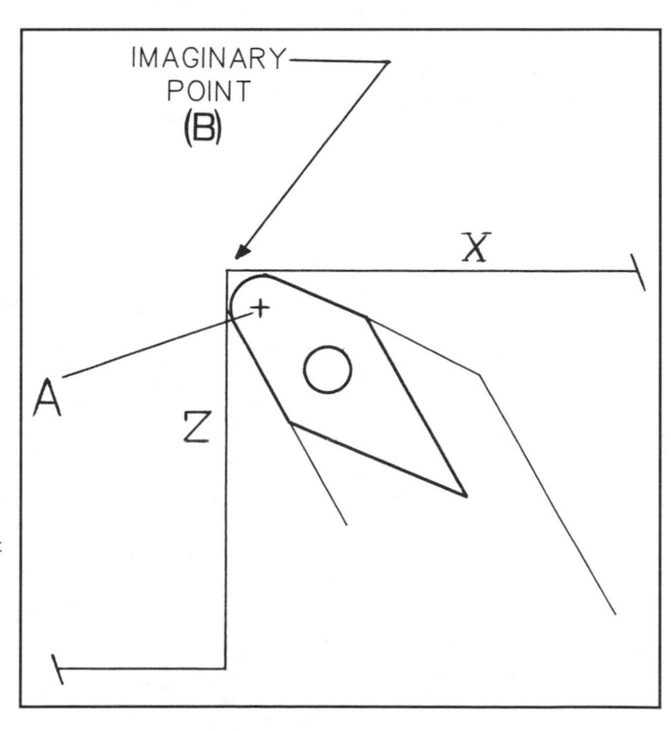

Fig. 21-8. The control assumes the imaginary point is real until it is told the tool radius and shape.

Explanation of Program Commands

N002 T0103 M06

Tool Codes

The first entry needed to initiate tool compensation in the example program must be the tool code. The tool code indicates two facts to the controller:

1. *The tool being used.* In the example above, this is Tool #1. If the machine has an automatic tool changer or tool turret, this entry tells the machine which tool to load for machining. This means that Tool #1 will position for machining when the tool change code (usually M06) is read.
2. *The set of tool offset facts to be used.* The cutter radius, cutter shape, and axis offsets are stored in tool storage register number 03, as indicated in the tool code. This information is put in the control during the setup.

N007 G41 G17 X1.5

Line N007 has a great deal of information in it. This command line completes the necessary information to begin compensation.

G41 Compensation on "Tool Left"

This code indicates the side of the following program line command to which the cutter is to compensate.

G17 Compensate in the X-Y Plane

Remember, this code may not be needed. Most machines have a default condition (a normal compensation plane). In this example, if the machine was a vertical mill, the compensation plane would automatically be the XY plane.

X1.5 Initial Move

At the X1.5 move command, the machine will execute two moves sequentially.

First, it will move up in the Y axis by the radius amount found in tool register #03. This means that since the first move after entering compensation is to the right along the X axis, that the control knows it must first move the cutter away from the line by .25″ in the plus Y direction. The control now has the sense of the program line.

Second, the control will then execute the X1.5 move, but tangent to the part line. The control will not stop the cutter at the 1.5″ position because it sees ahead to the next cutter move in a minus Y direction. The control will add the cutter radius to the X1.5 position. In the command block to execute the .375″ arc, the control will calculate the shift in the tangent points and also calculate the new arc radius for the cutter path. It will continue to do this around the part.

N016 G00 Z0.0

This line is a safety move before cancelling compensation. *You must be aware that when compensation is cancelled, the cutter will move.* It will move in the reverse direction of the last compensation adjustment. You must always remember to get the cutter away from the work before cancelling compensation.

N017 G40 X0.0 Y0.0 M02

This line ends the use of compensation and moves the cutter back to the PRZ at rapid travel. The M02 ends and rewinds the program. It usually turns off any M functions such as spindle and coolant.

Automatic Compensation

1. Refer to the example program on page 241. Assume that we changed line 17 and added line 18:

   N017 G00 X0.0 Y0.0
   N018 G40 M2

   Describe what would happen as the control executed line N018.
2. Line N009 takes the program through the .375″ arc. Write the remaining program lines to complete the shape up to the lower right corner of the part. You may use either the IJ or Radius method for the remaining arcs. You may shift to the G91 incremental mode, if you prefer.

The answers to question 1 are given in the Reference Point that follows In Your Lab.

Custom Sheet #20

For this exercise, you will need to know the compensation words used on your machine.

3. Write the program lines to initiate either a lathe or mill CComp program for a shape your instructor assigns. You may choose to use the drawings in the Appendix. Make sure that your program contains the:

   A) Tool Number Code or Word.
   B) Tool Left or Right command.
   C) Compensation-On command.
   D) Compensation Plane.

1. The changed program lines would cause the following actions:

   N017 — Tool would rapid position to PRZ but would still be in compensation so the tool perimeter would be tangent to the PRZ, not centered.

   N018 — Cancels tool compensation. Now the tool will park directly over PRZ at the tool center.
2. Test plot or run your program for verification.

COMPENSATION HINTS

The following hints will make your programs run more safely and smoothly.

1. Always start compensation off the part or in a move toward the part. Never attempt to touch the part with the cutter and then commence compensation. Notice that, in the previous example, we positioned the cutter to the left of the part. Then we entered the compensation. Then we lowered the cutter to the part. One technique to accomplish this is the "dummy move" mentioned earlier. This is a move of a very small increment in the exact direction as the first intended move when the cutter is on the part.

   Consider this example. When machining the part shown in Fig. 20-1, you wish to enter compensation while parked over the PRZ so you could start machining the left edge of the part. Line N003 could have been written:

   N003 M14 G00 G17 G41 Y.0001

   With this command, the cutter sense would have been a positive Y axis move of .0001″. The control would first move to the left a distance equal to the cutter radius. Then it would move the .0001″ Y distance. It now has a sense of the part path even though the cutter path called out was very small. The cutter could then be moved down in the Z axis without touching the part. The cutter would just be tangent to the part.
2. Compensation will be carried out for rapid as well as feed rate moves. Once the control has been locked into compensation, it will continue to compensate for moves in rapid travel as long as they are in the comp plane.
3. Moves in a third axis will not be compensated. Any comp moves will occur in the plane called out.
4. It is poor policy to reverse directions while in compensation. Think about it. The control is following to one side of a line. When the direction is reversed, the control must skip over to the other side of the line to stay on the correct side for the direction traveled.
5. Never end compensation while the cutter is touching the part. There will be movement toward the part when compensation is turned off. This movement will be normal to the last direction the tool traveled.
6. Compensation values may be changed without changing tools. This trick allows the roughing and finishing of a part with a single tool.

   For example, in the 20-1 example program, we could have roughed out the part by calling out T0102 where the cutter radius was stored in the tool register as .300″ instead of .250″. This would have tricked the control into staying away from the part line an additional .050″. Then switch to T0103 where the correct cutter radius was stored in tool register Number 03. The path would be followed again but with the correct compensation allowances for a finish cut.

   Once the cutter had made a first pass, a new tool code would be entered as T0103. This #03 set of offsets would have the true cutter size. A second pass could then be taken. To do this trick, we enter the same tool number. However, an M06 does not need to be given because the actual tool does not change. Only the stored operating parameters in the compensation logic tool registers change.
7. Use ramp compensation. It is possible to approach a part line that is not parallel with a machine axis. The technique used is that of making a dummy move or using an extended surface line that has exactly the same angular slope as the surface you intend to machine. This line is given as the final compensation parameter — an initial move in the program. This is known as *ramp compensation.*

   The control will sense this line direction and move away at 90° to the slope of the line given. Once the control has locked on to this "fake" line that has the same slope as the part, the cutter may then be moved to the part and machining may commence.

21-3 Tool Length Offsets and Zero Presets

There are two different techniques in common use in industry for correcting tool length differences. These are the use of offsets and the use of zero preset. We will look briefly at both.

OFFSETS

Tool length offsets are used during tool changing in a single program. Tool length offsets are used to correct the length and position of different tools so that they perform as though they had equal length and positions. The correcting offsets are called up from the tool offset registers (tool page) in the controller by the tool code.

Offset Length Example

Refer to Fig. 21-9 on page 245. A good example would be three drills in a lathe turret. Each has a different working length. Ideally, as each tool touches the workpiece, the Z axis register would read zero. That way each tool could be easily programmed from the PRZ. Since each tool has a different length, the Z axis offset is entered as a positive or negative adjustment. Thus, each can be programmed as though they had the same length as Tool #1.

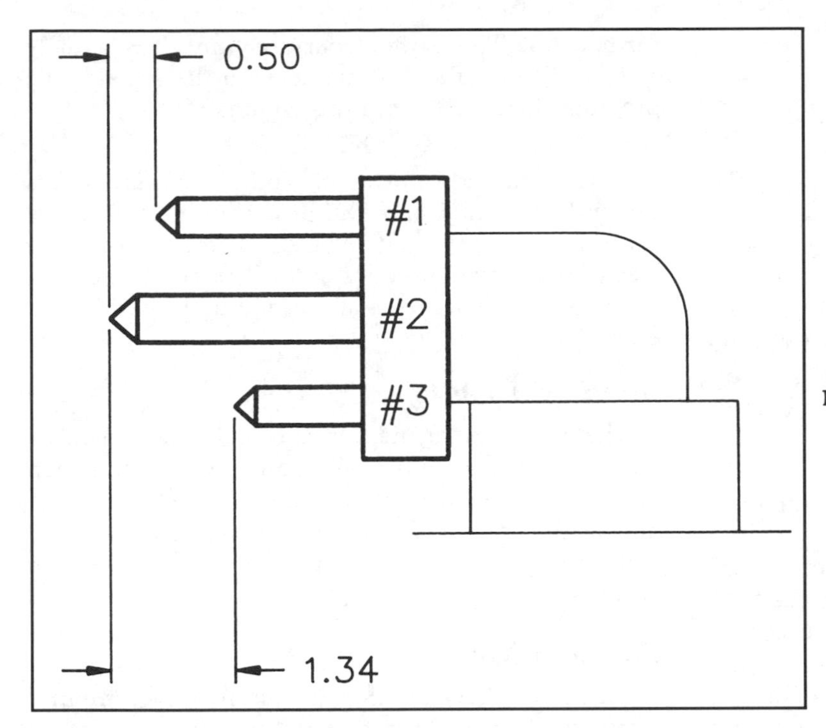

Fig. 21-9. A lathe tool turret loaded with three tools.

Tool #1 Is the Standard

In Fig. 21-9, Tool #1 is the standard with which Tools #2 and #3 are compared. Tool #2 is .50″ longer than Tool #1 along the Z axis. Therefore, the Z axis offset register would contain the offset of Z–.50″.

When Tool Code T0202 is read along with an M06 for a tool change, the machine will automatically move the tool to the compensated position. This CComp position will put Tool #2 in the same position as Tool #1 in its uncompensated position. The offset will apply until it is cancelled by a new tool code.

Tool #3 is shorter than Tool #1. Therefore, its Z axis offset would be a positive 1.34″ to bring it up to the Tool #1 position.

Safety Note. *Be careful when cancelling or changing compensation because there will be movement of the tool.*

In a similar fashion, the X axis would be adjusted to compensate for any difference in Tools #2 and #3. The lathe control tool register would have an X and Z offset for each tool.

Mill Offsets

Length offsets in the Z axis and X and Y positional offsets are also used on milling machines.

Setting Up Offsets

There are several ways to determine the axis offsets for each tool. The most common way is to park the tool at the machine home (lathes) or tool change position (mills). Each tool in turn is jogged up to the work or setup position and touched off, usually against a piece of paper. The absolute distance of the tool from the park position is written down in the setup person's tie-in sheet. Each succeeding tool is touched off and then recorded. The difference in each tool's position, as compared to Tool #1, becomes the offset value.

A second method is to actually machine a part or a test piece of material. The results are measured. The corrections needed to make the tools perform correctly are then entered into the registers.

A third method is to machine a small test part and simply enter the axis position of the tool while it is in line with the machined part. The tool does not need to be moved to the PRZ, but the diameter and length of the test part will tell the control the actual position of the tool. This information will then be used by the control to set its own offsets.

Breaking or Replacing a Tool

If any tool other than Tool #1 is broken, the new tool is simply touched off the part again and the new offset value is entered to correct the new tool size.

If Tool #1 is replaced, it must also be touched off. The difference between the old Tool #1 and the new Tool #1 is recorded. This difference is then entered into the tool offset register as the correction that will enable the new Tool #1 to perform as though it was the original Tool #1.

There are a variety of innovations on this general procedure when using new controls. Some controls have sensing probes or detectors to determine tool position offsets. After automatic probing of the tool position, some controls have the ability to automatically store the new tool offset information. However, the "touch off" tool teaching method outlined above remains the standard in industry. Even sensing-type machines can be set up with this "touch off" type of offset determination.

ZERO PRESET

Presetting a Program Reference Zero — G50

Commonly, on lathes, there is a way to set the PRZ from the departure point without actually resetting the machine axis registers. This is done by a code or word that causes the control to reset the zero point from the present location of the tool. The PRZ is established by the G50 code followed by the distance it lies from the present tool location. This is a useful code. A setup person needs to coordinate the tool to the part but not necessarily to the PRZ. The control reads the G50 code followed by the coordinates of the new PRZ. The present location of the tool is taken as the temporary reference.

For example, a piece of raw 1.00″ diameter stock that protrudes from the collet 3″ must have the PRZ at the collet face on the centerline. The setup person does not need to move the tool to that impossible position, which would be inside the stock. Instead, during the setup, he or she coordinates the Z axis by touching the tool to the outer face of the stock and then coordinates the X axis by touching off the diameter. These actions position the tool at the outer corner of the stock. The setup person then moves the tool back 3.0″ further in Z. This is the departure point. It is actually 6″ from PRZ at the 1.0″ diameter in the X axis. The program then starts with this command:

G50 X1.0 Z6.0

This command informs the control that the tool is presently 1.0″ in X and 6.0″ in Z from the PRZ. The control then knows that the PRZ for this operation is located 6″ toward the chuck in the Z axis and on center in the X axis. (Diameter programming is used in this example.)

Multiple Tools

The zero present method is useful when qualified or nonqualified tools are being used. Each tool is then touched off the stock and positioned in the same position as Tool #1. A G50 code is given for that particular tool. In this way, you can compensate for tool length differences.

Use with Mills

The zero preset method is a useful programming technique that can also be used on mills, but not all programmers use it. If a G50 technique is used, it should be indicated in the setup document. This technique is also used on mills and on conversational machines.

Modern manufacturing is interwoven with and controlled by computer technology. Although CNC is a historical core of computer-integrated manufacturing (CIM), the power of the microprocessor is used in every aspect of the manufacturing process. In this unit, we will look at some of the other uses of computers in the C/NC process.

Our main objective is to gain an overview of the entire computerized process of manufacturing. After studying this unit, you will have a sense of the other computer operations around the C/NC process and how to use them to your advantage. The chapters in Unit 7 also provide career planning information.

CONTENTS

An *acronym* is a word formed from the first letters of a term with several words. Some of the topics discussed in the chapters of this unit are commonly referred to by their acronym. All of these acronyms begin with the letter C. In these acronyms, what do you think C stands for?

Chapter 22. Computer-Integrated Manufacturing (CIM)—This is the umbrella of data-based management in a factory. All aspects of the manufacturing process are directed through host, or master, computers.

Chapter 23. Flexible Manufacturing Systems (FMS)—In a fully computer-integrated FMS, several machines run themselves. Robots load tools, delivering parts from machine to machine. Inspection machines check the work. All of these operations are monitored by a powerful computer.

Chapter 24. Computer-Aided Programming (CAP)—Computer-aided programming is sometimes called CAM programming. C/NC programs are generated by either a CAD unit or a special offline programming computer. The shape of the part is fed into the computer as an image. The computer generates the program and then translates it into the language of a particular machine.

Chapter 25. Computer-Aided Design and Drafting (CAD/D)—This chapter discusses the design and drawing of products at a sophisticated computer terminal using special software to assist engineers, drafters, and programmers.

CHAPTER 22

COMPUTER-INTEGRATED MANUFACTURING (CIM)

A discussion of CIM, could easily fill another text of this size. CIM is the overall umbrella of computer management in the manufacturing process. There are many aspects to CIM. The following is a partial list of the factory duties that can be managed by CIM:

- Inventory control.
- Order processing.
- Product data base maintenance.
- Product inspection.
- Materials ordering.
- Tooling design.
- C/NC programming.
- CAD Drawing.
- Manufacturing scheduling.
- Personnel payroll.
- Job progress reports.
- Inspection and statistical product control.
- Preventive maintenance scheduling.
- Parts collection.
- Production cost control.
- Material handling.
- Parts warehousing.
- Inventory reduction.
- Production cost estimates.
- Process planning.
- Control of heating and energy use.
- Waste recycling.
- Shop load schedules.

OBJECTIVES

After studying this chapter, you will be able to:

- Identify the two kinds of costs in CIM.
- List the benefits of a CIM system.

COST CALCULATION

Much of the data identified in the previous column is used in CIM to calculate two kinds of costs.

1. The **shop burden rate.** This is the actual cost per hour for operating a particular machine. The cost might also refer to a group of related machines (sometimes called cells) or to flexible manufacturing systems (FMS, see Chapter 25). It can also refer to the overall rate of manufacturing.
2. **Part unit pricing and quantity pricing.** This is the product cost of manufacturing. Until CIM, this cost and the burden rate were very difficult to calculate correctly.

HISTORY IN THE DATA BASE

Future costs can be accurately forecasted in CIM because information on similar products and situations has been stored in the data base.

This can positively affect design costs, inventory costs, and the manufacturing process. One large cost in manufacturing is the part and supply inventory that must be stored. The oversight of all aspects of the CIM factory allows a new concept in inventory control and manufacturing scheduling. This is called *just in time manufacturing.* Each component, supply item, or tool is not held in inventory. Instead, it is ordered just in time to assemble the product. This puts a tremendous burden on CIM to accurately schedule incoming supplies. However, it reduces the inventory cost and the manufacturing cost. Any change in need is immediately detected by the CIM. Adjustments are then made to increase or decrease the flow of parts from suppliers.

Every job completion adds valuable data to the history base. This will make future decisions even more accurate. In effect, a CIM factory is like your body. Many functions are controlled by a central brain.

22-1 CIM Computers

From the list of possible duties in a CIM factory, you can see that there are many transactions that might need attention at any given moment. A CIM system is like an orchestra. Each component duty must work with the others through a single data base, central software, and an open communications network. Each task is an independent process that must work with all others. The number of duties included in the CIM is chosen by the system managers and limited by the CIM software writers.

Each individual duty requires data and a RAM loaded with software to handle the transaction. This means that as a CIM unit becomes more capable (receives more duties) or has more users (more transactions per hour) or must contain more data, it must be a larger computer. At a certain point, a single computer becomes incapable of handling the transactions and holding the necessary programming in RAM. Small CIM systems run on single computers. As more is asked of the CIM system, more equipment is needed.

CENTRAL HOST COMPUTERS

A network of computers can be developed. This is controlled by a single central computer. A major feature of a large CIM system is that each component of control is integrated through a single data base. The data base contains the stored information about the products being produced. This data might be the part image from the CAD. It might be the inspection records of the parts previously produced. This data is the record of the duties performed or about to be performed.

Single duties or groups of similar duties are assigned to smaller computers by the central master unit. Each smaller unit may share information through the central computer and its data base.

22-2 How CIM Affects You

Exactly how CIM is used from factory to factory is dependent upon the individual needs of the facility. However, they all exhibit similar characteristics. We will look at these outer characteristics, what they can do for you and what action will be required by you, the C/NC operator, or the setup person.

THE CENTRAL COMPUTER

The heart of the system is the central computer and its associated data base and programming.

SECONDARY COMPUTERS AND INPUT STATIONS

The units you will use are secondary computers assigned a single duty or a limited group of duties. These computers are similar to the ones you use in your lab for CAD or for C/NC program writing and storage.

Often, lab data such as *work in progress* is entered at a terminal in the lab environment. This terminal is linked to a secondary or host computer. The actual transaction or data is dealt with at another location. This shop floor entry of progress is called *work in progress (WIP).*

Work in progress falls into two categories: *productive*—work that is making progress and *nonproductive*—work that is "down," or not making progress. By overseeing this N/P WIP, a process can be set up by the CIM to avoid as much down time as possible.

BENEFITS TO THE MACHINIST

The services you might realize from a CIM system are as follows:

1. Computer generated work orders. These contain the information to set up and make a certain part:
   - A) The part number.
   - B) The drawing number, including the revision level for the parts ordered.
   - C) Quantity to be made.
   - D) Tooling numbers needed for operations and programs.
   - E) Cutting tools required.
   - F) Special customer requirements.
   - G) Start and due dates.
2. Computer generated drawings. These are the beautiful part drawings you have seen. They were plotted on a CAD system.
3. Inspection data for part quality. Statistical analysis for shop control of part quality.
4. Information about related jobs throughout the factory. Instant information about the status of other jobs, parts or tooling needed to keep the production schedule.

22-3 What You Must Do as a Machinist in a CIM System

Several major responsibilities are assigned to the machine setup person, machinist, or C/NC operator.

WORK IN PROGRESS

As a job is started, the operation, part number, and location of the individual work station (machine) must be sent to the data base through a terminal. The entry of a single job into the data base makes the job visible to the computer and adds facts to the data base. This individual job entry and work in progress is called WIP. This is your responsibility. This information is given to the secondary computer, terminal, or host computer in one of three ways:

General Entries

A) These can be made with a bar code reader. The scanning wand is moved over a code printed on the work order. This code contains tracking information about the job. The part information and order number are sent by the bar code.

B) A card or magnetic code strip is fed into a reader, or an MDI entry is made by the machinist.

Specific Entries

C) The name of the employee is usually fed in by the use of an identification badge.

D) The name of the particular machine is fed in by a machine card in a rack of cards. There is one card for each machine in the shop.

E) The specific operation to be performed must then be fed in either through a card or MDI entries at a keypad at the terminal.

F) Clocking in and out on a specific job.

As a review, answer the following questions about CIM.

1. Define CIM. Be sure to include the following terms in your definition: host computer, terminal/secondary computer, duties, transactions.

2. List at least ten duties that might be monitored by a CIM computer.

3. List two benefits that you as a machinist might realize from a CIM shop.

4. What is your primary responsibility as a machinist in a CIM system?

5. Define WIP.

CHAPTER 23

FLEXIBLE MANUFACTURING SYSTEMS

There exists some controversy over the terms flexible manufacturing systems (FMS) and CNC work cells. The difference will be explained later. For now, they both stand for a group of CNC machines connected to, coordinated by, and controlled by a computer. The total system, or cell, contains more equipment than simple machine tools. The FMS or cell system may feature any of the capabilities discussed below, in Section 23-1.

OBJECTIVES

After studying this chapter, you will be able to:

- Identify the features of an FMS system that are not found on a CNC machine.
- List the ways in which efficient use of the work cell and FMS can reduce costs.
- List the features that are characteristic of a work cell.
- List the characteristic features of an FMS.
- List the career opportunities in cell/FMS technology.

23-1 Features and Goals of Cells and FMS Systems

FEATURES

The following features of cells and FMS systems are not found on standard CNC machines.

- Automatic part loading—robots or part loaders. Fig. 23-1 on page 253.
- Automatic part handling—robots, smart conveyors, and programmed shuttle carts. Fig. 23-2 on page 253.
- Tool changing for dull or broken tools, plus sensors to detect dull tools. Machining sound, pressure, and visual inspection are used to "see" dull tools.
- Automatic tool storage and delivery.
- Automatic machine setup for multiple part runs. This requires some form of part recognition. Bar codes, light reflection, silhouette vision, and weight are a few of the ways a smart machine recognizes the part to be made.
- Integration into the CIM system.
- Automatic part finishing and offload work.
- Communication and coordination with CIM master host computer.
- In-process and finished part inspection.
- Assembly of products.
- Palletizing and preshipping.

Fig. 23-1. A typical part shuttle. The wire-guided vehicle is part of a flexible manufacturing system. The Boeing Company

Fig. 23-2. A part and pallet transfer area in a working FMS. The part is fixed on the pallet, then moved to the machine automatically. The Boeing Company

GOALS

Efficient use of the work cell and the FMS can reduce costs in two ways. They can reduce costs through more efficient part production and through lower labor costs.

More Efficient Part Production

Either the cell or FMS is capable of round-the-clock production with little, if any, wasted time and less human intervention. The timing in an automatic production facility is crucial. Parts must not be bunching up at a single station. A computer must be able to make sure that all the components in a system are synchronized with the whole system. When parts have entered the system (have been taken by robot from the raw stock area), they must be kept moving through the cell. If they sit, not being machined or moved, they become unprofitable. This is called *nonproductive* work in progress (NP-WIP).

A cell or FMS is coordinated by a single computer, which may or may not be integrated into the CIM host computer. One of the major jobs for the supervising computer is to time the system to avoid NP-WIP. This supervising computer oversees the system or cell. It supervises all part movement and machine start times. Once the raw stock is loaded into an individual machine, the onboard CNC control takes over until the part is completed. Some waiting time is inevitable. When this does occur, there is usually a holding area for transfers.

Once the CNC part program is completed, the CNC control signals the supervising computer. If possible, the loading robot is brought up to unload the part and reload another part. Fig. 23-3. If the robot is busy or the next station is not ready, the robot is told to wait. Shuttle robots may be used to load the part. These are moving carts that obey direction commands or follow wires implanted in the factory's floor. The shuttle cart may unload the part in a waiting area or simply hold the part until the machine is ready.

Palletizing

One system designed to avoid NP-WIP is the concept of universal pallets. A metal board is loaded with a production part or tooling fixture. This pallet is designed to fit each machine or system station in exactly the same orientation. Tooling or actual parts may fastened to the pallet and shuttled from machine to machine. In general, pallets are an FMS tool. Parts can be quickly loaded from pallets.

Less Human Intervention

As you can see, all of the additions listed in the introduction are designed to move, machine, and inspect parts with little or no involvement from the machinist. Labor costs in part production amount to a large part of the price. By some estimates, that cost can be as high as 80%. The use of robots and CNC machines reduces labor costs. This is not a totally negative message for a student in C/NC training.

Your Future in FMS-Cell Technology

While the tedious work is being automated, exciting technical jobs are being created. Overall productivity is increased. Studies have shown that automating a manufacturing process does displace less skilled workers. However, it creates work for more highly trained persons. Your present studies are a step in the right direction.

Fig. 23-3. The wire-guided shuttle delivers the part and fixture pallet to a machining center. The Boeing Company

23-2 Differences in FMS and Work Cells

Both an FMS and a work cell are able to move parts, machine them, and inspect them. There are, however, some differences. The definitions of these two terms overlap. The following is a generalization of the industry accepted definitions.

WORK CELL

A work cell can be any single unit that produces parts to which a cost per part unit may be assigned. This cost per part unit is called the *burden rate.* In a broader sense, a cell is a small group of machines that produce a single part or very similar parts at a predictably low burden rate. The following features characterize a work cell.

- A work cell is usually thought to be a dedicated more rigid automatic machining facility. It is less comprehensive than an FMS. A work cell usually is dedicated to a single part for a very long time. Sometimes it is dedicated to a group of similar parts.
- A work cell has no mobility or adaptability in part transfer. Usually the part movement is accomplished by smart conveyors or similar devices. Robots are present, but they are used only to load and unload parts. The robots are fixed. They do not shuttle nor have they any flexibility in their tasks.
- A work cell is a lesser version or lower version than an FMS. A cell costs less, is less complicated, and is simpler.
- In its simplest form, a work cell may not need a supervising computer. Simple trip switches and gate functions in the control and loading devices can signal progress of the production parts.

FLEXIBLE MANUFACTURING SYSTEM (FMS)

An FMS is a more complex and adaptable machining facility. Refer to Fig. 23-4 on page 256. It is characterizeby the following features.

- Fast and/or automatic turnaround from one type of part to the next. An FMS is, as the name implies, "flexible."
- Some FMS systems can even recognize random parts delivered by shuttle, conveyor, or robot. As the part is recognized, the correct CNC machining program is called up and the correct tooling and holding are applied. This advanced feature requires part recognition but makes the timing of the system much more efficient. Parts requiring different machine run times may be mixed in a matrix to reduce NP-WIP.
- An FMS requires a supervising computer. It may even require one or more secondary computers. The FMS is usually part of a CIM system. Management of an FMS within the factory structure would be too difficult by standard means.
- An FMS requires a reserve of tooling. Tools that wear or break must be replaced immediately. The system often tracks tool wear and schedules tool replacement. This feature is also found in cells.
- An FMS is often very costly, although school versions have been recently placed "on line."
- An FMS may contain cells within. A large FMS could control and load a group of machines completing similar parts.

23-3 How FMS/Work Cells Affect You

As stated earlier, the C/NC process is undergoing change. As a factory moves into cell or FMS technology, the jobs, job titles, and responsibilities expand from those in a C/NC shop.

To stay ahead, you will need to develop new skills. But first let me assure you that a good C/NC machinist will be in demand for a long time. Not all shops will go to full computer management or work cells. The skills you already possess are valid and will remain useful for a long time. Finally, the new jobs and professions are directly related to the C/NC skills you have just learned.

JOBS IN CELL/FMS TECHNOLOGY

The manufacturing process using cell/FMS technology requires the same stages as a standard C/NC job. These three stages are the creation of the idea, the planning, and the production of the product. The tasks, however, will be different. The dividing line between job titles will be changed. For example, programmers may be more closely related to drafters because programming will be accomplished at the terminal where the design is drawn. We will briefly examine the process and see if we can predict the new job titles. Naturally most of the change is due to the computer involvement. Thus, some of the new C/NC process will be done by the computer itself.

Fig. 23-4. An industrial FMS. Cincinnati Milacron

Design Engineer

The design engineer is responsible for the product idea. In FMS/CIM, his or her role remains unchanged in that the product idea and product plan start with engineering. There are, however, the following exceptions:

- To be economical, product design must be planned more closely around the automated production equipment. The cell or FMS may dictate design features in an automated factory.
- The engineer may draw the design on a CAD terminal that has the programming ability that will be discussed in Chapters 24 and 25. In such a case, the work of the drafter and programmer are eliminated. The C/NC program is created directly from the CAD data stored in the data files.
- In a CIM factory also, the design is fed into the data base to initiate tooling, work orders and material purchasing.

Shop Planner

The shop planner's job will change in that the computer data base will sharpen and quicken his or her ability to schedule and organize. The planner will assume more of the role of an expediter. The computer is able to schedule production and monitor shop load.

There will be no need to request programs. The CAD design, along with program software, virtually provides the program data. An additional job responsibility may be troubleshooting cell failures. With automation, this new job becomes critical.

Toolmaker

The skills of the toolmaker will still be needed. The emphasis on universal tooling will change some of his work. However, there will be additional needs such as the need for terminal grippers on robots or the need for fixtures that can transfer from station to station. He or she must be able to understand the FMS to make functional tools.

Programmer

Programmers may or may not be eliminated. In some shops, their job responsibilities will remain unchanged. In other automated systems, their work will be radically modified, making them more of a troubleshooter. Programs in these systems will be generated by design data.

Setup Person

The work of a setup person may be similar to that of a systems troubleshooter. Of course, not all setups will be automatically changed. Also, flexible tooling will not work for every part run. Ideally, the designer will strive to create parts that fit the system and can be integrated with little or no customizing.

Machine Operator

The job of machine operator will be eliminated in many automated systems. Human intervention adds much of the cost to production products. There will still be shops that specialize in part production that does not require automation. In some jobs, a fast C/NC machine will never be outdated.

THE NEW JOBS

In addition to the new versions of the jobs listed above, some new jobs have emerged from CIM, Cell, and FMS technology.

Robotics/Automation Technician

The job of robotics/automation technician may be several jobs. Each will concentrate on keeping the system up and working. An autobot-tech will need to have a working knowledge of physics/mechanics, electronics, electricity, pneumatics, and hydraulics. This new job will require a minimum of two years training beyond high school. These programs are available in community colleges and technical institutes. A four-year college degree, or master's degree may be necessary in the more technical aspects of autobot technology. Specialization will become necessary as the field expands.

Systems Programmer/ Communications Specialist

The new job of systems programmer/communication specialist requires a solid working knowledge of computers, intercommunication protocols, and automation principles. This person will be a setup person for new systems and/or a troubleshooter for systems in use. This job will require a minimum of two years of college. A four-year degree in computer sciences will become necessary as systems become more sophisticated.

WHAT YOU MUST DO

There are three things you can do to help ensure job opportunities.

1. Get a solid background in the C/NC process. Operating a C/NC machine is a good start.
2. Set a particular career goal. Decide upon one of the above job titles. Do you plan to move into automation beyond C/NC? Watch the requirements and nature of the job while your C/NC career develops. Find formal and informal educational opportunities. On-the-job training is still a good way to learn, as is self training.
3. Stay informed. Find out what training you will need beyond that which you are now receiving. Look for openings to gain knowledge and experience in your own factory or shop. Attend trade shows. Subscribe to publications that feature the latest automation techniques. Join professional associations that promote technological growth and awareness. The Society of Mechanical Engineers (SME) is one such organization. Stay in touch with educational institutions that are training people for the new jobs.

As technology develops, the jobs will continue to change. Who knows what wonderful new jobs will emerge?

CHAPTER 24

ALTERNATIVE PROGRAMMING METHODS

This chapter examines the C/NC programming method that is similar to CAD drawing. You may hear it referred to as computer-aided programming (CAP) or computer-assisted machining (CAM). Both words refer to a system that uses software to create an image of a part that can then be translated into a C/NC program. This can be done on a dedicated computer or on a microcomputer with moderately large RAM.

IMAGE DATA

The image data created by the CAP unit may become part of the total CIM shared data. This means that the image description may be uploaded from the CAP into a CAD unit from which a part drawing may be developed. The CAP image may also be postprocessed (translated) into a specific machine control language C/NC program. Also, through sharing data, a CAD image can be used through additional CAP software to create a C/NC program. We will investigate this in the next chapter.

CAP Is Compatible with CAD

Although less comprehensive than CAD, CAP programming is similar. The image is developed as a series of lines and circles that the machine can connect to create a continuous part profile. Several languages available today are compatible with specific CAD units. These languages are capable of using the CAD developed data for the part image.

OBJECTIVES

After studying this chapter, you will be able to:

- List the ways CAM program images are created.
- Understand how a CAM image is prepared for a C/NC program.
- Visualize raw image data and finished C/NC part drawings.

24-1 The CAP-CAM Background

PROGRAMMING LANGUAGES AND SYSTEMS

The earliest CAP systems were created to deal with the mathematics created in writing C/NC programs. The first widely accepted NC programming language was created by the government and aerospace industry to deal with the impossibility of calculating all the significant points on three dimensional parts.

This language is called APT (automatic programmed tooling). APT is a powerful programming tool that has been improved and made available for microcomputers. The major disadvantage perhaps is that it is too universal. Thus, it requires too many entries for two dimensional part programs. Newer programming languages and systems have been developed as a response to the APT complexity. These new language systems require less RAM capacity. Thus, they are more easily used by microcomputers. Also, C/NC control manufacturers have developed specific systems tailored to their control. Other languages have been written to fill the need for a less complicated offline programming system; Compact is the most common of these. There are several other good languages on the market.

An APT program was originally handwritten and then loaded into a main frame computer through telephone connections. The computer would process the raw data into a program. Today, the APT system resides in a hard disk in the user's shop, as do all other leading systems.

24-2 Writing A CAP Program

Creating a CAP program requires the user to enter two types of input data. This data is entered in response to prompts given by the CAP software.

MACHINING PARAMETERS

The machining parameters include work material, machine process (lathe or mill), cutter shape, part position relative to PRZ and the machine axis frame, fixture requirements, clamp locations, and any special conditions. The desired tool path may be entered or the tool path may be chosen by the computer.

GEOMETRIC LAYOUT

A user at a CAP computer creates an image of the part similar to a part layout. To lay out a part, you would draw a series of lines and circles from known information. All CAP systems take advantage of geometry. For example, you could lay out the shape in Fig. 24-1 using the facts given on the print. But to program that line, you must have the coordinates for points P4 and P5 in Fig. 24-2. You may not know the coordinates of every significant point on the part geometry. However, you could lay out the complete image. Once the raw layout is complete, the CAP system can calculate the significant points that are not directly supplied in the input data.

POINTS, STRAIGHT LINES, AND CIRCLES

Refer again to Fig. 24-2. To lay out this shape, you would draw the location of points P1, P2, and P3. You would also draw circle C1. Lines L1 and L2 are tangent to C1. They pass through points P1 and P3 respectively. Also, line L3 passes through both P1 and P3.

You could lay out this shape because you know the exact location of points P1, P2, and P3. You also know the radius of circle C1. You do not know the location of tangent points P4 and P5, but you do know that lines L1 and L2 are tangent to C1. You could then calculate points P4 and P5.

A CAP program requires the user to describe the raw image of the part. This image is a series of straight lines and circles fixed in the grid by known points. These raw lines connect to become a continuous part image. The significant points needed to develop the C/NC program are either directly in the image or are calculated by the CAP software. This saves much time and helps prevent errors.

USE WITH CAD

If the system is being used with a CAD system, the geometric data may be derived from the CAD drawing. The CAD drawing must be prepared with a technique called overlaying. *Overlaying* means that the complete image of the part is drawn as though on a clear overlay sheet. The image is drawn without dimensions or nonimage lines. Any notes, dimensions, and other information that would confuse the geometry are drawn on a separate overlay. When plotting the CAD drawing, all overlays are used. When writing a C/NC program, only the part image overlays are called upon. Often, these overlays are shown in different colors on the CAD screen.

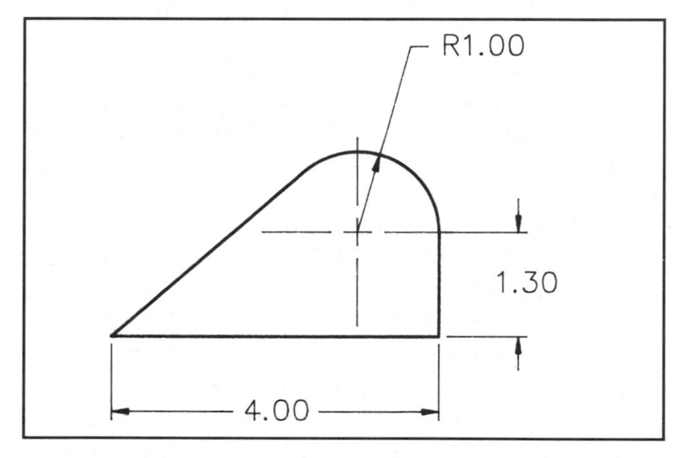

Fig. 24-1. This part is to be programmed using CAP/CAM software.

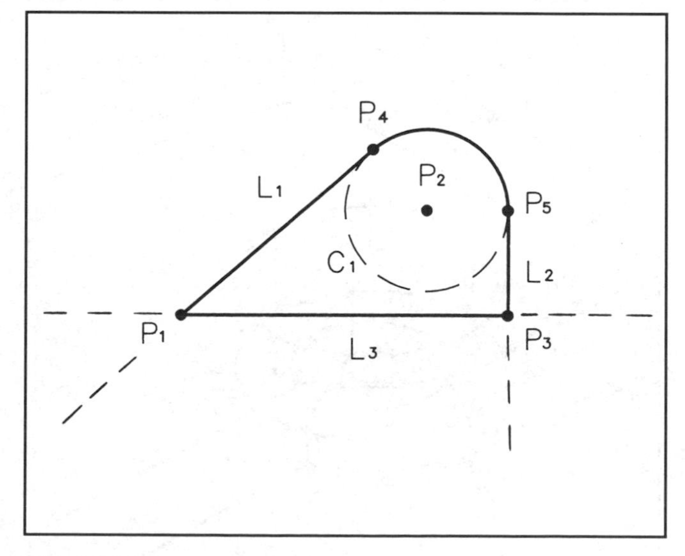

Fig. 24-2. The points and lines are identified for Fig. 24-1.

24-3 Industrial Example of CAM Program Software

The following examples are for specific industrial CAM software packages. We will look at SmartCam® by Point Control company of Oregon and Mastercam® by CNC Software of Connecticut. We will examine the process and the differences in the process for the simple profile mill job illustrated in Fig. 24-3. This example is offered as a learning tool only. No comparison of product superiority has been made, nor is this an endorsement.

SMARTCAM® PROTOCOL

All CAM softwares have particular operating protocols. A *protocol* is the way the software package works from start to completion of the program. SmartCam® begins by completing a *Job Plan* page on the screen. On this planning page the user sets up the job parameters. These include items such as machine type (for example, lathe, mill, or punch), cutter size, speeds and feeds, and tool numbers. This is type-one information as outlined above. It is saved in the *Job Plan File.*

Type-two data, the part drawing, is entered next. The user can draw the part geometry at the screen. This is similar to making a CAD drawing. If the part drawing already exists as CAD data, with some preparation and through the use of CAM Connection® software, the drawing may be uploaded into the CAM computer. At this stage, the user must have the intended tool path and tool changes planned as the geometry is being set up. Tool path lead-in and lead-out lines as well as tool change points must be drawn when creating the part shape.

Once the geometry has been completed and saved in the *Shape File,* the user enters the *Show Path* stage. Here a trial of the intended program is displayed on the video screen. At this point, the program follows the tool data in the Job Plan. However, the program is a generic version, not usable on a particular control. Once the tool path is shown to be acceptable, it is saved with the shape file. Figure 24-4 shows a tool path at this stage.

Code generating is the final stage. This requires selecting a particular control for the program. The program is prepared by selecting a specific *Template* and a *Machine File* stored in the software. These files make the commands acceptable to the user and customize the program for a specific control. The user has access to these files and can change them to suit his or her needs. With SmartCam®, all four files work together to generate the CNC code program. The operations in Figures 24-5 and 24-6 will produce the part shown in Fig. 24-3, but each is for a different control. Notice that most of the program is simply EIA codes. Line 009 in both programs is a G02 curve command with the same characters but in slightly different form.

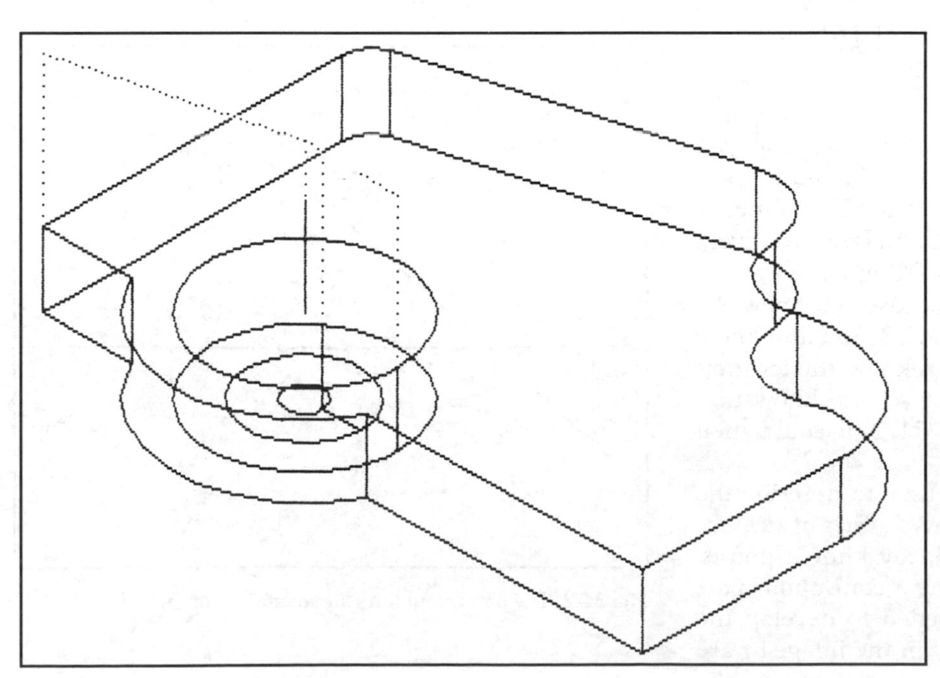

Fig. 24-3. Plotted part shown in 3-D.

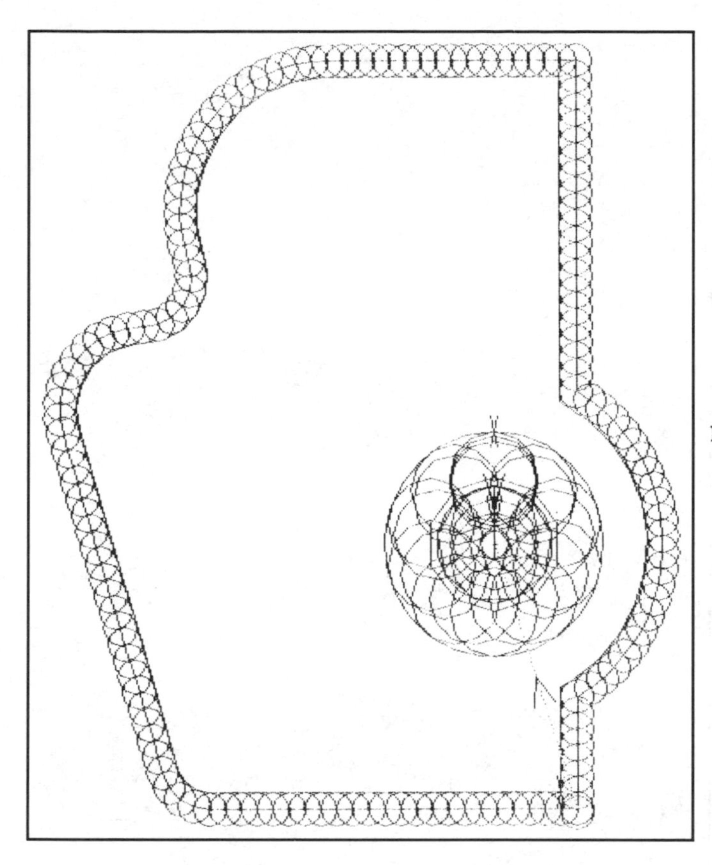

Fig. 24-4. A plot of the cutter path.

MASTERCAM® PROTOCOL

To use Mastercam®, the user starts by creating the part geometry, using a CAD image data transfer into the CAM computer, or calling the geometry out of memory. This geometry is not prepared for the tool path at this time. It is simply the part drawing. This drawing is saved in the CAM geometry file.

Next, the tool path is chained on the screen. *Chaining* is the process of selecting the order and direction of drawing elements to follow. A drawing element is a line segment, either straight or curved. The user indicates the point at which the cut is to begin, how the tool approach to the work is to be made, and the direction in which the work is to proceed. Tool change or depth change points must also be selected during chaining. Once the tool path has been created, the job parameters must be established on the parameters page of the software. This is type-one data, as listed previously. Once completed, the tool path contains information such as tool and depth of cut changes, number of passes, and finish material. The user then saves this tool path into a generic intermediate (NCI) program file. In the final stage a particular control template is chosen. The NCI file is postprocessed to a custom machine-usable program.

These examples are stand-alone CAM packages. They do not represent the range of protocols. Both software packages complete the same work, but each approaches the job differently. There are other systems that work differently yet. There are also CAM systems that reside within CAM software. NC Programmer® is a good example. It works within AutoCad®. NC Programmer® takes advantage of the drawing and geometry capabilities of the AutoCad® software. These units have more capability in drawing the part shape due to the full CAD software. However, the user must then have the full CAD software as well as the CAM package.

LEARNING CAM

The use of any CAM software will require practice and training, especially if you are unfamiliar with CAD drawing. Many CNC and CAD classes are teaching CAM. Another common way to learn is with factory or sales representative training. On-the-job training or off-hour courses are also options.

After the training, you will need practice. Hands-on application is the only way to become efficient. There are many skills to be learned. This brief development was intended to explain the concept of CAM programming only. CAM programming is rapidly growing in manufacturing. The programming skills you already have will make you a better user of CAM software as there is still a need to edit and improve CAM programs. CAM provides a quick way to write CNC programs. However, CAM programs are not necessarily the fastest programs on the machine. Often a skilled CNC technician will see improvements in CAM-generated programs.

Figure 24-6 shows the same part program. Here it is posted for a FANUC. Follow the commands as the cutter progresses around the part.

How many lines can you read?

```
%
N1G90G00G75X-2.9875Y-0.9375T2M6
N2Z2.0
N3G00X-2.9875Y-0.9375
N4Z0.0
N5G01Z-1.0F0.5
N6G41X-2.9875Y-0.9375
N7G01X-2.9375F1.0
N8Y3.25
N9G02X-2.5339Y3.7896I-2.375J3.25
N10G01X1.2352Y4.8993
N11G02X2.4375Y4.0I1.5J4.0
N12G03X3.1875Y3.4697I3.0J4.0
N13G02X5.4375Y1.8787I3.75J1.8787
N14G01Y-0.9375
N15X1.6956
N16G02X-1.6956I0.0J0.0
N17G01X-2.9375
N18G00Z2.0
N19G90G00G75X-2.9875Y-0.9375T1M6
N20X0.0Y0.0
N21Z0.0
N22G81Z1.3004F0.4
N23G80
N24G90G00G75X-2.9875Y-0.9375T3M6
N25Z2.0
N26G00X0.25Y0.0
N27Z0.0
N28G01Z-1.0F0.2
N29G03I0.0J0.0F0.4
N30G40G01X0.75
N31G03I0.0J0.0
N32G00Z2.0
N33G00X0.7Y0.0
N34Z0.0
N35G01Z-1.0F0.2
N36G41X0.7Y0.0
N37G01X0.75F0.4
N38G03I0.0J0.0
N39G00Z2.0
N40G40G90G00X-2.9875Y-0.9375M2
%
```

Fig. 24-5. Part program for Bridgeport Boss CNC control.

Fig. 24-6. Part program post processed for FANUC CNC control.

```
%
N1 G49 G4Ø G9Ø G17
N2 GØØ X-2.9875 Y-Ø.9375 T2 MØ6
N3 MØ3 S1Ø19 F1.Ø
N4 G43 H12 Z2.Ø MØ8
N5 ZØ.Ø
N6 GØ1 Z-1.Ø FØ.5
N7 G41 D2 X-2.9375 F1.Ø
N8 Y3.25
N9 GØ2 X-2.5339 Y3.7896 I-2.375 J3.25
N1Ø GØ1 X1.2352 Y4.8993
N11 GØ2 X2.4375 Y4.Ø I1.5 J4.Ø
N12 GØ3 X3.1875 Y3.4697 I3.Ø
N13 GØ2 X5.4375 Y1.8787 I3.75 J1.8787
N14 GØ1 Y-Ø.9375
N15 X1.6956
N16 GØ2 X-1.6956 IØ.Ø JØ.Ø
N17 GØ1 X-2.9375
N18 GØØ Z2.Ø
N19 G4Ø
N2Ø MØ9
N21 GØØ G91 G28 ZØ
N22 G9Ø XØ.Ø YØ.Ø
N23 G49 T1 MØ6
N24 MØ3 S382 FØ.4
N25 G43 H2 Z2.Ø MØ8
N26 G81 Z1.3ØØ4 RØ.Ø
N27 G8Ø MØ9
N28 GØØ G91 G28 ZØ
N29 G9Ø XØ.25 YØ.Ø
N3Ø G49 T3 MØ6
N31 MØ3 S382 FØ.4
N32 G43 H13 Z2.Ø MØ8
N33 ZØ.Ø
N34 GØ1 Z-1.Ø FØ.2
N35 GØ3 IØ.Ø JØ.Ø FØ.4
N36 GØ1 XØ.75
N37 GØ3
N38 GØØ Z2.Ø
N39 XØ.7
N4Ø ZØ.Ø
N41 GØ1 Z-1.Ø FØ.2
N42 G41 D3 XØ.75 FØ.4
N43 GØ3
N44 MØ9
N45 GØØ G91 G28 ZØ
N46 G49 G9Ø X12.Ø Y8.Ø
N47 M3Ø
%
```

CHAPTER 25

CAD/D COMPUTER-AIDED DESIGN AND DRAFTING†

At this point, you must have heard of a CAD unit. CAD units are common in industry and in modern tech schools. As stated earlier, the drawings in this book have all been produced by a CAD unit. The host computer was an NEC APC III. The software used was AutoCad by AutoDesk. The plotter was a Hewlett Packard 7570.

C/NC and CAD overlap in several areas. As you expand your C/NC career, it is more than likely that you will work at a computer console that generates C/NC programs from images produced on a screen. Your knowledge of absolute and incremental polar and Cartesian coordinates will be invaluable.

Chapter 25 will acquaint you with computer-aided drafting (CAD). A CAD system consists of the following basic components:

1. A computer.
2. Drafting software.
3. A graphics/system.
4. An input device (usually a digitizing pad with a mouse or pen and also a keyboard).
5. An output device (usually a plotter for drawing prints).

With these components, lines, circles, arcs, and points may be produced. These are combined into part shapes, letters, numbers, and lines.

Originally, CAD was accomplished on large, costly mainframe computers. However, since 1982, low-cost software that runs on micro-computers has become available. Even at this affordable level, CAD software and equipment costs many times more than a manual drafting station. Therefore, to be cost effective, a CAD station must be more éfficient than a manual drafting station.

OBJECTIVES

After studying this chapter, you will be able to:

- Present an overview of the CAM process.
- List the components of a CAD system.
- List the advantages of CAD.

†This chapter was prepared by Steve Monson

25-1 What Can CAD Do?

CAPABILITIES OF CAD

CAD drawings are created using basic geometric forms. However, this is not all that CAD can do. On a CAD drawing, the software can make the following changes:

- Rotate shapes.
- Stretch or compress shapes.
- Copy entire shapes or parts of shapes.
- Move shapes or part features.
- Evaluate mating part fits and functions.
- A CAD drawing may be instantly mailed electronically to another location via phone lines and modems.

CAD software can be customized for the particular discipline in which it is being used. For example, in electronic circuit drawings, the special symbols used may be inserted as part of the menu of options. More importantly, the data base created by the CAD drawing can be used by other software programs to evaluate the strength, acceptabiltiy, safety and other design features of the product before it is actually produced.

There are two other advantages to CAD drawings.

1. CAD drawings are produced at full scale, which eliminates the scaling errors found in manual drafting.
2. CAD drawings are *extremely accurate.* The plotted image is as accurate as the plotter resolution. The data image is accurate up to fourteen decimal places. This makes the CAD data image highly acceptable for CAM programming.

25-2 CAD Process

THE THREE CAD FUNCTIONS

CAD has two functions in the design lab and one more for the manufacturing facility. We will look at these two functions first. Then we will look at the way a CAD unit works for you in the machine shop.

Function 1—Drafting

A CAD unit is a computer with special software. It has dedicated internal programming that allows the drafter to draw prints using a screen, keypad, and—usually—a digitizing board. Fig. 25-1. These prints are stored in the computer's memory. They are then called up as needed for revisions or redraws. The print is drawn as a hard copy when needed by a machine called a *plotter.* A plotter is a C/NC machine of sorts.

Function 2—Designing

Product design can be assisted through the use of a CAD computer. This assistance comes in two forms:

1. *Better product visualization.* Fit and function can be seen from several angles. Design rotation and color enhancement are possible.
2. *Product analysis.* A simple drafting CAD unit cannot perform engineering analysis. However, with the addition of more software, large CAD computers can perform engineering tests and analyze CAD-designed products. They can simulate physical forces on the design and predict actual part failure. Such tests can identify areas where improvement is necessary.

Function 3—C/NC Programming

C/NC programming is the CAD function that most nearly fits your interests. After a part image is created by CAD, it may be analyzed by the computer as a solid object. In a fashion similar to the way the controller

Fig. 25-1. Using a digitizing pad and CAD unit to draft part prints.

undertakes radius compensation, the computer sees the shape. It then writes a set of point coordinates that translates into a program that will machine the part.

The object lines are drawn on a single overlay. The dimensions are then drawn on another overlay. These overlays are similar to a series of clear sheets—one on top of the other. Notes might be on yet another overlay.

Once complete, the engineer or aid can run the overlay that contains the actual part through a second software process. This process breaks the shape down into basic lines, circles, and points. This shape breakdown may be used for engineering evaluations or for C/NC programming. The programming software is identical to that which was described in Chapter 24. The only difference is that the image used for the program is provided by the CAD data, rather than manually input by the user.

Following the geometric rules used to write a C/NC program, the lines join at significant points. These are the same points you would need to program the part in the conventional way. The computer contains the image as a set of coordinates. Thus, it is not difficult to convert this information into data for C/NC programming.

POSTPROCESSING

Another software program now writes the data into a C/NC program for an individual control. This procedure is called *postprocessing*, or *post*. Many postprocessors can translate the image into several different types of programs.

The postprocessor would need to know the operating parameters of the intended machine. These parameters are entered in response to prompts. They include parameters such as material type, and cutter type. Usually the tool path is indicated manually. Some CAM softwares, however, will actually select the path. With this information, the post software can produce a usable C/NC program. This C/NC program may then be viewed as a first stage dry-run test of the program.

After the post, the program may need minor debugging. Obviously, the smarter the post, the fewer the problems with the program. Even if the program is not perfect, most of the hard work is done by the computer. Certain software programs called *mastery* postprocessors learn from the debugging suggested by the machinist. Each time an improvement is made to the program, it is added to the software.

CNC to CAD Interface

First, we looked at a CAD image becoming a C/NC program. A second version occurs in reverse of this process. The program is written at the CNC control, using either conventional programming or interactive graphic programming. Interactive graphic programming is very similar to CAD in that it creates an image by describing lines and circles to the control. Once the image is complete, the control processes it into program language.

This program may then be fed into the computer through a reverse postprocessing to create a CAD drawing of the part. CADs and CNCs can communicate.

25-3 CAD and You

BENEFITS TO THE MACHINIST

A CAD unit benefits you in the shop in the following ways:

- It produces accurate up-to-date drawings. CAD drawings are nearly perfect, save for errors in actual input data. The lines and shapes are drawn with computer accuracy. Because an update is completed on a video screen, drawing revisions are easy to keep up to date.
- It provides fast access to part shape data for C/NC programs. Until CAD or CAM programming, the part shape was being described twice. It was being described first by the engineer/drafter and a second time by the programmer. This is no longer necessary.

WHAT YOU MUST DO

Unless you are the drafter, you will not actually operate a CAD unit. You may, however, be called upon to enter part shape in a CNC control or at an offline CAM-CAP computer. You will be the end user of the CAD-CAM link. As these improve, fewer improvements will be needed once the post is complete. The best advice I could offer is to take a class in CAD drawing at a local school. The skill is very complementary to C/NC and will become more valuable as CAD and CAM become more common.

Since 1982, CAD usage and capabilities have developed more than in the previous twenty years. Today, as this growth and improvement continue, the limits seem unbounded. The CAD technician needs to be in constant training. He or she need to be aware of the newest products and capabilities on the market and in industry.

The predictable trend is for the job responsibilities of the CAD technician, the CAM technician, and the C/NC machinist to merge. These combined responsibilities will evolve into a new job responsibility. This responsibility will be assumed by a technician whose job title has not yet been defined. Education and information will be the keys to success.

APPENDIX A

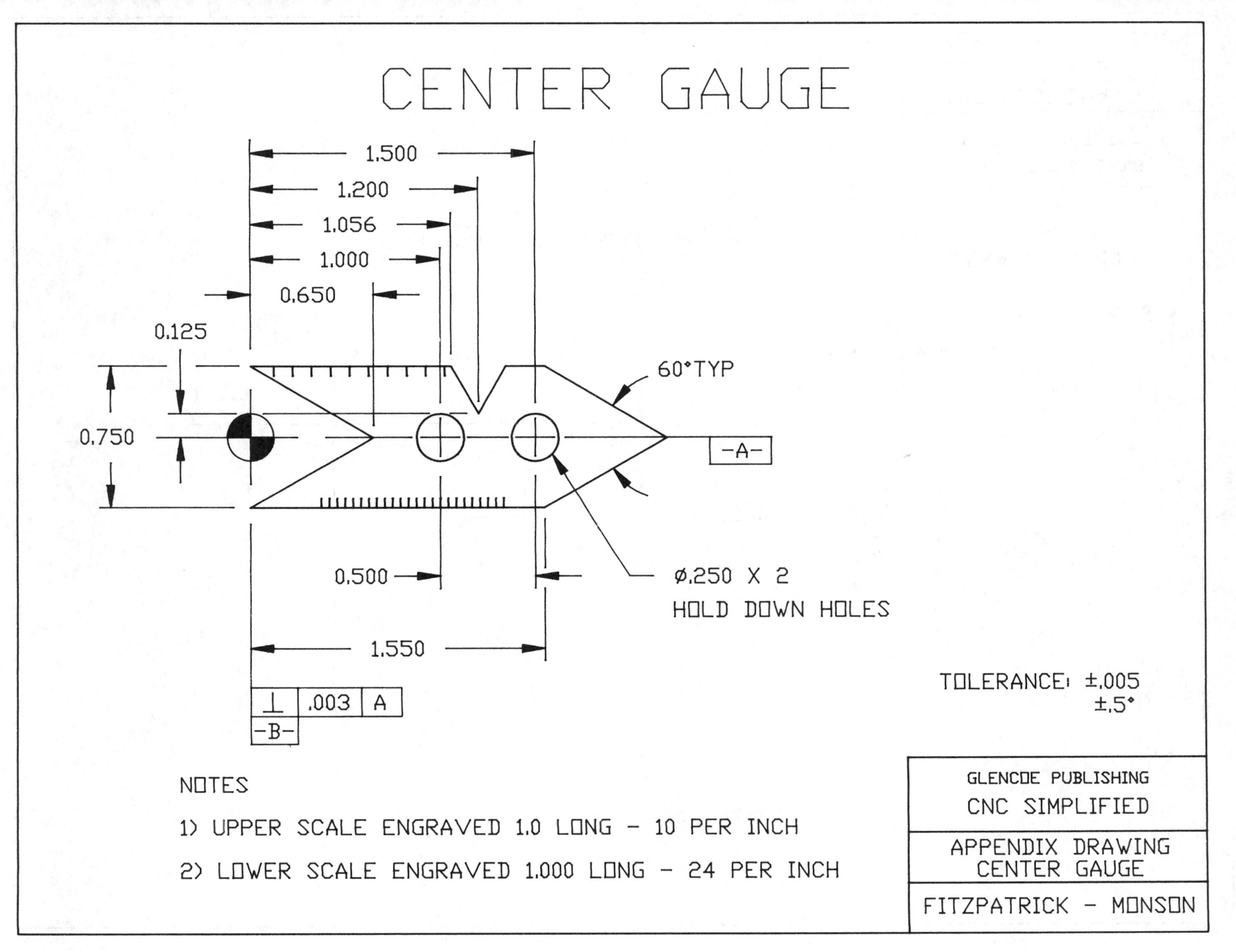

Drawing 1. Center Gauge.

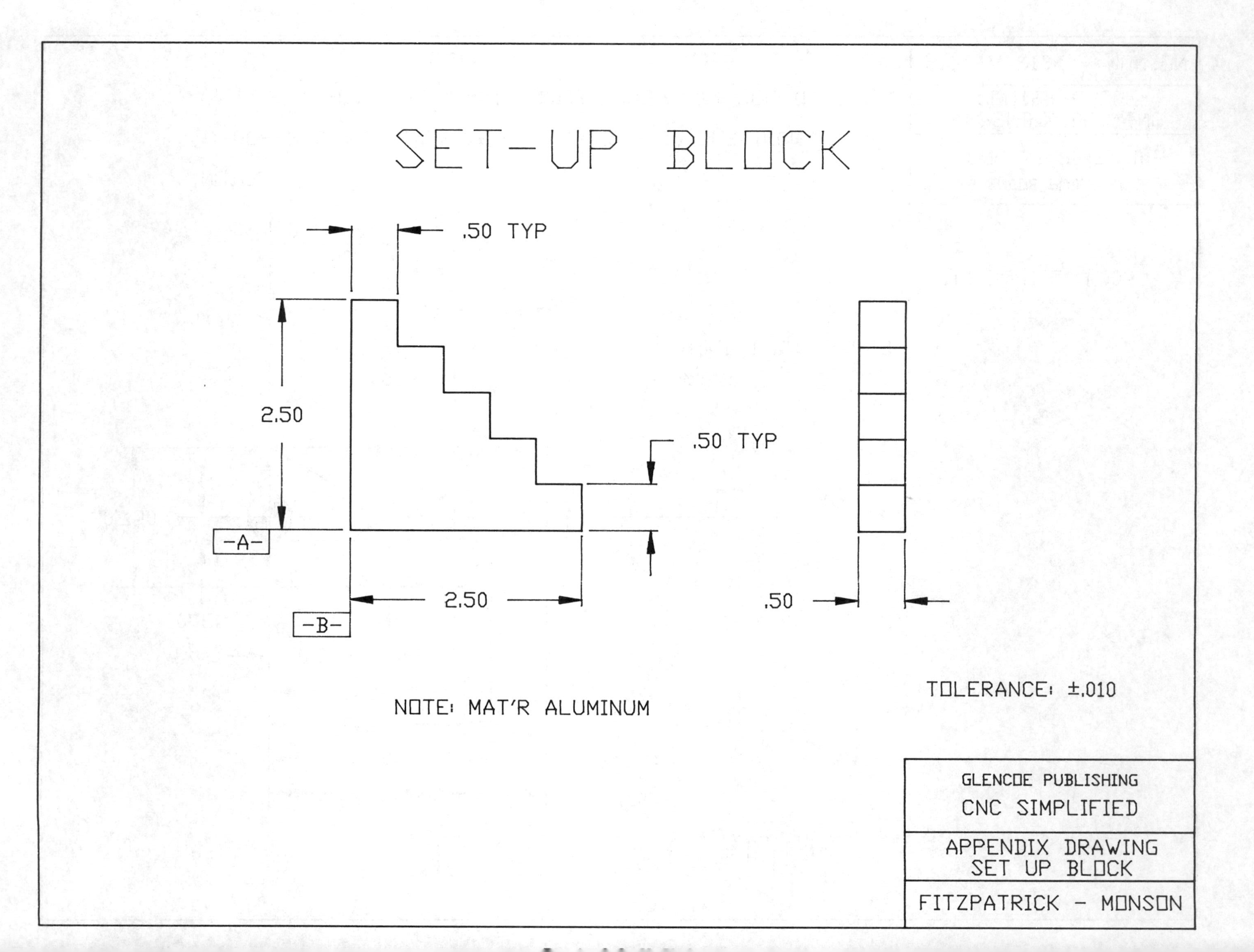
SET-UP BLOCK
.50 TYP
2.50
.50 TYP
-A-
2.50
-B-
.50
NOTE: MAT'R ALUMINUM
TOLERANCE: ±.010
GLENCOE PUBLISHING
CNC SIMPLIFIED
APPENDIX DRAWING
SET UP BLOCK
FITZPATRICK - MONSON

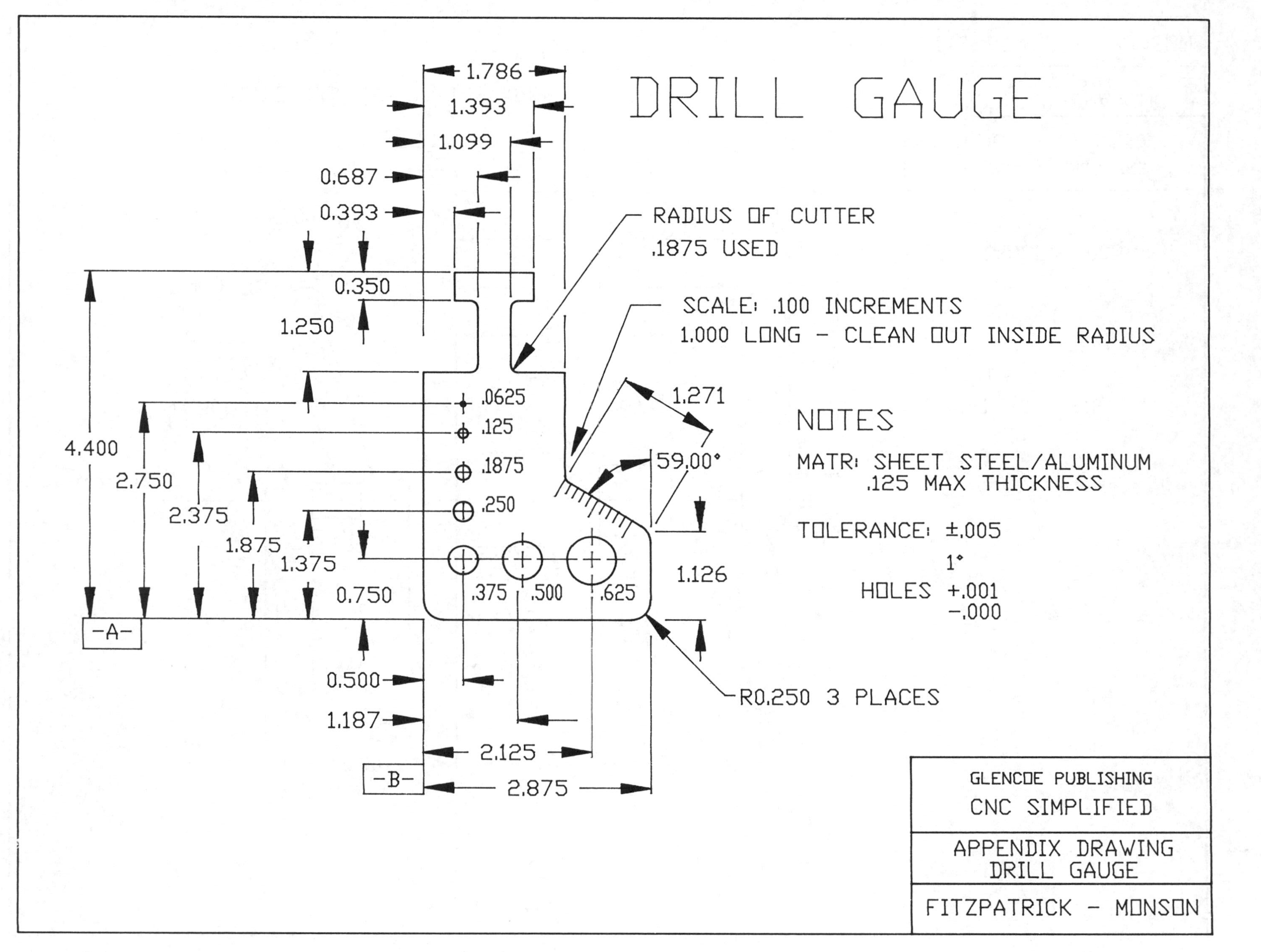

Drawing 3. Drill Gauge.

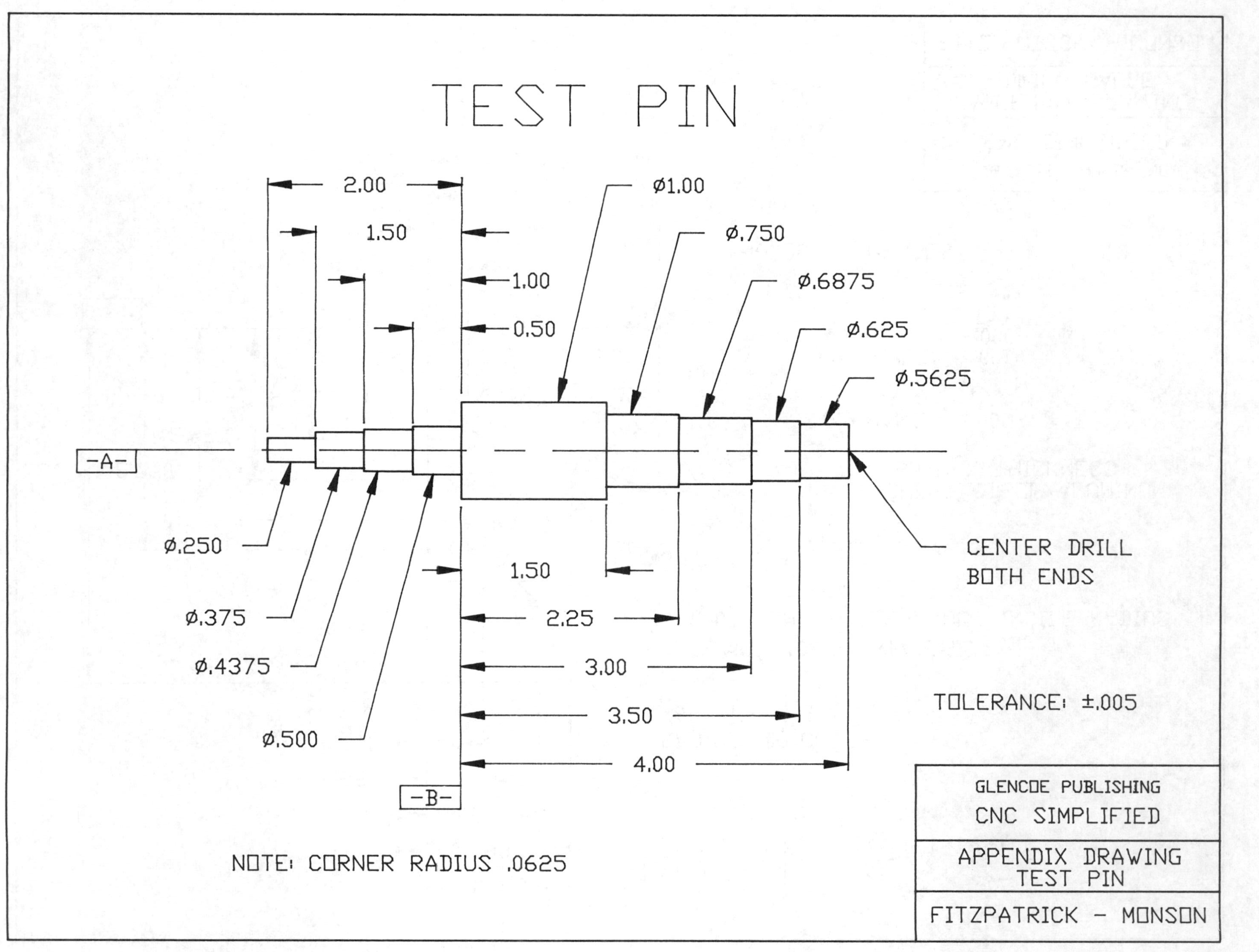

Drawing 4. Test Pin.

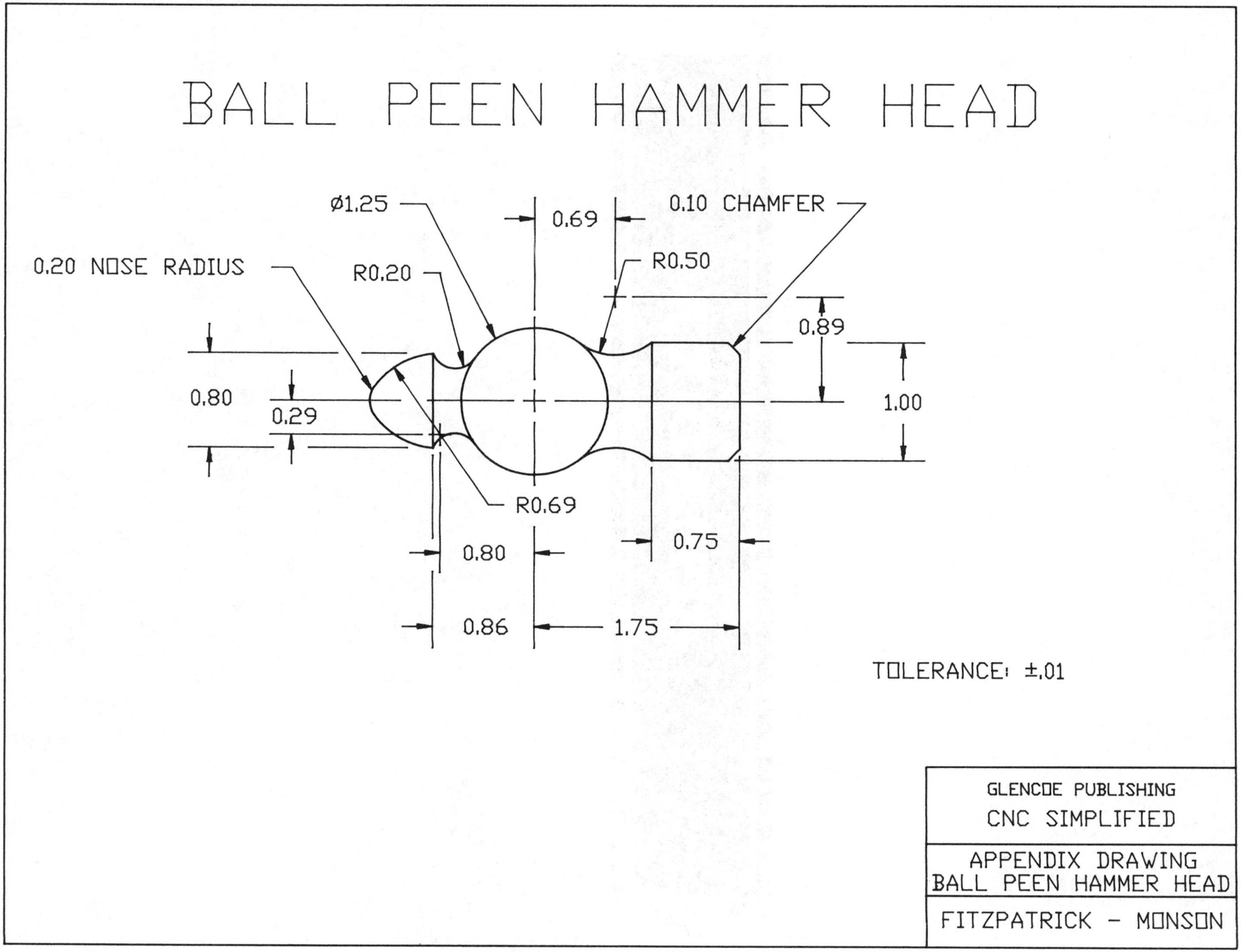

BALL PEEN HAMMER HEAD

Ø1.25
0.69
0.10 CHAMFER
0.20 NOSE RADIUS
R0.20
R0.50
0.89
0.80
0.29
1.00
R0.69
0.80
0.75
0.86
1.75

TOLERANCE: ±.01

GLENCOE PUBLISHING
CNC SIMPLIFIED
APPENDIX DRAWING
BALL PEEN HAMMER HEAD
FITZPATRICK - MONSON

Drawing 5. Ball Peen Hammer Head.

APPENDIX B

APPENDIX B
PREPARATORY AND MISCELLANEOUS FUNCTIONS‡

B.1 PREPARATORY FUNCTIONS (For Explanation, see Section B.3)

Code	Function Retained until Cancelled or Superceded by Subsequent Command of the Same Letter Designation	Function Affects only the Block within which It Appears	Function
G00	A		Point to Point, Positioning
G01	A		Linear Interpolation
G02	A		Circular Interpolation Arc CW (2 Dimensional)
G03	A		Circular Interpolation Arc CCW (2 Dimensional)
G04		X	Dwell
G05	*	*	Unassigned
G06	A		Parabolic Interpolation
G07	*	*	Unassigned
G08		X	Acceleration
G09		X	Deceleration
G10-G12	*	*	Unassigned
G13-G16	B		Axis Selection
G17	C		XY Plane Selection
G18	C		ZX Plane Selection
G19	C		YZ Plane Selection
G20-G24	*	*	Unassigned
G25-G29	*	*	Permanently Unassigned
G30-G32	*	*	Unassigned
G33	A		Threadcutting, Constant Lead
G34	A		Threadcutting, Increasing Lead
G35	A		Threadcutting, Decreasing Lead
G36-G39	*	*	Permanently Unassigned
G40	D		Cutter Compensation/Offset, Cancel
G41	D		Cutter Compensation-Left
G42	D		Cutter Compensation-Right
G43	D		Cutter Offset, Inside Corner
G44	D		Cutter Offset, Outside Corner
G45-G49	*	*	Unassigned
G50-G59	*	*	Reserved for Adaptive Control
G60-G69	*	*	Unassigned
G70	I		Inch Programming
G71	I		Metric Programming
G72	A		Circular Interpolation—CW (3 Dimensional)
G73	A		Circular Interpolation—CCW (3 Dimensional)
G74	J		Cancel Multiquadrant Circular Interpolation
G75	J		Multiquadrant Circular Interpolation
G76-G79	*	*	Unassigned
G80	E		Fixed Cycle Cancel
G81	E		Fixed Cycle No. 1

‡Reprinted with permission of the Electronic Industries Association from EIA 274-D.
*The choice of a particular case must be designated in the Format Classification Sheet.

Code	Function Retained until Cancelled or Superceded by Subsequent Command of the Same Letter Designation	Function Affects only the Block within which It Appears	Function
G82	E		Fixed Cycle No. 2
G83	E		Fixed Cycle No. 3
G84	E		Fixed Cycle No. 4
G85	E		Fixed Cycle No. 5
G86	E		Fixed Cycle No. 6
G87	E		Fixed Cycle No. 7
G88	E		Fixed Cycle No. 8
G89	E		Fixed Cycle No. 9
G90	F		Absolute Dimension Input
G91	F		Incremental Dimension Input
G92		X	Preload Registers
G93	G		Inverse Time Feedrate (V/D)
G94	G		Inches (millimeters) per Minute Feedrate
G95	G		Inches (millimeters) per Spindle Revolution
G96	H		Constant Surface Speed, Feet (meters) per Minute
G97	H		Revolutions per Minute
G98-G99	*	*	Unassigned

*The choice of a particular case must be designated in the Format Classification Sheet.

NOTES:

1. Permanently unassigned codes are for individual use and are not intended to be assigned in the next revision of the standards.
2. For future selection of "G" code pairs for on-off functions, the higher number shall initiate the function and the lower number shall cancel the function.
3. Assignments in previous revision: G72, G73, G74, G75 unassigned.

B.2 MISCELLANEOUS FUNCTIONS (For Explanations, see Section B.3)

Code	Function Starts Relative to Commanded Motion in Its Block: With	Function Starts Relative to Commanded Motion in Its Block: After Completion	Function Retained until Cancelled or Superceded by an Appropriate Subsequent Command	Function Affects only the Block within which It Appears	Function
M00		X		X	Program Stop
M01		X		X	Optional (Planned) Stop
M02		X		X	End of Program
M03	X		X		Spindle CW
M04	X		X		Spindle CCW
M05		X		X	Spindle OFF
M06	*	*		X	Tool Change
M07	X		X		Coolant No. 2 ON
M08	X		X		Coolant No. 1 ON
M09		X	X		Coolant OFF
M10	*	*	X		Clamp
M11	*	*	X		Unclamp
M12		X		X	Synchronization Code
M13	X		X		Spindle CW & Coolant ON
M14	X		X		Spindle CCW & Coolant ON
M15	X			X	Motion +
M16	X			X	Motion –
M17-M18	*	*	*	*	Unassigned
M19	*	*	X		Oriented Spindle Stop
M20-M29	*	*	*	*	Permanently Unassigned
M30		X		X	End of Data
M31	X			X	Interlock Bypass
M32-M35	*	*	*	*	Unassigned
M36-M39	*	*	*	*	Permanently Unassigned
M40-M46	X		X		Gear Changes if Used; Otherwise Unassigned
M47	*	*	*	*	Return to Program Start
M48	X		X		Cancel M49
M49	X		X		Bypass Override
M50-M57	*	*	*	*	Unassigned
M58	X		X		Cancel M59
M59	X		X		Bypass CSS Updating
M60-M89	*	*	*	*	Unassigned
M90-M99	*	*	*	*	Reserved for User

*The choice of a particular case must be designated in the Format Classification Sheet.

NOTES:

1. Permanently unassigned codes are for individual use and are not intended to be assigned in the next revision of the standard.
2. For future selection of M codes, the higher number shall initiate the function and the lower number cancel it.
3. Assignments in previous revision: M12, M46, M47, M58, M59 Unassigned.

B.3 EXPLANATIONS OF FUNCTIONS

Code	Function	Explanation
G00	Point to Point Positioning	Point to point positioning at rapid or other traverse rate.*
G01	Linear Interpolation	A mode of contouring control which uses the information contained in a block to produce a straight line in which the vectorial velocity is held constant.
G02	Arc Clockwise (2 Dimensional)	An arc generated by the coordinated motion of two axes in which curvature of the path of the tool with respect to the workpiece is clockwise, when viewing the plane of motion in the negative direction of the perpendicular axis.
G03	Arc Counterclockwise (2 Dimensional)	An arc generated by the coordinated motion of two axes in which curvature of the path of the tool with respect to the workpiece is counterclockwise, when viewing the plane of motion in the negative direction of the perpendicular axis.
G02-G03	Circular Interpolation (2 Dimensional)	A mode of contouring control which uses the information contained in a single block to produce an arc of a circle. The velocities of the axes used to generate this arc are varied by the control.*
G04	Dwell	A timed delay of programmed or established duration, not cyclic or sequential; i.e., not an interlock or hold.
G06	Parabolic Interpolation	A mode of interpolation used in contouring to produce a segment of a parabola. Velocities of the axes used to generate this curve are varied by the control.
G08	Acceleration	A controlled velocity increase to programmed rate starting immediately.
G09	Deceleration	A controlled velocity decrease to a fixed percent of the programmed rate starting immediately.
G13-G16	Axis Selection	Used to direct a control to the axis or axes, as specified by the Format Classification, as in a system which time-shared the controls.*
G17-G19	Plane Selection	Used to identify the plane for such functions as Circular Interpolation, Cutter Compensation, and others as required.
G33	Thread Cutting, Constant Lead	Mode selection for machines equipped for thread cutting.
G34	Thread Cutting, Increasing Lead	Mode selection for machines equipped for thread cutting where a constantly increasing lead is desired.
G35	Thread Cutting, Decreasing Lead	Mode selection for machines equipped for thread cutting where a constantly decreasing lead is desired.

G40	Cutter Compensation/Offset Cancel	Command which will discontinue any cutter compensation/offset.
G41	Cutter Compensation-Left	Cutter on left side of work surface looking from cutter in the direction of relative cutter motion with displacement normal to the cutter path to adjust for the difference between actual and programmed cutter radii or diameters.
G42	Cutter Compensation-Right	Cutter on right side of work surface looking from cutter in the direction of relative cutter motion with displacement normal to the cutter path to adjust for the difference between actual and programmed cutter radii or diameters.
G43	Cutter Offset-Inside Corner	Displacement normal to cutter path to adjust for the difference between actual and programmed cutter radii or diameters. Cutter on inside corner.
G44	Cutter Offset-Outside Corner	Displacement normal to cutter path to adjust for the difference between actual and programmed cutter radii or diameters. Cutter on outside corner.
G50-G59	Adaptive Control	Reserved for adaptive control requirements.
G70	Inch Programming	Mode for programming in inch units. It is recommended that control turn on establish this mode of operation.
G71	Metric Programming	Mode for programming in metric units. This mode is cancelled by G70, M02, and M30.
G72	Arc Clockwise (3 Dimensional)	An arc generated by the coordinated motion of 3 axes in which the curvature of the tool path with respect to the workpiece is clockwise when viewed from a certain angle.
G73	Arc Counterclockwise (3 Dimensional)	An arc generated by the coordinated motion of 3 axes in which the curvature of the tool path with respect to the workpiece is counterclockwise when viewed from a certain angle.
G72-G73	Circular Interpolation (3 Dimensional)	A mode of contouring control which uses the information contained in a single block to produce an arc on a sphere. The velocities of the axes used to generate this arc are varied by the control.*
G75	Multi-Quadrant Circular	MODE Selection if required for Multi-Quadrant Circular, cancelled by G74.
G80		Command that will discontinue any of the fixed cycles G81-G89.

G81-G89 Fixed Cycle** — A preset series of operations which direct machine axis movement and/or cause spindle operation to complete such action as boring, drilling, tapping, or combinations thereof.

Fixed Cycle			At Bottom			
Number	Code	Movement In	Dwell	Spindle	Movement Out to Feed Start	Typical Usage
1	G81	Feed	--	--	Rapid	Drill, Spot Drill
2	G82	Feed	Yes	--	Rapid	Drill, Counterbore
3	G83	Intermittent	--	--	Rapid	Deep Hole
4	G84	Spindle Forward Feed	--	Rev.	Feed	Tap
5	G85	Feed	--	--	Feed	Bore
6	G86	Start Spindle, Feed	--	Stop	Rapid	Bore
7	G87	Start Spindle, Feed	--	Stop	Manual	Bore
8	G88	Start Spindle, Feed	Yes	Stop	Manual	Bore
9	G89	Feed	Yes	--	Feed	Bore

G90 Absolute Input — A control mode in which the data input is in the form of absolute dimensions.

G91 Incremental Input — A control mode in which the data input is in the form of incremental data.

G92 Preload of Registers — Used to preload registers to desired values. No machine operation is initiated. Examples would include preload of axis position registers, spindle speed constraints, initial radius, etc. Information within this block shall conform to certain specified character assignments.

G93 Inverse Time Feedrate — The data following the feedrate address is equal to the reciprocal of the time in minutes to execute the blocks and is equivalent to the velocity of any axis divided by the corresponding programmed increment.

G94 Inches (Millimeters) Per Minute Feedrate — The feedrate code units are inches per minute or millimeters per minute.

G95 Inches (Millimeters) Per Revolution — The feedrate code units are inches (millimeters) per revolution of the spindle.

G96 Constant Surface Speed Per Minute — The spindle speed code units are surface feet (meters) per minute and specify the tangential surface speed of the tool relative to the workpiece. The spindle speed is automatically controlled to maintain the programmed value.

G97 Revolutions Per Minute — The spindle speed is defined by the spindle speed word.

M00	Program Stop	A miscellaneous function command to cancel the spindle and coolant functions and terminate further program execution after completion of other commands in the block.
M01	Optional (Planned) Stop	A miscellaneous function command similar to a program stop except that the control ignores the command unless the operator has previously validated the command.*
M02	End of Program	A miscellaneous function indicating completion of workpiece. Stops spindle, coolant, and feed after completion of all commands in the block. Used to reset control and/or machine. Resetting control may include rewind of tape to the end of record character or progressing a loop tape through the splicing leader.*
M03	Spindle CW	Start spindle rotation to advance a right-handed screw into the workpiece.
M04	Spindle CCW	Start spindle rotation to retract a right-handed screw from the workpiece.
M05	Spindle Off	Stop spindle in normal, most efficient manner; brake, if available, applied; coolant turned off.
M07-M08-M09	Coolant, On, Off	Mist (No. 2), flood (No. 1), tapping coolant or dust collector.*
M10-M11	Clamp, Unclamp	Can pertain to machine slides, workpiece, fixtures, spindle, etc.
M12	Synchronization Code	An inhibiting code used for synchronization of multiple sets of axes. For example, see Appendix C.1.
M15-M16	Motion +, Motion –	Rapid Traverse or feed direction selection, where required.*
M19	Oriented Spindle Stop	A miscellaneous function which causes the spindle to stop at a predetermined or programmed angular position.*
M30	End of Data	A miscellaneous function which stops spindle, coolant, and feed after completion of all commands in the block. Used to reset control and/or machine. Resetting control will include rewind of tape to the end of record character, progressing a loop tape through the splicing leader, or transferring to a second tape reader.*
M31	Interlock By-Pass	A command to temporarily circumvent a normally provided interlock.*
M47	Return to Program Start	A miscellaneous function which continues program execution from the start of program, unless inhibited by an interlock signal.

M49	Override By-Pass	A function which deactivates a manual spindle or feed override and returns the parameter to the programmed value. Cancelled by M48.
M59	CSS By-Pass Updating	A function which holds the RPM constant at its value when M59 is initiated/cancelled by M58.
M90-M99	Reserved for User	Miscellaneous function outputs which are reserved exclusively for the machine user.

*The choice for a particular case must be defined in the Format Classification Sheet.
**This command initiates a sequence of events which will be repeated at the appropriate times until cancelled or changed.

NOTE: Additional commands may be required on specific machines. Unassigned code numbers should be used for these and specified on the Format Classification Sheet. On certain machines the functions described may not be completely applicable; deviations and interpretations should be clarified in the Format Classification Sheet.

INDEX

T

V

W

Z